P9-CDL-995

The
Silicon
Boys

The Silicon Boys

and Their Valley of Dreams

DAVID A. KAPLAN

Perennial

An Imprint of HarperCollins*Publishers*

Excerpts from "The Tinkerings of Robert Noyce: How the Sun Rose on Silicon Valley" by Tom Wolfe, which originally appeared in *Esquire*, December 1983, are reprinted by permission of International Creative Management, Inc.

A hardcover edition of this book was published in 1999 by William Morrow and Company, Inc.

THE SILICON BOYS AND THEIR VALLEY OF DREAMS. Copyright © 1999 by David A. Kaplan. All rights reserved. Printed in the United States of America. No part of this book may be used or reproduced in any manner whatsoever without written permission except in the case of brief quotations embodied in critical articles and reviews. For information address HarperCollins Publishers Inc., 10 East 53rd Street, New York, NY 10022.

HarperCollins books may be purchased for educational, business or sales promotional use. For information please write: Special Markets Department, HarperCollins Publishers Inc., 10 East 53rd Street, New York, NY 10022.

First Perennial edition published 2000.

Designed by Bonni Leon-Berman

The Library of Congress has catalogued the hardcover edition as follows:

Kaplan, David A., 1956–
The silicon boys and their valley of dreams / David A. Kaplan.
p.   cm.
Includes bibliographical references and index.
ISBN 0-688-16148-0 (hardcover)
1. Microelectronics industry—California—Santa Clara County.
2. High technology industries—California—Santa Clara County.
3. Businessmen—California—Santa Clara County.   4. Santa Clara
County (Calif.)—Economic conditions.   I. Title.
HD9696.A3U56284   1999
338.4'7621381'0979473—dc21   99-21947
                                                  CIP

ISBN 0-688-17906-1 (pbk.)

00 01 02 03 04 RRD 10 9 8 7 6 5 4 3 2 1

For my parents

Contents

Of all the possible values of human society, one and one only is the truly sovereign, truly universal, truly sound, truly and completely acceptable goal of man in America. That goal is money.

—C. Wright Mills, *The Power Elite*

Prologue
Woodside 94062

There's rich, there's filthy rich—and then there's Woodside. In Silicon Valley, there are dozens of affluent towns where the overlords of tech build their megalo-mansions and try to prove there's no such thing as too conspicuous construction. Palo Alto and Portola Valley, Atherton and Los Altos Hills—these are among the places that the Siliconillionaires call home when they're not at work counting their stock options or sporting about the globe in their Gulfstream 5s.

But if you really want to know Silicon Valley, as I tried to do during the year of 1998, you have to visit its seat of power—bucolic Woodside, the symbol of an era's accumulation of ineffable wealth, the Beverly Hills of high-tech. This little enclave—thirty miles south of San Francisco, astride the San Andreas Fault—is tucked into the fogbound redwoods, where no property's too modest to have an electronic Fort Knox gate at the driveway. You'd think Patton was invading. Has it occurred to anybody that intruders can defeat the purpose of these hulking barriers by . . . well, just walking around them? Over in neighboring Portola Valley, Scott McNealy, CEO of Sun Microsystems, has figured this out and instead wants to hire a security force and erect guardhouses around his property. If the 1980s had the "Barbarians at the Gate," the 1990s in Silicon Valley has got them just behind the gates.

Woodside: It's sedate, it's rural, and according to the horse census, the ponies practically outnumber the people. There's actually a five-page publication available at town hall titled "Keeping a Horse in Woodside," which includes regulations on hitching racks, and "hay and feed sales," and definitions of "corral" and "pasture," and "accessory uses—horses." When county officials tried to restrict equestrian access to some of the backwoods trails—bikers and hikers had complained—the horsefolk kicked up a fit. Same thing happened when someone suggested pooper-scoopers.

More than any other of the well-hoofed communities around the Valley, Woodside tries to say, "We're loaded, but we're not telling." For instance, there are the pickup trucks. "Why does Woodside have lots of rich people driving pickup trucks?" asks Larry Ellison, the CEO of Oracle, who's moving to town next year. Ellison already understands the narcissism of the place. "Late-model, clean pickup trucks with leather seats have cachet, as in 'I'm not a materialistic person and I can prove it—I drive a pickup truck.' Trouble is, for me, I need a backseat for my kids, my briefcase, my topcoat, and, you know, the four million dollars in cash that people think I drive around with." So Ellison drives a Jeep—that is, unless he's driving his McLaren, Acura NSX, or Bentley turbo convertible.

Houses are an entirely different matter than wheels. Everybody seems to want a bigger, better, cooler, hotter one than the next guy. And the prices are pornographic. Average cost for a home in 1998: $1,514,065. There's faux-château, faux-maritime, faux-Japanese, faux-Tara, and authentic late-twentieth-century American Dermatologist. Zoning for homes is typically three-to-five acres, and always has been. That's how Woodside stays rural and also how it keeps the middle-class riffraff outtatown; when the zoning police in the fall of 1998 found out some property owners were using barns for human habitation, they cracked down. "Fabulous half-acre lot with beautiful heritage trees, winding paths take you back in time to lovely garden parties and tree climbing," reads a typical real-estate ad. "Offered at just $847,950. *Two-bedroom house included.*" Some of the listings cut right to the chase: "Potential rebuild situation." This is short for: Come scrape our house. "To scrape: *v.* To reduce one normal house to rubble and erect something much bigger. Scraper. *n.*"

The nouveaus here have more than they can possibly spend in a lifetime, but they're trying anyway; if you can't take it with you, you might as well spend it on eighteen-dollar-a-pound ostrich salami at the upscale market. Didn't the French have a revolution over something like this? The ex-wife of one Oracle engineer has forty-eight rack *feet* of clothing. It took Carnegie and Vanderbilt lifetimes to get rich—and they didn't have the luxury of doing it poolside in California. The Valley's best foodery, Draeger's, carries not just specialty meats, but an orgy of cigars and brandy, a locked cabinet of tiny $1,500-a-bottle balsamic vinegar infused with truffle, and more brands of olive oil than shampoo. Draeger's sells more Rosenthal than any store in America outside Manhattan and Miami. The proprietor, of course, lives in Woodside, eking out a respectable living off the tekkies. Let the plebes eat, well, Oscar Mayer. Woodside keeps its money quiet, but within the place, everybody's keeping count. There is no better, insular repository of gossip in Silicon Valley than Woodside 94062.

Gordon Gekko may seem like a bad memory from another go-go decade, and the Digital '90s may seem to reflect a less vulgar worldview. But you need to tour Woodside to get a real glimpse of what the Valley is becoming. A. C. "Mike" Markkula, Jr., the reclusive cofounder of Apple Computer and its former chairman of the board (as well as founder of the Markkula Center for Applied Ethics at Santa Clara University), is busily putting up a house that's 7,840 square feet, not including his parking garage for twenty motorcars. Problem, though, for the baronial Markkula is that he initially wanted twice that space—maybe something large enough that the *Sojourner* spacecraft could see it, even if that meant losing some of his privacy.

Markkula said the reason he needed the space was the towering redwoods, some of which are 120 feet tall. "If you put a small house in that location," he explained to the planning commission, "it gets dwarfed by the trees. It looks funny." Woodside said no way. Markkula also argued he was entitled to a zoning variance, since, after all, he was providing a "substantial public benefit" by setting up his caretaker's quarters as "affordable housing." Woodside said no way. (This was the funniest request Woodside town hall had seen since the wife of venture capitalist John Doerr asked for a variance to keep a pet pig in the backyard.) Markkula already has spent more than $1 million to restore a twelve-thousand-square-foot 1920s theater set in the groves. Outside the fortified construction site, somebody with a sense of humor posted a sign that said WILL CONSULT FOR FOOD.

At the planning commission brouhaha, Markkula was so mad at having his picture taken by the local newspaper that he lunged for the photographer's camera. In the late summer of 1997, his construction manager was fined $11,000 by the California Fair Political Practices Commission for what it termed "money laundering." It seems the manager used other people's names to make contributions to the campaign of a Woodside mayoral candidate—who, upon his election, would likely consider changing the town's zoning code, allowing Markkula to put up a much bigger house. The candidate apparently knew nothing of the scheme, but when asked if Markkula was behind the contributions, said: "It would take a very dense person not to figure that out." Woodside may see itself as quaint, but its politics are strictly big-city.

It's even got its brand of class warfare, an Edith Wharton comedy of manners. Some years ago, in what was dubbed "The War of the Garage Lights," a councilwoman got mad that her neighbor's outdoor halogens were shining into her bedroom window. She couldn't very well install curtains or shades, so she did the next reasonable thing: As a town elder, she drafted a law that lights could not be visible beyond the boundaries of a homeowner's property. Then the Woodside flatlanders (the ones with the flashy estates) joined in, squawking that their western vistas were being marred by the lights in the hills, and the hoi polloi up there—the old guard who got there first—replied that they wanted all the lights down below switched off. Sanity prevailed when one resident pointed out that the laws of physics superseded those enacted by men.

Gordon Moore, cofounder of Intel and California's richest man, also has a home in Woodside. Gracious and unassuming, he and his $10 billion or so are what pass for normal here. "I live in the unincorporated part of Woodside," notes the seventy-year-old Moore, "so they can't tell me what color my roof has to be." Even he gets the Woodside psyche. So does T. J. Rodgers, the outspoken libertarian CEO of Cypress Semiconductor. He's got a beautiful modern house in the hills, complete with Pinot Noir winery and a sock drawer perfectly organized by color and size, each pair in its little square nest. T.J. makes confrontation a sport. His own dog bites him. When the Woodside building department refused to give him a certificate of occupancy because the ledge of the shower was *two inches too high,* T.J. refused to budge. Instead, he installed an additional nozzle and got it classified as a "Roman tub." See, it's okay for a *tub* to have a two-inch ledge, though you have to be pretty small to get much of a bath. The civil servants in town hall like to torment the richies, and T.J., who says "money is the root of all good," likes to return the favor. EAT ME, says his weekend T-shirt (an ad for edible surfboard wax). T.J. once took on a nun who dared to criticize him for not having women and minorities on the Cypress board of directors. "Bluntly stated," he told her, "a 'woman's view' on how to run our semiconductor company does not help us, unless that woman has an advanced technical degree and experience as a CEO of an important technology company."

Steve Jobs of Apple has two places off Robles Drive—this is where Bill and Hillary stay while visiting Chelsea—and Jobs doesn't even live in Woodside (preferring his Tudor in Palo Alto). But Neil Young and Shirley Temple Black do. Joan Baez, too (once an item with Steve Jobs). And Koko the Gorilla, world renowned for her five-hundred-word sign-language vocabulary. Where else in America can you find a town with a computer genius, two music legends, and a primate who can order latte no-fat, no-foam?

Or a junior-high curriculum that teaches basic skills in "How to Be a Millionaire." Every year the first math assignment for seventh-graders is spending one million hypothetical dollars and plotting it on a spread-sheet. The rule is that you're limited to fifteen purchases, so that you can only buy expensive stuff. One of them can be a house for up to $700,000. (Forget about Woodside—this is how the kids learn to pro-nounce "San Jose.") Two can be cars. No more than $25,000 can go to charity. And for every expenditure other than the charity donation, the students have to find an advertisement of the item, with a photo and listed price. Talk about practical education: Woodside is teaching its kids—the sons and daughters of the nation's leading entrepreneurs and venture capitalists—how to be smart shoppers! Where else must all seventh-graders study the pages of *Town & Country*? Only in Woodside.

Woodsiders know how to throw a party. IPO parties, dog parties, helicopter parties, divorce parties (rich husbands doing the divorcing refer to their pretty new wives as "up-grades"), bar mitzvah parties, villa-warming parties, birthday parties for little kids that re-create entire children's stories in a backyard (and come with portable eight-foot hedges just in case less fortunate neighbors want a peek), Microsoft-bashing parties, and just-because-you-have-an-extra-$50,000-to-spend-on-them parties. Barb Ellison, the richest divorcée in town, a woman who's bought out so many neighbors she no longer has to have any, puts on the party of parties every two years—a party organized, naturally, around her own name. There was Barbstock, then Barbarella, and then Barbi Gras in 1998, which came with its very own professionally produced videotaped invitation for each of the three hundred guests. "An exotic adventure filled with spicy food, hot rhythm, and daring pleasures!" Barb exhorts her friends in the two-minute video. Did she mention the bare-breasted dancers, too? "When the sun goes down, the spirits awaken and you've got your pass to the Voodoo Lounge . . . a place to rid yourself of burdens. And friends, I need your help to rid me of my biggest burden—my name and that damned doll!" Barb instructs her guests to "bring your own sacrificial voodoo Barbie doll. Decorate your voodoo Barbie, place it on our altar, and our witch doctor will perform his magic." As she pokes her own doll, Barb says without a hint of humor, "Take revenge on a cheating spouse!" (More on him in a moment.)

On a cool spring night, there were eight hours of Barbi Gras '98—complete with musicians, aura reader, masseuse, inflatable Barbies, the witch doctor, and those dancers, along with all the chocolate truffles and raspberry mousse you could eat, plus the Voodoo Lounge down by the barn and tented tennis courts, where the fiery Barbie exorcism was performed to chants and drums. The only thing the guests groused about was those four Porta Potties, even if they were the flushable high-end models, with lights, mirrors, and sinks. Barb thinks the world of her Woodside guests, but she locked the doors to her stately Tara-like mansion, just in case; the last time, they made a mess of the master bathroom (the one with the color TV built into the tub and the naked-Greek-god-in-tile on the floor). Barb hates having the locals gossip about her life, but she doesn't get the irony of the stack of *People* magazines scattered throughout the house.

Barb Ellison's parties are fun. However, if you really want to understand Woodside today, you have to attend its annual charity auction for
the public elementary school. But don't expect to get in through your
checkbook alone: You've got to be pals with the party police, otherwise
known as the million-dollar Woodside School Foundation. If you don't
seem discreet, if you're not from Woodside, or heaven forbid, if you don't
seem part of the club, they'll have you stopped in the parking lot and
taken away by the sheriff.

I know, they tried with me. I asked some friends if I could attend the
ball, and the next thing I knew, the foundation mailed me a frilly invitation and a request for a donation. But at the last minute, when the
foundation decided I had been spending too much time in Woodside,
they reneged. There were concerns that I was going to write where families keep their silver, how many kids they have, and, as a sarcastic
message to me said, "Please come and abduct our children." And what
if I got a list of the guests? It would be a veritable Handbook for Kid-
nappers, I was told. The foundation then distributed a description of me
to the party police—"glasses, beard, about 150 pounds." I only found
out about this when I asked one of the cops why he was giving me the
evil eye on Grand Auction Night 1998. "We're supposed to be on the
lookout for someone who looks *just like you*," he tells me as I walk by.
"And then I have to have him arrested. Let me know if you see him,
okay?"

It's a good thing I don't look just like me.

Woodside had done this to visitors before. Two months before the
auction, poor Ray Piecuch decided to ride through town on the last day
of an equine journey across the United States. Carrying $200 and a
guitar, Piecuch had set out from New Hampshire for the Golden Gate—a
dream to become the first person to ride a horse across the country in
one year. He and his faithful companion, Bo (named for John Wayne's
horse in *True Grit*), slept under the stars and were befriended by all
manner of strangers. Then he stopped for a few beers at the Pioneer
Saloon in Woodside, parking Bo outside. Next thing he knew, the cops
showed up with the local animal-control officers; it seemed that someone
reported that Bo looked too travel-weary. Piecuch went to the pokey for
the night; his horse was taken in for questioning.

The school charity auction is the best party of the year. Forget the bake sales and car washes of Smallville USA. Think of, well, the cars themselves. Or boats—really big boats. Here we are at the 1998 benefit, themed after *Casablanca*. On a starry Saturday night in May, a five-peaked big top has been pitched across the school's soccer field together with two smaller tents. It's a fitting setting for the carnival of excess about to begin. Huge movie stills of Bogie and Bergman adorn the walls. There's a piano on the dance floor for Sam to play again. Genuine Moroccan art and clothing serve as decor, courtesy of one of the parents who made a special trip to North Africa to get them. The menu: roasted top sirloin of lamb with charmoula, potato gratin, and honey lemon gâteau. Outside is an ersatz Moorish marketplace of baskets, fruits, beans, belly dancers, and three huge pythons; right nearby are a cage of live chickens, leading someone to suggest they were some venture capitalist's idea of a Darwinian joke. If only they had replaced the fifty-nine Mercedes in the parking lot with camels, the *Casablanca* atmosphere would be complete.

The big-bidder fat cats and hot dogs are all here—rock stars by any other name. Many of the men come in white dinner jackets and red turbans; the women appear to have raided the *I Dream of Jeannie* wardrobe department. A few people have their Palm Pilots; more brandish cell phones and pagers, which don't do much for the Mediterranean look. Scott Cook, cofounder of Intuit, the software company that makes Quicken, is dressed in a towering gold lamé headdress topped with a feather. Normally quite modest, Cook this evening also wears Halloween-colored harem pants and a skimpy red vest that bares his chest; only the beach sandals and eyeglasses tucked in his waistband belie the look. John Doerr, the much-hyped venture capitalist behind Netscape, Amazon.com, and other companies of the new Wild West, skips the costume and shows up in chino pants and baseball cap.

By contrast, one of his partners, Brook Byers, is the best-dressed sheikh at the party, even if he still looks like the Georgia preppie who graduated from the Stanford Business School. There are also Gettys, Quists, and other Old Money among the four hundred guests—the *Up-*

stairs, Downstairs crowd that's been in Woodside since the Coolidge Administration and so resents the new cyber-wealth. And at the center of the event there's blond and bejeweled Barb Ellison, the latest ex-wife of Larry Ellison—the Valley's most celebrated, mediagenic womanizing rogue, the anti-Gates, and the second-richest person in the Golden State. Larry's completing an exotic $40 million samurai-style retreat in Woodside.

Larry Ellison didn't come tonight. A few days before the auction, always the sport, he tells me he's tempted to show up and bid more on items if I'm there—the better to show off. "They don't want to be conspicuous in their largesse?" he snickers. "If I were you, I'd just bring a camera crew." He gets it. "Rich people no longer throw parties here," he says. "They throw *fund-raisers*. It's hilarious."

I try to convince Ellison to take me as his date for the evening. My wife said it would be okay. "If you came as my date," he says, "then someone would write about *that* and I don't think it would be a very pretty story for either of us."

I can't convince him, but Ellison is there in spirit. He's offering his boat—his really big boat—to be the grand prize of the auction. She's Item No. 18, *Sakura*, a 192-foot, five-deck aluminum-and-teak megayacht that can crisscross the Atlantic without refueling. For Larry, size matters and this is a pretty good vessel through which to display it. *Sakura* is longer than any pickup truck, and handles better. What would you pay for you and nine friends to spend seven days along the Riviera aboard "one of the world's finest and most elegant motor yachts," as the auction brochure breathlessly proclaims? *Sakura* comes with other smaller runabouts, a crew of eleven (give or take a pastry chef)—and a chance to sleep in Larry's bed.

Take your time deciding how much to bid on this one because there's other stuff to buy first. "Male Bonding"—a chance for fourteen grown men to smoke cigars, eat meat, and tell jokes—goes for $6,200 to Joe Lacob, another of Doerr's partners and a co-owner of the San Jose Lasers, a women's basketball team. Not to be outdone, Barb Ellison pays $6,600 for "Ladies' Night Out," which includes a tattoo artist and a tarot reader. (Presumably it will be on a different night than the guys' outing; her daughter, a seventh-grader, can put it on her spreadsheet.) Doerr buys a PTA Comedy Club for $5,250, while Byers's wife shells out $3,900 for a Lobster Feast flown in from Maine. (She should've been so generous when she tried to steal Barb Ellison's housekeeper away; the housekeeper knew better, being the only housekeeper in Woodside assisted by an outside cleaning service.) The Doerrs also donate a party in their backyard—complete with three-legged races, an old-fashioned BBQ (maybe with that pig?), and "pool fun that includes a wet T-shirt relay and no-hands diving for pennies and a greased watermelon." "Outdoors at the Doerrs' " goes for $1,600. Ski lodges, soapbox-derby racers, deluxe Broadway weekends, Super Bowl tickets, tee-time at California's best links, a week at NASA Space Camp—each goes for $3,000 to $8,000. (While astronaut camp sounds like fun, it's not nearly as cool as the "Fighter Pilot for a Day" back in 1994, which included instruction for an eight-year-old in a Marchetti SF 260 air combat trainer used by NATO.) A family of Beanie Babies fetches $1,000. "You know the market's good when . . ." chirps a bemused investment banker by the name of Duff, filling in the sentence each time with the latest sky-high bid. Duff eventually buys a reserved parking space in front of the school for $800. The lowest bid of the evening comes for a baby-sitting contract, proving that the law of supply-and-demand operates even at charity events: Why buy a service that your live-in house staff already provides?

Along with smaller items like hay and alfalfa and other Mr. Ed accessories, as well as artwork made by the 480 students of the elementary school—there have to be some items under $1,000 for the rabble who aren't on the Forbes 400—the auction raises $439,000 in "private" funds for the school ($99,000 of which goes to pay for the event itself). That $340,000 in proceeds will go to programs like drama and art, and this year the purchase of a concert-sized tuba. (Luckily, Koko the Gorilla, IQ of 95, lives outside the boundaries of the school district, lest she lower the median test scores.) The $340,000 take is a pretty good end run around state law, which prohibits wealthy communities from having higher tax rates than poor ones.

The auctioneer, a pro flown in every year from Arizona just for Woodside, announces that bidding for *Sakura* will open at $10,000. Several paddles go up. "Fifteen! Twenty! Thirty thousand!" the auctioneer cries, taking bids from an engineer, a banker, and a ridiculously successful real-estate broker. Roger Sippl enters at $65,000, a virtual rounding error in his centi-millionaire portfolio. Sippl, at twenty-four, was a founder of Informix, which makes database software for large companies; ironically, Informix has been eaten alive by rival Oracle, the company started by Ellison. The surprise of the evening comes at $70,000. That's when Bruce Thompson, the Woodside school principal, leaps in. "We must be paying him too much," says one mother. A "VC wife"—what they call a venture capitalist's spouse—bids $75,000. Sippl's in for eighty, Thompson for eighty-five. The VC wife drops out. It's a two-tycoon race.

"Rog—er! Rog—er! Rog—er!" chant the couples at one table in the back, as if they were at a boxing match. They haven't had this much fun since watching good ol' boy Brook Byers display his prowess at "Body Shots" a few years back. (The game involves applying lime and salt to somebody else's body part, then licking it off and taking a shot of tequila. After the merriment in 1995, the sheriff had to escort one guest home after she was found stumbling in a roadside ditch. The next year, party planners with a sense of humor put out a gift basket of lime, salt, shot glasses, and fifty-dollar-a-bottle tequila for the frolickers.)

Roger responds to the encouragement. He bids $90,000. The band whoops it up.

Thompson and Sippl go back and forth in $5,000 bumps until Thompson wins at $125,000. He gets a standing ovation, discovering he has more friends in the town than he realized. Sippl still gets to go home with a "Five Star Family Campout" at Scott Cook's six-acre hideaway, which comes with use of the host's private observatory. It's a bargain at $3,900. If the guests are better behaved than Barb Ellison's, maybe he'll let them see the super-duper toilet in the guest bathroom in the front hall. Cook saw the model in Japan and was so impressed he arranged to ship one back to Woodside. The toilet, a conversation piece if ever a bathroom fixture could be one, was heated, padded, and, best of all, had a remote control, just like on *Ally McBeal*. If you could read the posted Japanese instructions, who knows what other tricks it could perform?

Meanwhile, everybody wants to know what stock options Thompson has lucked into for an oceanic vacation that will cost twelve dollars per minute. One story has it that Bill Gates was behind the bid, as a way of getting back at Ellison, who's made a vocation of making fun of Gates and who's been trying for years to buy a MiG warbird to strafe Bill's new Xanadu up in Seattle. Another rumor claims that Ellison put Thompson up to it so nobody would actually get the boat. And another says that

Brook Byers was using Thompson because he feared publicity about his own finances. Since Woodside has just 5,250 residents, and since almost everyone with a big mouth has attended the auction, it takes all of about three minutes to figure out that Thompson was a surrogate for one J. Taylor Crandall. Not exactly a household name in Silicon Valley, but he's the guy who gives financial advice to the Bass family, some of whom also live in Woodside.

Managing money for a Bass means he gets to keep a few minnows himself. A few Halloweens ago, the Crandalls put on a themed $5,000 birthday party for their daughter that the other thirty kid guests still talk about. (*Raising Self-Reliant Children in a Self-Indulgent World* isn't a best-seller in Woodside.) The Crandalls hired Rick Herns, the Valley's premier party planner, to erect in their yard a facade of a haunted house, with fake fire in the windows and a cast of witches and wizards to entertain; Herns is the same special-effects impresario who helps put on the auction. Upon returning home to her own, more modest surroundings, one young girl exclaimed to her mother: "They even have a man who works for them just to carve the pumpkins!" When you've already got that, how can you not buy a cruise on *Sakura*?

Crandall and his wife are incensed that their attempt to stay in the background at the auction has failed. On their behalf, Byers writes a nasty note to the gossip columnist of the local paper, asking why she named J. Crandall in a story. Like the rest of the Woodside aristocracy, the Crandalls like the revelry of money, but are ambivalent about showing it off to just anybody. Five days later, the Woodside School Foundation even issues a statement that Crandall isn't the real buyer. Maybe he just got around to reading the fine print in the auction catalog. It seems that Larry Ellison only included three hours of fuel per day. J. Taylor is on his own for the rest of the gas ($100 an hour), along with food, booze, and dockage fees. But why worry? This is Silicon Valley— Valley of the Dollars. Either you've hit it big, or you believe you're just about to.

Chapter I

Dreams

Once upon a time, but not that long ago, Silicon Valley was just a dry, sleepy patch of orchards between San Francisco to the north and San Jose to the south. Among the willow thickets near the shore and sunbaked chaparral in the foothills, apricots and cherries blossomed here, not a technological awakening that changed mankind. The Valley was brown, not the color of money. It wasn't even called Silicon Valley until the rise of the microprocessor in the early 1970s. It was just the Peninsula or South Bay. Now, "the Valley"—as those who know it call it—is an American icon, a Sutter's Mill for our time.

If the Valley were a nation, it would rank among the world's twelve largest economies. Back during the Cold War, the area ranked near the top of the Soviets' list of nuclear targets. But far more than the mother lode of economic miracles, fount of overnight millionaires, and international symbol of high-tech know-how, the Valley competes with Hollywood as a place in the culture of money and celebrity, success and excess. Washington and Wall Street have been left gasping for air in the rearview mirror.

The Valley is otherworldly—a foreign land we know little about, populated by the gearheads we used to tease at recess. Now they're the ones who own the playground. The meek didn't inherit the earth—the geeks did. They live in the Valley so they can work there; you'd think it might be the other way around, given the spectacular setting. But the Valley is one giant company town. Anthropologists are studying it the way they used to look at Papua New Guinea; they know that the more rarefied the culture, the more it develops its own values. What do they find? Narcissism. Stuff like a hard-driving mother in one company—we'll call her Compulsive Cathy (not her real name)—whose two kids complained she wasn't spending enough time with them. One morning she had a few hours free, so she called for a "team meeting" with the children. They looked at her as if she were from another planet. Another sees herself as "project manager" for her kids. They wear pagers—just in case Mom's looking for them. At his therapist's suggestion, an engineer is writing a "mission statement" for his personal life. The newest-fangled anthropologists have taken to calling themselves "entrepreneurialologists" (which manages to use all five vowels even if it's not actually a word).

In Hollywood, everyone's got an almost-finished script in the top drawer; in the Valley, it's a business plan for a start-up they're convinced will make investors drool. Hollywood has agents, the Valley has venture capitalists; and between them, their attention span still isn't fifteen minutes. Every season, Hollywood has a new list of releases, while the Valley has new IPOs. Each loves its reborn stars: John Travolta and Steve Jobs. Commuting stinks in both. Bigger, better, faster, richer. The best dream wins—unless Disney or Microsoft gets involved and mucks it up.

Hollywood has Morton's, the Valley has Buck's (in Woodside, naturally); Spago has a chichi restaurant in both places. (Explaining why he opened in Palo Alto, rather than cuisine-happy San Francisco, Wolfgang Puck explained unpuckishly, "It's the demographics.") Eat a meal at any of them and you'll hear deal-making blather at a nearby table. In Hollywood, percentage "points" are the coins of the realm; in Silicon Valley, it's options. *Variety* runs the weekly box-office scorecard; *The Wall Street Journal* publishes twelve pages of stock quotes every day. Just as everybody in Hollywood knows *Titanic* is the No. 1 box-office hit in history, everybody knows that a dollar invested in 1995 in Cisco Systems (the plumbers for the Internet) octoplied three years later and that during one stretch of time in 1998, Yahoo did the same thing. Hollywood has two Lamborghini dealers, Silicon Valley has four. The buzzword "convergence" isn't about the Internet and cable and Ma Bell. It's about Silicon Valley and Hollywood: The animated movies out of Pixar now feature Steve Jobs as an executive producer and the premieres boast as many digerati as glitterati.

Big money, instant money: The industries may change, but entrepreneurs have always been after it. "Money is life's report card," is the saying of W. Jerry Sanders III, CEO of Advanced Micro Devices, Intel's semiconductor arch-nemesis. One particularly good fiscal year, Sanders distributed profit-sharing in doubloons. Another time, when asked why he kept a black Rolls-Royce convertible in San Francisco and a white one at his Malibu beach house, he cracked, "So I can tell where I am in California." Next to Larry Ellison, Jerry Sanders is the nattiest titan in the Valley. The stripes in his pinstripe suits say, "Jerry Sanders Jerry Sanders Jerry Sanders."

Even George Lucas, the creator of *Star Wars*, whose 2,000-employee empire north of San Francisco represents a kind of antithesis of Silicon Valley, has recently conceded a bit to the Valley money culture. Lucas is the sole owner of a multibillion-dollar, entertainment-technology conglomerate that consists of a film company, video-games division, and special-effects studio. He may buy top-of-the-line computers and software from the Valley, but he's steadfastly tried to remove himself from its ways (just as he's kept out of Hollywood). He's never considered a public stock offering, he doesn't pay his people top dollar, and he stays away from the Valley's high-profile, high-adrenaline ways. Yet Lucas has now been forced to give senior employees stock options that would materialize if he ever sold out. Darth Vader would be proud.

Hollywood and Silicon Valley both think they're so savvy they can affect the national political agenda: Hollywood for social issues, the Valley on technology. Politicians humor the respective self-proclaimed expertises of Kim Basinger and Larry Ellison, but there's a simpler explanation for their frequent visits to California. And in the northern half of the state, when the Washington politicians need spending money, they've stopped bothering with the old biddies on Nob Hill in San Francisco. They go to the Valley where the cash is—money trees they can regularly shake down. Woodside and Atherton and Palo Alto have as many $1,000-a-plate political dinners as Georgetown and the Upper East Side of Manhattan. The tekkies think they're getting their dollar's worth on Capitol Hill—favorable bills on securities litigation, mergers, and trade barriers; often the goal is just to get Washington to stay *out* of high-tech regulation—that is, unless it involves antitrust and Microsoft. Whatever their successes, if nothing else, the tekkies get great seats at

White House dinners. Do you really think John Doerr is invited because Bill Clinton wants to hear his views on the arcana of encryption?

Say this for Hollywood: Most of its leaders recognize, most of the time, that they're in the entertainment business. Sure, *Titanic* taught us "timeless lessons about human nature" (yes, and also about the importance of counting the lifeboats *before* leaving shore); but unless you're James Cameron, you know it's just a movie. In Silicon Valley, things are a bit more grandiloquent. "The largest legal creation of wealth in the history of the planet!" is how moneyman Doerr describes the fabulous economic engine he's helped build. Unemployment is virtually zero (prized programmers go like real estate, bid up past any rational number), half the workforce are minorities (though few are black), a quarter are foreign-born, wages are the highest in the United States. If you can spell "cat" on a keyboard, there's probably work available. Doerr again: "We're the cornerstone of the New Economy . . . a cradle of prosperity . . . a beacon to every other nation . . . the purest expression of capitalism." Fair enough, but have you ever heard General Motors in Detroit claiming it was "changing the world"? Einstein and Salk, Roosevelt and Churchill, Ruth and Ali—they, too, changed the world, but had the misfortune of not becoming billionaires in the process. If the Valley isn't about money but about changing the world, why do so many dweebies retire to a life of Ferraris and ski holidays in the Alps before they're thirty?

Then there are the giddy comparisons to the Mesopotamia of antiquity, Florence in the fifteenth century, Paris in the Twenties. So, that would make Doerr, Medici and Jobs, Hemingway? Silicon Valley is not just epicenter of high-tech, but also soul of the Zeitgeist, attracting everyone from everywhere, triggering a latter-day Renaissance. Maybe it's the weather. Maybe it's the good press. But almost every greedy, driven twenty-three-year-old here somehow thinks he (or, on rare occasions, she) has the idea that will transform the way we live, think, play, even govern ourselves—when, in fact, most of the ideas just have to do with yet another way to sell jelly beans on the Internet. On top of that, most of the ideas will fail anyway. God bless their efforts—but are they about changing the world, or the more modest goal of striking it rich? At least P. T. Barnum didn't delude himself.

Perish the thought, but could money really be part of it? "You just don't get it, do you?" I was told in conversation after conversation. Actually, I just don't *have* it.

Depending on yesterday's Nasdaq close, roughly a quarter-million millionaires live in the Valley, give or take an ostrich salami. By one count, the Valley on average produces sixty-four new millionaires every twenty-four hours, including Christmas. Given that a lot of them will retire young, the Valley stands to become the Boca Raton of the West—though it'll need a lot more golf courses. Silicon Valley exports more goods than any other region in America. The market value of its companies is approaching a trillion dollars, which dwarfs the numbers of Detroit and Hollywood; the federal government makes more money, but that doesn't count, because it literally makes the money. The Forbes 400 includes more engineers than movie moguls. Glossy magazines are as likely to profile Doerr as Tom Cruise (who in 1997 was reported to be moving to Woodside, erroneously as it turned out). The president of the United States makes more trips to Northern California than to any other part of the country, and it isn't just to visit his daughter at Stanford. Boys used to dream of growing up to be Mickey Mantle. Or, in the bygone Age of Milken, to become Masters of the Universe. Today, they want to rule cyberspace by "writing code" and becoming Jerry Chih-Yuan Yang of Yahoo! Inc.: thirty years old, a leader of the Internet revolution, with several billion on account. Silicon Valley is the Newest New Thing, the spot to be to change the world (or, at least, to get rich trying).

It is a location on a map, the flatland between two coastal mountain ranges that gets lots of blue sky and the occasional eight-point-something earthquake. It has its recognizable images—Intel's dancing bunnies, Apple's apple, Netscape's nerds, rows of Dilbertian cubicle farms filled with unkempt hackers. Important things were invented or created here: the integrated circuit, the first commercial radio broadcast, video games, minicomputers, microprocessors, gene-splicing, 3-D computing, Internet commerce. Yet these miss the point. The Valley represents, for better and for worse, part of the American imagination of the late twentieth century. Like the West of earlier generations, it is a state of mind, combining real-life drama and unadulterated myth. Our adoration of its culture, or lack of it, tells us something about ourselves and offers a hint of the future. The Valley has its admirable moments, its venal moments, and, best of all, its absurd ones. After all, there aren't many places where a boat ride costs $125,000.

This book is the story of that place and the people who'll gladly pay it. The least they could do is buy a copy.

How did Silicon Valley happen and why in that location? It's not happenstance, but intrinsic to the character of California—the state's DNA. The techno-entrepreneurial revolution belongs here as much as palm trees and sunshine. For 150 years, ever since the Gold Rush, every California story has begun with a dream.

An Oracle engineer named Roger Bamford likens the Valley to a "vein of gold," a river that "anybody can reach down into and strike it rich." Most will come up empty, for the gold is hidden in a roaring river that constantly changes course and fortune. "No matter how big your hand is," he says, "if you reach down in the wrong spot, you don't get anything." But those like Bamford, who find the vein, will be awash in gold and become part of the essential folklore of California. Bamford, of course, has a home in gilded Woodside.

Bamford is one of Larry Ellison's chief software programmers, who can write code the way Shakespeare wrote sonnets. Microsoft covets Bamford so much that it periodically opens the vault for him; last year, a key executive toted him around Redmond, Washington—at two thousand feet in the air. They toured the scenery, visited houses for sale, and were granted an audience with Chairman Bill himself. But Bamford decided to stay with Oracle. He has a reputation as one of the best programmers in the Valley, but also has a rare sense of the absurd. He started working at Larryland in 1984 and is well enough off to own two helicopters that, depending on the outcome of his divorce, he hopes to keep in a hangar Frank Gehry will design for him. At forty-three, Bamford knows he's one lucky engineer. "What's the difference between the real smart and real personable guy like Larry Ellison and the real smart and real personable guy who's in some loser start-up?" he asks. "Being in the right place at the right time. It's random."

El Dorado is a recurring metaphor in Silicon Valley. Books, headlines, business plans all talk of a latter-day Gold Rush. It's easy to see why:

Today, the very alchemy of the Valley is turning silicon into gold. In another time and place, gold gave birth to modern California. Gold symbolized a new economic belief—that anybody could "strike it rich." Expectations changed, almost overnight, even if the odds of a windfall were staggeringly high. Before, subsistence farming ruled the day—earning a living from the earth was hard work—and great wealth came slowly, amassed over a lifetime of labor. This kind of "earned" riches had long been a part of the country's conservative heritage. The wrong pedigree, limited education, crushing debt—these could now be instantly transcended. It all changed on the morning of January 24, 1848, in a town called Coloma, on the western slope of the Sierra Nevada.

Along the South Fork of the American River, in a narrow valley backed by sugar pines, thirty-eight-year-old James Marshall was supervising construction of a sawmill. Nine years earlier, his Swiss partner—Captain John Sutter—had arrived in California in search of wilderness he could turn into an agrarian empire. The Mexican government, still in control of the area and eager for a buffer from the approaching United States, gave him a huge land grant. On fifty thousand acres, Sutter planted wheat fields, orchards, and vineyards. Settlers would stay at his adobe-walled fort while waiting out the Sierra snows.

Forty-five miles to the east, where the foothills provided abundant trees and the swift waters of the river offered power, Marshall was putting up the mill that Sutter needed for lumber. By New Year's 1848, all that remained to be finished was the tailrace, where the spent water would make its way back to the river. To accommodate the millwheel, the race needed to be deepened. On January 24, Marshall went to inspect the digging and found the yellow nugget—the proverbial "flash in the pan"—that would profoundly transform the California landscape and American values. Sutter and Marshall vowed to keep their find quiet— they had a mill to complete. But, as one writer put it, that would be "like trying to contain a sunrise in a bottle."

Rumors of a gold strike began to spread out of the Sierras. The editor of the *California Star* heard them and scrawled "HUMBUG" in notes he took at a river camp. Seven weeks later, his newspaper ceased to print, after everyone in the shop had left for the mines. It was a San Francisco huckster who struck the Promethean match. Sam Brannan was both a publisher and owner of a trading post at Sutter's Fort. He visited Coloma upon hearing about Marshall's find and returned to San Francisco in May, hoping to stir interest in the goods sold by his store. "Gold! Gold! Gold from the American River!" he declared in Portsmouth Square (what would become Chinatown), brandishing a bottle of gold dust in one hand and waving his hat with the other. Beforehand, Brannan had bought up every digging implement he could find, cornering the market in a way Bill Gates would surely have admired. (One could think of the pan and the rocker as a sort of "operating system" for the Gold Rush.) On May 29, one newspaper reported: "The whole country . . . resonates to the sordid cry of gold, gold!, GOLD! while the field is left half-planted, the house half-built and everything neglected but the manufacture of shovels and pick-axes." Dreams and greed were a potent combination. And the world rushed in.

The 48ers, and then the miners of '49, came by all means—on foot, on horseback, over land, and along the Sacramento River. By the middle of June, three-quarters of the men in San Francisco had headed for the hills. The Bay was clogged with abandoned ships (which then sank and were used as the landfill on which the present-day Embarcadero financial center is built). Gold seemed available for the taking, literally a free-for-all. Ten million dollars' worth was mined in 1848 alone. Sutter and Marshall could do nothing to keep the miners out. By the end of the following year, there were fifty thousand fortune-seekers and adventurers in new towns like Rich Bar, Hangtown, Dry Diggings, and Rough and Ready.

They came first from other states and elsewhere in California, then Mexico and Hawaii, China, South America, Europe, and Australia. They came around the Horn, over the Plains, across the Panama isthmus, traveling thousands of miles in pursuit of sudden wealth. Covered wagons and clipper ships came to be rituals of passage. The Gold Rush reined in the Western frontier and set off the largest voluntary migration in world history. In the meantime, California, earlier ceded to the United States by a Mexican government that was unluckily unaware of gold's discovery, became the thirty-first state in the Union—the Golden State—on September 9, 1850.

Before Sutter's Mill, the nonnative population of California was 12,000. Six years later, it had rocketed to 300,000 (with women making up fewer than 10 percent). The little, sand-duned maritime village of

San Francisco, after its virtual desertion in 1848, became the portal to Gold Country, a hundred miles away. At one point in 1849, the city's population was doubling every ten days. Real-estate values soared, commencing a great Bay Area tradition; a single lot purchased for $16 in 1847 sold for $6,000 the next year, and seven times that in 1849. Gambling thrived, along with booze, even with saloon whiskey costing $30 a quart. San Francisco's reputation as a gastronomic mecca got its start during this period. As immigrants from everywhere poured in, and with little in the way of family life, "eating out" was de rigueur, as were ludicrously high prices. Freedom and independence reigned—go as you wish, dress as you please. The wide-open lifestyle was 150 years ahead of its California time. In a different epoch, Steve Jobs might have struck gold.

The romance of a lucky strike like Marshall's was evanescent. The easy placer gold quickly disappeared, replaced by one technological innovation after another—a hint of the California to come. The pan and the rocker gave way to the sluice, and then to the diversion of entire rivers and a hydraulic assault upon the mountains themselves. These new commercial mining technologies of land and water—a far cry from James Marshall—in turn found other uses. Irrigation reclaimed desert land, and aqueducts led to the urbanization of Los Angeles and San Francisco. From 1860 to 1960, California's population doubled every twenty years. Rapid improvements in mining techniques required infrastructure that only companies could offer. Individual miners all but vanished; who in 1848 and '49 would have thought of a way of life called "subsistence mining"?

The money to be made was no longer in the ground or the riverbeds. Gold might be found just around the bend and might not be, but miners needed pants either way. A Bavarian immigrant named Levi Strauss began providing them and the rest is blue-jeans history. Merchants and shopkeepers made a buck regardless of what the next pan bestowed. A miner's wife could make more money doing laundry than her husband did in the streams. Knives, pickets, tents, blankets, boots, beef, barley, brandy, rum—these were the staples peddled by entrepreneurs like Brannan. He bought a pan for 30 cents, then sold it for $15 (or, today, close to $400). Brannan became California's first millionaire—and didn't have to discover a thing; his gift was exquisite timing. Banks, with plenty of gold already in their coffers, were happy to finance the new mining ventures that needed vast capital. The "Big Four" who built the Central Pacific Railroad in the 1860s—Charles Crocker, Mark Hopkins, Collis Huntington, and Leland Stanford—were Sacramento merchants who started out selling supplies to the miners. It was Stanford who, in memory of his son, would found the university that became the wellspring of talent in Silicon Valley.

The Rush ended in the 1850s, after 24.3 million ounces had been taken out of the earth. But the state's genetic code had been irrevocably altered. This was a place to find one's fortune. Other entrepreneurial booms followed: first, the silver of the Comstock Lode; then agriculture, the railroads, oil, real estate, motion pictures, the aerospace industry; and, finally, Silicon Valley. No matter that James Marshall and John Sutter died broke, more than three decades after their discovery. Every year, even now, prospectors come, panning and pining for a little piece of the boundless dream.

The Gold Rush happened because of luck—and the vicissitudes of tectonics. So, too, in its own way, did Silicon Valley, never resting on solid ground.

At the beginning of the Jurassic period, 180 million years ago, the oceanic crust of the Pacific began to spread eastward. Bumping into the North American coast, it swept under the continent in a geologic process called subduction. Back then, there was no San Francisco, no San Jose, no Silicon Valley—these areas were underwater. Sacramento was the seacoast, a beach for the dinosaurs. Cracks within the ocean crust contained water, and as the crust dove one hundred miles deep, it heated rocks to their melting point. The resulting magma rose to the surface, heating other material along the way and fueling the great volcanoes that covered the landscape of the modern Sierra Nevada.

Volcanic activity stretched from the Andes in South America to the Sierra Madre in Mexico and into the cordillera of the United States. Within the rocks caused by volcanic eruptions were such elements as magnesium, iron, zinc, lead, silver, platinum—and gold. As the magmatic fluid cooled, the various elements congealed, with some deposited in large veins of rocks and others in the minute interstices of other minerals. Many of the large veins—the Mother Lode—eroded and were swept away down rivers toward the plains of the Sacramento Valley. Because gold is so much denser than other minerals, it's the first to settle out of a turbulent river. Dense flakes of gold accumulated in pockets along the floor of streams. These are the deposits that the miners panned for in riverbeds, sandbars, and the tailrace at Sutter's Mill. The gold had been there for the taking for eons, but the native populations attached little value to it.

At the same time the Pacific and North American plates were conspiring to make volcanoes and create the Mother Lode, they were building California. The colliding plates brought in massive amounts of oceanic debris—crustal flotsam and jetsam—that became attached to the western flank of the continent, as desperate castaways might to a life buoy. Wind and rain loosened additional material from the continent itself and this material wound its way to the oceanfront, piling up as mineral detritus. In concert, these two processes doubled the size of central California, extending the coastline to its current location.

Subduction, volcanism, accretion—this medley of geological M.O.'s continued over the course of 150 million years—"assembling California," as the writer John McPhee describes it. Only then did a new era of upheaval commence; geology, so it seems, abhors telling the same old story. Instead of continuing great collisions, the Pacific and North American plates started to grind laterally past each other. This sliding motion mimicked a zipper, opening first in Southern California 12 million years ago and proceeding northward an inch or two a year. Every now and then, the zipper zipped and an earthquake happened. It still does. The zipper—where the crustal slabs meet and pass each other by—is called the San Andreas Fault; in time, the city of Los Angeles, sitting on the Pacific side of the zipper, will march right by San Francisco on the North American side. It is the San Andreas that's responsible for creating the chief boundaries of the peninsula containing Silicon Valley.

To the west are the Santa Cruz Mountains; to the east, beyond the lower part of San Francisco Bay, which peters out at San Jose, is the Diablo Range. Both chains, running parallel in a roughly northwest-to-southeast direction, are the product of the plates crunching away. That compression periodically generates upthrusts and wrinkles—mountains by any other name. In a large earthquake, such as Loma Prieta in 1989, a mountain could grow by two feet in an instant. It's true enough that the waters of San Francisco Bay currently bound the eastern side of Silicon Valley, and the Diablo Range is then beyond. But that's just coincidence. As recently as five thousand years ago—when the continental Ice Shield was still large—the Bay was dry, and camels and mastodons roamed its floor. The modern San Francisco Bay is merely an extension of Silicon Valley that happens to be wet this geologic epoch. In another time, if the continental glaciers of Antarctica were to melt, sea level would rise and Silicon Valley itself would be underwater. Here today, gone tomorrow.

Geology surely was destiny in the Sierra Nevada; without it, there would've been no Gold Rush. But Silicon Valley? Its location has no strategic value. Gold came from the land itself, just as oil and soil and water were physical gifts. The phenomenon of the Valley isn't tied to any such thing—indeed, silicon is Earth's second-most abundant element—resting instead on intellectual resources. The newest frontier isn't in a mine, but in a more ethereal place called cyberspace. How so, then, the significance of geology?

The answer lies, perhaps, in another analogy. Over the last half-century, Silicon Valley has managed to create its own critical mass of ingredients for entrepreneurial explosions. Academics, financiers, lawyers, scientists, engineers, even the low-tech rabble who run the restaurants—all find themselves bouncing around hyperkinetically in a relatively small area. There are thousands of companies squashed into a small space, like so many transistors on a chip. The total population of 2.3 million ranks larger than thirty-two states; more than a million people work in an area of fifteen hundred square miles and there's not much room to expand. The mountains are utterly inhospitable to development, filled with ravines, covered by poison oak, prone to burn, laced with earthquake faults, and susceptible to slides. The city of San Francisco constitutes a terminus to the north—some folks in the Valley now derisively refer to it as a bedroom community, where Old Money is a relative trifle, where a scion like Will Hearst gave up the family's publishing business to become a venture capitalist in the Valley—and San Jose is the nexus in the south. The center of gravity is neither city, but a force field in between.

In theory, the Valley corridor could keep growing out over the Golden Gate or into Oakland or in the more open spaces beyond San Jose. And there are outposts, notably companies like George Lucas's special-effects empire (*Star Wars*) and Steve Jobs's Pixar Animation Studio (*Toy Story*), both a short drive north of San Francisco. But most start-ups incubate within the confines of Santa Clara and San Mateo counties in flatland towns like Palo Alto, Redwood City, Menlo Park, Cupertino, Milpitas, Sunnyvale, and Mountain View (a town that has neither), the high office rents notwithstanding. Major companies like Apple and Intel and Hewlett-Packard have always been located here, exerting a gravitational pull on newer businesses. The venture capitalists are along *one road,* afraid to be away from the action.

For all the talk of the virtual office and telecommuting, being *in* the Valley is vital. Five states—California not among them—rank high-tech as the leading private-sector employer, but nobody thinks of them as anything other than wannabes. In states like Oregon and Vermont, there are more software engineers than plaid-shirted lumberjacks. Silicon Fen, Silicon Glen, Silicon Tundra, Silicon Snowbank, Silicon Wadi, Silicon Dominion, Silicon Gulch, Silicon Plantation, Silicon Bog, Silicon Prairie, Silicon Mesa, Silicon Plateau—the names sprouting up worldwide are ridiculous, as if the nicknames create the culture. Bob Metcalfe, founder of 3Com and an industry pundit, likes to say Silicon Valley itself is "the only place on Earth not trying to figure out how to become Silicon Valley."

Four hundred miles down the freeway from the Valley, Southern California has more high-tech start-up companies in a given year, yet no one would mistake it for Silicon Valley. (Witness L.A.'s earnest attempt to market itself as the "Digital Coast"—Silicon Beach and Silicon Alley were taken, and they lacked the courage to go for HollyWeb or Mo Betta Media—leading detractors to call it "Digital Toast.") The Silicon Valley locals will repeatedly tell you they're laid-back, but the chatter at lunch every day isn't about sprouts. Computer code, business plans, who Microsoft just did in—this is the gossip that undergirds the place. If Larry Ellison has a new girlfriend or Steve Jobs had any friends, word would be around in a Menlo Park minute. It is in this round-the-clock buzz on the periphery of the Valley where much of the action is. The Valley is the ultimate exercise in networking. Where you live, where you eat, where you work—whether or not you get rich may be decided by these variables as much as by your great idea. You can never have too many breakfasts at Il Fornaio, no matter how much the cappuccino costs.

For all the talk of a global economy and multicultural labor pool, Silicon Valley works because it's small and incestuous—the Hollywood of the North, an island of sorts that has allowed an entrepreneurial species to evolve. If it weren't for maddening round-the-clock traffic jams on the main drag, Highway 101 (I wrote part of this chapter between two exits), you might hear the locals sweetly whistling "It's a Small World, After All." In this tightly packed terrain, overrun by ugly, low-slung tilt-ups, it's no wonder that so much business gets done. Everybody's a professional, because the factory workers who actually assemble the chips and package the software all live in the Far East, where labor costs twenty-five cents an hour, not twenty-five dollars. If you mix together obsessed engineers and venture capitalists flush with billions to invest, along with real-estate developers, marketers, and public-relations specialists, how can this Brownian motion *not* help but form a new company or two? (In 1999, the CEOs of @Home, a high-speed Internet service, and Excite, a popular Internet site, supposedly concluded their merger in a parking lot between their buildings in Redwood City.) Nobody's in charge, except rampant capitalism. "In any restaurant today, you can't swing a cat and not hit an entrepreneur," says Al Alcorn, who cofounded Atari in the 1970s.

Greed obviously fosters intense competition. But the players in the various fiefdoms—hardware, semiconductors, computer networking, software, and Internet commerce—are smart enough to know that cooperation has its advantages. Notions are continually shared, teams formed, favors exchanged (in return for some preferred stock down the road). The traditional "vertical integration" of companies just doesn't exist in the Valley. Instead, it's a "social architecture" of interacting competitors, according to AnnaLee Saxenian, a Berkeley professor of urban planning who wrote a book, *Regional Advantage,* arguing that the monolithic nature of the high-tech companies along Boston's Route 128 structurally doomed them. The Valley's ecosystem prizes time, which, in the Age of the Internet, is the most precious commodity. Geology and geography didn't predestine the economic fecundity of Silicon Valley—it didn't have to happen here and Microsoft, for example, did not—but they sure helped. The Valley is a small place, where everybody knows each other and everybody's keeping score.

The physical underpinnings and surroundings of the Valley mean something else. Much of the geological narrative has already been told, but tectonic drama remains. As a chunk of real estate, the Valley is remarkably unstable. While that makes owning a house next to the San Andreas Fault an adventure, it also produces a different outlook on life. Instability and change pervade one's existence. Taking risks becomes commonplace. Ask David Howell of the U.S. Geological Survey in Menlo Park, a kept government operation and bureaucratic jungle in this land of a thousand start-ups. Howell's been with the USGS for a quarter-century and is known as "Dr. Mud," for his expertise on landslides. His office is filled with colored maps and charts illustrating the tectonic havoc lying in wait throughout Northern California. The dangers notwithstanding, he savors its arrival.

I went to visit Howell one afternoon simply to learn about the land on which Silicon Valley rests. I wound up getting an education on Valley subculture, of which he is more observer than participant. As much as ancient rocks and crustal plates, "geopsychology" fascinates Howell. He talks philosophy more than science. "The earth affects people differently here," he says. "Maybe people who come to the Valley are predisposed to be a certain way or maybe the land does something—it's a chicken-and-egg thing. But you live your life here with no sense of permanence. In the East, the Appalachians are bedrock, you were rooted in them, secure. Out here, you wake up every day knowing the whole thing could come crashing down in a moment. It produces anxiety, but also a subliminal force that drives people to keep overachieving."

Howell's a geologist, but he understands the shaky psyche of Silicon Valley as well as anybody: Get it today, because it may be gone tomorrow. And if you happen to fall down, get up and start over. Failure exists on as grand a scale as success: Marshall and Sutter proved that. Howell even has a few ideas for starting a private company someday. Maybe we could set him up with Roger Bamford.

Chapter II

Genesis

Silicon Valley doesn't care much for its own history, high-tech or otherwise. It should.

In a place that was once called the "Valley of the Heart's Delight"—full of orchards, canneries, and drying sheds—the Santa Clara County Fair is being taken over by the techno-weenies. There used to be pigs and goats and clowns. Cow milking used to suffice. But now the fair tries to lure the nerds with modem yodeling (as in birdcalling) and a Bill Gates look-alike contest. The fair's 1998 theme was "Hayrides and Hard Drives" and its logo sported a bespectacled carrot using a cell phone. In Sunnyvale, they've actually protected the fair as an historical site, perhaps so parents can show their kids what an apple tree looks like.

Instead of history, the Valley prefers to bask in its most recent triumph or the one just round the bend. The past may connote a legacy and experience, but it also turns up mistakes. Nobody in the Valley likes to admit a blunder. In the great semiconductor bust of the mid-1980s, headlines proclaimed the place the Valley of Death, but you won't find a hint of that in the advertising-rich sections of the San Jose *Mercury News*, which bills itself as "the newspaper of Silicon Valley."

The second-most-popular canon in the Valley—"We tolerate failure" (following right behind "I'm not in it for the money")—flows from an entrepreneurial spirit that reveres risk-taking. But it also comes from the fact that nobody much remembers what happened before—if you don't recall the failures, what's not to tolerate? John Doerr, for instance, is always happy to sing the praises of his venture capital firm's lucrative investments in Netscape or Amazon.com. But mention Dynabook, a laptop computer company from the late 1980s that resulted in gargantuan losses and an embarrassing front-page profile in *The Wall Street Journal*, and Doerr will barely blink. Some of that is disingenuous public-relations shtick, part is the Valley's wonderful ability to forget.

In downtown Palo Alto, at the corner of Channing and Emerson, is a large plaque cemented in sidewalk stone. Chipped and decaying, it reads:

> ELECTRONIC RESEARCH LABORATORY
> Original site of the laboratory and factory of Federal Telegraph Company, founded in 1909 by Cyril F. Elwell. Here, with two assistants, Dr. Lee de Forest, inventor of the three-element radio vacuum tube, devised in 1911–13 the first vacuum tube amplifier and oscillator.

In a neighborhood of electronics bazaars, coffee bars, and a hardware store that sells "private label" Acme Doggie Biscuit Mix for $6.95 ("handsomely boxed, complete with bone-shaped cookie cutter!"), the plaque is about as close as the Valley gets to honoring its past. Between all that shopping and the eternal search for parking, who's got time to notice some history?

Lee de Forest, big-eared and always tinkering, was one of the Valley's first eccentrics and a relentless self-promoter. To celebrate his seventieth birthday in 1943, he went to the Sierras and climbed Mount Whitney, tallest peak in the lower forty-eight at 14,494 feet. Just before undergoing cancer surgery as an old man, he overheard a physician mention that they would be removing his tumor by a process called electrodesiccation. "Commonly known as the hot wire," he said from the O.R. table, "I invented it in 1907." De Forest worked under trying conditions—he lived with his mother and for a time was on bail for stock fraud for telling investors it would be possible someday to transmit the human voice across the Atlantic (he was acquitted). But it was de Forest's vacuum-tube amplifier—resembling large incandescent lightbulbs with electrical parts inside—that gave birth to the electronics age. By making it possible for the first time to regulate electric current without a mechanical switch, de Forest's "audion" facilitated the development of modern radio communication, long-distance telephony, radar, and television—despite a telephone executive once telling him, "You could put in this room, de Forest, all the radio telephone apparatus that the country will ever need." Not until the discovery of the transistor nearly four decades later was the importance of the vacuum-tube amplifier transcended. Palo Alto, where Stanford University was just a fledgling, became a center of experimentation in radio.

The Federal Telegraph Company was the area's first start-up, and unless you liked growing fruit, it was the most interesting place to work. Able to staff itself with the best of Stanford's electrical engineering graduates, Federal provided communications for the Navy during World War I. Stanford's first president, David Starr Jordan, helped to start it with a $500 investment. Two early employees quit to form another company—the first in a chain reaction of corporate defections that continues to the present era. This new company invented the loudspeaker and came to be known as Magnavox.

It was also in the Valley that the first continuous radio broadcasts began in 1909. *San Jose Calling* was the forerunner of today's KCBS in the Bay Area. One of the listeners was Frederick Terman, the first of so many Westerners and Midwesterners who would create the Valley, and the son of a Stanford psychology professor. As Tom Wolfe later pointed out, he was part of a rich intellectual lineage of small-town pio-engineers whose very distance from the conservative ways of the Atlantic seacoast seemed to nourish their imaginations. Thomas Edison before him came from Port Huron, Michigan. Thereafter, the twentieth century produced such men as de Forest, from Council Bluffs, Iowa; William Shockley, from Palo Alto; John Bardeen, from Madison, Wisconsin; Walter Brattain, from Tonasket, Washington; Jack Kilby from Great Bend, Kansas; Lester Hogan from Great Falls, Montana; Gordon Moore from Pescadero, California; and Robert Noyce, from Denmark, Iowa. These were technologists as much as scientists.

Terman, born in 1900, came from strong academic stock. His father, Lewis, developed the famed Stanford-

Binet IQ test, and Terman raced through grade school in half the usual time. Like other boys in Palo Alto, he caught the radio bug, and with the son of Herbert Hoover, he built an amateur radio transmitter. Later, working occasionally at Federal Telegraph, Terman studied chemistry at Stanford and obtained his degree in 1920. He then left for MIT for a Ph.D. in electrical engineering under Vannevar Bush, who went on to run the Office of Scientific Research and Development in World War II; it was pretty much accepted that those in the backwater West went east for professional training, just as those in the East often went to Europe. Terman might have stayed in Cambridge, but while home on the West Coast he contracted tuberculosis and had to stay. The dean of engineering at Stanford hired him in 1924 to run a new radio communications lab, housed in an attic on campus. Within a few years, he wrote the leading textbook on "radio engineering," before "electronics" was even in the lexicon.

Leading little of the high life and taking little advantage of Northern California's sun and fun, Terman was nothing like de Forest. Stiff and severe, he worked Sundays, never took a vacation, and considered bridge to be a walk on the wild side. In some ways, he was progenitor not just of many modern-day businesses, but the image of today's propeller head (at least until the stock options come in). Professor Terman didn't wear T-shirts or flip-flops, and with his horn-rimmed glasses and social awkwardness, he never was mistaken for a movie star.

As good a scholar as he might have been, Terman's faculty was as talent scout and electronics evangelist. Drawing from his own practical experience at Federal Telegraph, as well as from the example of MIT's ties with such inventors as Thomas Edison and Alexander Graham Bell, Terman encouraged students to work for nearby companies and to consider starting their own, even during the Depression. As part of his courses, Terman took his students to visit start-ups like Philo Farnsworth's television lab in San Francisco. "As you can see, most of these successful radio firms were built by people without much education," Terman told his tuition-paying students, without a trace of irony. After World War II, he brought in massive amounts of federally funded projects and lobbied for top salaries for the engineering faculty, "creating a modern community of technical scholars" who roamed freely between the ivory tower and industry workshop. Terman called his philosophy of recruiting stars "building steeples of excellence." In so doing, he made Stanford the incubator of high-tech talent it remains today—an essential ingredient in the Valley's critical mass. The university's scientists knew that any experiment might someday become a gold mine. In one stretch of the late 1970s and early 1980s, three different Valley high-tech gi-ants—Sun, Cisco, and Silicon Graphics, Inc.—were all being spawned in Margaret Jacks Hall (an engineering building named for a cheese heiress whose family invented "Monterey Jack"). Silicon Graphics' foun-der, Jim Clark, became the greatest entrepreneur of his time, finding gold again in 1995 with Netscape. Cisco made the high-tech plumbing that links up computer networks, which would be so critical for the Internet. Sun, short for "Stanford University Network," made extremely powerful desktop computers known as "workstations." The University of California at Berkeley—up in the East Bay, across the bridge from San Francisco—could never match Stanford in start-up scoring.

Terman chafed at the notion that so many of his best graduates went into "exile in the East," where the jobs were. For the next quarter-century, before becoming engineering dean and provost and acting pres-ident of Stanford, Terman prodded and nurtured a generation of engineers, whom he described as "electronic nuts, these young men who show as much interest in vacuum tubes . . . as in girls." Though more recent entrepreneurs and scientists may be better known, Fred Terman is still regarded as the "Father of Silicon Valley." His two most suc-cessful offspring founded a company and an institution called Hewlett-Packard.

In the spring of 1933, Terman invited a six-foot-five Stan-ford junior named David Packard to take his radio-engineering graduate course. Packard was hardly an electronics prodigy-in-waiting: Besides being a ham radio enthusiast, he played end on the varsity football team and served chow at his fraternity. But Packard also spent time at the university's amateur radio station and that's how Terman knew him; both were "hams," in radio parlance. In the course of his campus activities,

Packard befriended Bill Hewlett, another student. Among other things, both had shared a childhood interest in explosives; Packard's mangled left thumb was the result of a failed experiment and a sloppy surgeon.

They had met during freshman year and taken many of the same math and science classes. Packard, born in Pueblo, Colorado, had already decided to be an electrical engineer; Hewlett, the son of a San Francisco physician, wasn't so sure, but said he chose electrical engineering because he liked model trains. "I liked electricity," he explained elegantly enough. During their senior year, relying on Terman's constant encouragement, Hewlett and Packard talked about launching some kind of radio business "someday."

But Packard was offered a job at General Electric in upstate New York and took it. Hewlett went off to MIT (where he often took the long train ride to visit Packard), before returning to Palo Alto in 1936. Terman helped Hewlett get a job making equipment for a local M.D. At G.E., Packard initially worked in the refrigeration products division and grew bored very quickly. In the summer of 1937, on a visit back to California, Packard got together with Hewlett in what they later described as HP's first "official" meeting. They talked about high-frequency receivers, medical devices, and television. The following summer, Terman arranged a graduate fellowship at Stanford for Packard, who quit G.E. and drove west with his new bride, Lucile, and the used Sears Roebuck drill press that would be HP's first piece of equipment.

Reunited, Packard and Hewlett returned to their plan for a business. Hewlett had found a two-story house in Palo Alto for the Packards to rent, in a vintage part of town that came to be known as "Professorville." Hewlett himself lived in a bungalow on the same property. The two engineers decided to use the one-car unpainted garage as a workshop, giving birth to the legend of the high-tech "garage start-up." Half a century later, 367 Addison Avenue—just down the road from the de Forest plaque—was declared a state landmark, the "Birthplace of Silicon Valley," which isn't quite right, of course, because neither Hewlett nor Packard had a clue about silicon.

On January 1, 1939, Packard and Hewlett formally established their partnership. They flipped a coin to determine whose name would come first on the letterhead. Packard lost. The company had no master plan and $538 in the bank. "Our original idea was to take what we could get in terms of an order," Hewlett recalled years later. It was still 1939 and times were hard economically. Most of their jobs were lone contracts—like making a foul-line signal for a bowling alley or custom controls for air conditioners. It was Hewlett's idea to take the audio oscillator he had earlier designed and try selling it nationally. The oscillator was hardly the thunderbolt that Edison or even de Forest unleashed. But it did represent the first efficient way of testing the kind and range of audio frequencies needed in the emerging field of electronics. "Model 200A"—so designated, according to Packard, "because we thought the name would make us look like we'd been around for a while"—was featured in a sales brochure and sent to several dozen potential customers that Terman recommended.

One man wanted to use the oscillator to produce a tuner for harmonicas, but it didn't work. But luck struck next time: A young studio in Southern California run by Walt Disney placed an order for eight HP 200B oscillators—at $71.50 apiece—to be used in sound production for an experimental animated film called *Fantasia*. The audio oscillator became HP's inaugural electronics product, and some version of it would remain on the market until 1985. Hewlett made them, Packard got them to the customers. Said Hewlett: "He was the entrepreneur and I did the work." Their collaboration thrived exactly because their skills interlocked. Longtime Valley journalist Michael Malone described Packard's handshake as "huge and gentle," whereas Hewlett's was "small and firm." But the two got along. In the photograph for the annual report to shareholders, the Bunyanesque Packard was invariably shot sitting down, so he looked nearly the same size as Hewlett, who in reality didn't reach his shoulder. Founders don't always mesh so well. A generation later, for example, Steve Jobs and Steve Wozniak—cofounders of Apple who worked at HP as lads—became famous for their personal disagreements, which reached such a level that they now barely talk to each other anymore.

HP's only early rival was General Radio of Cambridge, Massachusetts. But General Radio's CEO seemed to think competition would help his business and, according to Packard, he actively encouraged HP. Without such benevolence, HP might not have had such an auspicious start. Terman, who invested his own money in his "boys" and wound up serving forty years on HP's board of directors, delighted in explaining how he could tell how his "Bill and Dave" were doing. "If the car was in the garage, there was no backlog. But if the car was parked in the driveway, business was good. They had some work soldering, wiring, painting, you name it."

HP soon introduced a line of instruments to measure audio frequencies, which would sustain it well before the advent of computers and printers and programmable pocket calculators. At the end of 1939—its first year in business—the company had turned a profit of $1,563, not bad considering this represented a 29 percent margin. There would never be an unprofitable year. By the end of the war, the company had more than a hundred employees and sales approaching $1 million. It didn't go public until 1957, when the two founders truly struck it rich in stock; for a time, Packard was one of the three wealthiest people in the country. By the 1990s, with Packard and Hewlett retired, HP was still a leader in electronics. The company's annual revenues surpassed $40 billion and it employed more than 120,000. More radical than a great bottom line, though, was *how* Hewlett-Packard did business. The "HP Way," as it came to be known, represented a kind of institutional morality unknown in American business culture, including at IBM.

Hewlett and Packard trusted their employees, respected them. The company tried to avoid hierarchy. "Bill and Dave" called their workers by their first names and expected the same in return. Money mattered, but not at any cost. HP assumed that each worker was worthy unless shown otherwise; according to the old company joke, the only way to get fired involved a revolver and your boss. In the early 1970s, during one of the periodic recessions to hit Silicon Valley, HP avoided layoffs by cutting pay 10 percent across the board and reducing the workweek by every other Friday—the "nine-day fortnight," as the phrase went. Managers didn't bark orders from afar, clueless about what was really going on; they were present in the trenches. "Management by walking around" was the credo. Goals were set and individuals were left to do their jobs. HP instituted profit-sharing long before it was fashionable in the Valley. Until the company grew too big, the cofounders would hand out the booty themselves at the annual Christmas party.

Treating employees right seemed to come from utilitarian rather than utopian ideals, but the result was the same. Employees were loyal; they dubbed the place a "country club." Hewlett and Packard made few enemies and were hailed as beacons of stability. And HP routinely ranked at or near the top of every pop-list of "Most Admired Companies in America." It still does, sixty years after its founding. The founders' philanthropic example is unmatched in Silicon Valley: Between them, Hewlett and Packard donated more than $300 million to Stanford, including funds for the main engineering building, which is named for Terman. That total, in inflation-adjusted dollars, rivals the founding bequest of Leland and Jane Stanford themselves. Packard bankrolled the Monterey Bay Aquarium, a children's hospital honoring his wife, and a host of causes ranging from population control to education to open spaces. On Packard's death in 1996, the David and Lucile Packard Foundation received most of his fortune and overnight turned into the second wealthiest private charity ever. Its assets of $10 billion or so exceed even those of the Rockefeller and Getty Foundations.

About the only downside to all the decency and teamwork at HP was that individual achievement didn't easily percolate up. Acres of egalitarian cubicles for virtually every employee except Hewlett and Packard—the early version of the mazelike "cube farms" so popular in the Valley today—mute ego. So does any monolithic corporate culture. Unfortunately, it also tends to stifle the kind of ambition that drives entrepreneurs. Jobs and Wozniak, after all, didn't stay at HP long. It's probably the case that a suit-and-tie environment like HP can't have it both ways, but the price HP paid for its paternalism is that it never quite had the impact on the Valley it might have. In the end, "most admired" doesn't necessarily mean "most important."

That accidental legacy belongs to William Shockley—egotist, eugenicist, Nobel laureate—one of the Valley's smartest, most vilified characters. As venerated as Hewlett and Packard were, it was Shockley who ignited the postwar scientific revolution and brought silicon to town. The transistor he helped discover turned out to be both a theoretical marvel and the guts of a new machine.

Every fifty years or so, a great discovery or invention comes along (and that's not including baseball's designated-hitter rule): electricity, the steam engine, automobiles, then the transistor that Shockley helped make. It's hard to think of an important device or company of the latter half of the century that didn't flow from the transistor.

Shockley grew up in a splendid Victorian
just blocks away from Federal Telegraph and the HP garage at 367
Addison. His father, a successful British engineer, came to the United
States to mine gold, though he kept London as a base of operations.
Shockley was born there in 1910. When he was three, the family moved
to Palo Alto, where Shockley's mother had attended Stanford. Little Billy
was quite an armful: At times he could just be mischievous—like when
he set up a hidden switch under the living room rug so that, every time
someone stepped on it, mysterious chimes would ring in the attic. But
in addition to collecting strange pets, playing with an imaginary train
set, digging up the backyard in search of treasure, and alienating most
of the other children in the neighborhood, Shockley was prone to intense
rage. He would turn out to be an only child.

His parents, usually the target of his outbursts, recognized that he was
special and they knew public school wouldn't be able to contain him.
They tried to educate him at home, but wound up enrolling him at the
Palo Alto Military Academy, where he first heard of radio and other
current scientific developments. After the family moved again, to Los
Angeles, Shockley enrolled at Hollywood High and then headed off to
study physics at Caltech and MIT, where for some reason he arrived with
a pistol in hand. Even when he was a teenager, Shockley's intelligence—
and hauteur—were apparent. "Our age is eminently mechanical," he
wrote in a high-school report in 1928. "We travel from one place to
another at relatively monstrous speeds; we speak to each other over great
distances; and we fight our enemies with amazing efficiency—all by the
aid of mechanical contrivances." He was correct, of course, and in time
he would help invent one of his own that would change the world.

After receiving his doctorate from MIT, Shockley, at twenty-six, joined
Bell Telephone Laboratories, the distinguished AT&T research facility
founded in 1925. Shockley was the first scientist hired after a
Depression-era freeze. To make its service more reliable, Ma Bell needed
a better device than the vacuum tube to boost the current in the network. As groundbreaking as vacuum tubes had been, they were bulky,
fragile, hot as a stove, and quick to burn out. "Nature abhors the vacuum

tube," quipped one of the AT&T engineers, who later coined the word "transistor." Then located in a twelve-story building in lower Manhattan's West Greenwich Village, Bell Labs was assembling a team of physicists to work on a possible successor to the tube. Shockley would become the leader, after directing antisubmarine research for the Navy during World War II. The team's chief focus was semiconductors—seemingly inert crystalline substances like silicon and germanium, which were elemental cousins that acted as both insulators and conductors.

Unlike traditional metallic conductors, these materials had curious traits that for several decades seemed to hold promise for electronic applications. Precisely because of their atomic ambivalence—straddling the fence between being an insulator like glass and a conductor like copper—physicists thought the semiconductors could be manipulated to do electronic tricks. They knew, for example, that adding different impurities to semiconductors ("doping"), and giving them positive or negative charge, would change the direction of electric current as it passed through the materials. Conducting one minute, insulating the next—this was the wonder of a semiconductor (which, presumably, could also have been called a semi-insulator). While various theories in quantum mechanics and even geometry offered hints about how to regulate this phenomenon, no physicist had broken through. It was in this realm of semiconductors—"solid-state" physics—that Shockley devoted his efforts.

Shockley loved being the company hot dog, relishing the attention he got from such stunts as scaling the lab cafeteria's stone walls. Ferret-eyed with a beetle-browed forehead, he would have been a looming presence even without his odious personality. His gifts included the ability to simplify science. Asked to explain amplification, for example, he offered: "If you take a bale of hay and tie it to the tail of a mule and then strike a match and set the bale of hay on fire, and if you then compare the energy expended shortly thereafter by the mule with the energy expended by yourself in the striking of the match, you will understand the concept of amplification." But while he appreciated the cerebral challenges, Shockley craved the limelight and wanted to reap the profits of his intellect. He wasn't shy about his intention of making a million dollars someday and trading up on the black MG convertible he drove to work. Shockley tried various methods of solving the semiconductor puzzle—forcing electrons to flow the right way along defined paths inside a solid block of semiconductor material—but his experiments with silicon never succeeded and he became involved in other solid-state research. Shockley handed the project to two of his younger researchers, John Bardeen and Walter Brattain.

From the fall of 1945 to late 1947, at the sprawling new campus of Bell Labs near the Watchung Mountains in northern New Jersey, they worked at the puzzle. Bardeen was the theoretician, Brattain the tinkerer. Bardeen liked the blackboard, Brattain liked to say he knew how to "wiggle" things "just right." Bardeen made the key intellectual breakthrough: Electric current couldn't pass through a semiconductor because its surface was like a Roach Motel; electrons could check in but they couldn't check out. Bardeen and Brattain had to figure out how to liberate the electrons and thereby amplify the electrical current. After tedious trial and error, some advice from Shockley, and some plain dumb luck, they triumphed. The solution was really quite simple: Electrical flow could be controlled by placing two fine wires on a sliver of semiconductor material in just the right place and using just the right combinations of doped material. When his experiment worked, Brattain exclaimed, "Eureka!" an accidental but fitting reference to another discoverer, Archimedes.

A week later, on a snowy December 23, 1947, Bardeen, Brattain, and Shockley demonstrated the device to Bell Lab executives. It was a small, inelegant contraption of germanium, batteries, a triangular wedge of plastic, a strip of gold foil, and a spring fashioned from an uncoiled paper clip. Jury-rigged on a workbench, the device was hooked up to a microphone and a headset. Jabbing the two electrodes into the germanium brick, Brattain then spoke into the microphone and his voice boomed into the headset. One by one, each executive listened. What came out was one hundred times greater than what went in. In short, electrical juice was being amplified without use of de Forest's vacuum tube—a perfect illustration of Arthur Clarke's observation, "Any sufficiently advanced technology is indistinguishable from magic."

The "transistor"—the very substructure of the future—had been born, though it took months before it had that name. "Transistor" was short for "transfer resistor." Compared with the vacuum tube, the transistor was faster, sturdier, cooler, and smaller—and it was susceptible to the remarkable miniaturization that became a hallmark of electronics. Eventually, transistors became microscopic—the tiny on-off, lightning-quick electrical switches that run all things digital. In the coming years, materials science would show that silicon was an even better semiconductor than germanium and had the advantage of being so plentiful. While it doesn't exist in nature independently, in other compounds and in dioxide form, silicon constitutes 90 percent of the earth's crust and is the principal ingredient of sand. Once engineers developed efficient ways to manufacture it with just the right mix of impurities, silicon became the chief semiconductor for transistors. This was a good thing for Northern California, lest it become the home someday to Germanium Valley.

Within a year of the Bell Labs discovery, the transistor was introduced for commercial use. In 1954, nearly a million were being sold to such companies as G.E. and RCA and Texas Instruments, as well as AT&T; Raytheon alone sold hundreds of thousands of transistorized hearing aids. Just in time for Christmas 1954, there was the Regency TR1 transistor radio, powered by batteries and small enough to fit in a pocket. At $49.95 (close to $350 today), it wasn't cheap, but Regency sold 100,000 of them the first year—the Nintendo Gameboy of its day—and made "transistor" part of retail. Then a Japanese start-up called Tokyo Tsushin Kogyo came along, later changing its name to Sony and dominating the market for TVs and radios. This was just the beginning. Transistors became elemental to the digital age, the invisible building blocks of anything electronic. Three hundred million billion of them allowed the development not only of rock guitars and a trip to the moon, but every computer on the planet today.

Few scientific achievements this century were as momentous. The silicon transistor would lead to the creation of companies like Apple, Intel, Oracle, Netscape, Yahoo, and a little start-up far away from the Valley named Microsoft. "My first stop on any time-travel expedition," Bill Gates once said, with rare humility, "would be Bell Labs in December 1947." (No word on whether his second stop would have been his own arrival in 1955.)

The public wouldn't learn of the transistor until six months after its discovery, and then just barely. At a large press conference in Manhattan on June 30, 1948, Bell Labs announced the invention—this "little bitty thing," as Ralph Brown, the lab's research director, described it. "We have called it the transistor," Brown explained, slowly spelling the word out, "because it is a resistor or semiconductor device which can amplify electrical signals as they are transferred through it from input to output terminals. [But] it has no vacuum, no filament, no glass tube. It is composed entirely of cold, solid substances." But the reporters there evidently were not intrigued. The next day, as subway fare went from a nickel to a dime, *The New York Times* buried the announcement on page 46 in "The News of Radio," after an item about CBS's summer replacements for its regular *Radio Theatre; Our Miss Brooks* would star Eve Arden as "a schoolteacher who encounters a variety of adventures." Lee de Forest understood the moment better. Invited to the Bell Labs campus for a demonstration of the transistor, he sent regrets, noting that he wouldn't be able to attend the "wake of my 42-year-old infant, the audion."

Shockley, too, appreciated what he had helped bring to life, all the more so at Bell Labs (now part of Lucent Technologies). "Hearing speech amplified by the transistor," he wrote nearly thirty years later, "was in the tradition of Alexander Graham Bell's famous 'Mr. Watson, come here, I want you.' " Nonetheless, Shockley was seething, because he knew Bardeen and Brattain were the real creators of the transistor. Shockley wasn't even present when the other two first got the transistor to work in the lab, a fact that forever haunted him. "My elation with the group's success was tempered by not being one of the inventors," he wrote. "I experienced frustration that my personal efforts, started more than eight years before, had not resulted in a significant inventive contribution of my own." Bell Labs' first transistor patent didn't include Shockley's name. He became obsessed with rectifying that, making himself a worthy member of the electronics trinity. On New Year's Eve, 1947, rather than bask in the discovery of the week before, he toiled away on diagrams for a refined transistor that would use a "sandwich" of differently charged semiconductor materials. This "junction transistor" eliminated the less reliable "point contacts" of the Bardeen-Brattain creation.

More than Bardeen or Brattain or anyone else at the lab, Shockley understood what the transistor might hold for the future, opening up new fields of electronics rather than merely helping Ma Bell. Ever the showman, he liked to hand out transistors at speaking engagements. Or, after glowing opening remarks by a host, he'd tell his audience that the only time he got a better introduction was when he delivered it himself, upon which a bouquet of red roses would magically appear in his palm. Shockley had the prescience to think about the marketplace. "There has recently been a great deal of thought spent on electronic brains or computing machines," Shockley said in a 1949 interview on G.E.'s *Science Forum* radio program. "It seems to me that in these robotic brains the transistor is the ideal nerve cell."

He was talking about computers. Postwar computers were maintenance nightmares that took up more space than a hippo convention. The University of Pennsylvania's ENIAC (Electronic Numerical Integrator and Computer) in 1946 was the first attempt at a large-scale digital computer that could help the military, federal government, and major corporations. But it was a capricious, thirty-ton beast. While the machine could perform calculations, it sucked up 150,000 watts of juice, was a morass of wire, and had 18,000 vacuum tubes that never all seemed to work at the same time. The warmth and light of ENIAC's tubes presented another problem: Moths liked them and would trigger short circuits. (Henceforth, "computer bug" meant a problem inside and "debugging" meant fixing it.) "Computers in the future," predicted *Popular Mechanics* in 1949, "may have only 1,000 vacuum tubes and perhaps only weigh one-and-a-half tons"—just one of many breathless prognostications that turned out to be wildly wrong, underestimating what the transistor would create.

Before the Bardeen-Brattain transistor was even disclosed to the public, Shockley was well along with his own new design. In time, the new "junction transistor" proved superior. But even though Bell Labs gave him his break back in 1936, it would not be the place to keep Shockley satisfied. Throughout the early 1950s, he was angry about not rising into senior management, not that anybody could blame the lab. Bardeen and Brattain detested him and he drove both away. (Bardeen left in 1951 to teach at the University of Illinois, going on to win another Nobel Prize in physics, for research on superconductors. Brattain switched to another department within Bell Labs.)

Shockley was also furious that he didn't share in the royalties from the patents he had helped win. In 1954, he started to prowl for a new job at another company or a university. But either the work or pay wasn't good enough. By 1955, he decided to venture out on his own. "After all," he wrote his soon-to-be second wife, "it is obvious I am smarter, more energetic and understand people better than most of these other folks." He was at least right on two of three.

California is where Shockley longed to return. It was where he grew up, his mother still lived there, it offered him the outdoor recreation he loved. He considered both the Los Angeles and San Francisco areas. The former was the location of Beckman Instruments, a burgeoning $20 million business founded by a former Caltech professor, which would provide Shockley seed money; the latter had Stanford and Fred Terman, now university provost, who aggressively courted Shockley. Terman knew that solid-state physics was the future and he wanted electrical engineering students to have the benefit of a Shockley-driven company in the neighborhood. Terman won out. Shockley chose the Bay Area, a pioneer returning to his boyhood home. In early 1956, he announced the creation of Shockley Semiconductor Laboratory, which would manufacture transistors and other semiconductor devices—and allow him a chance at El Dorado, like his father before him. It would be a spectacular failure, accomplished with breathtaking speed.

The peninsula south of San Francisco was nothing like the place Shockley left thirty-four years earlier. The rise of HP and aftermath of World War II had begun to transmute the region. From 1942 to 1945, the Golden Gate was the funnel for thousands of servicemen off to fight in the Pacific. Even a glimpse of the Bay Area's beauty and weather was enough. After the war, much of this young and skilled armada came back to settle in the West. The orchards and canneries began to give way to office parks and ranch houses, as well as to a manufacturing economy. Terman eagerly courted the government contracts and defense projects that he noted were "the beginnings of a great new era of industrialization." The Valley's critical mass was coming together.

Russell and Sigurd Varian, inventors of the klystron tube—a microwave variant of the vacuum tube that led to airborne radar during the war—set up Varian Associates. IBM, which had opened its first local facility in 1943, established a research lab in San Jose that went on to design the first disk memory for computers. G.E., Sylvania, and West-inghouse also arrived. The Ames Research Center—adjoining Moffett Field in Mountain View and a famous landmark with its enormous wind tunnels—became a leader in aeronautics research and high-speed flight, and, after *Sputnik,* a base for NASA. Lockheed Aircraft moved its mis-siles and space division from Southern California to Sunnyvale, in part to be near Ames, next door to which it built submarines for the Navy. For a long time, stoked by federal defense contracts, Lockheed was the peninsula's largest employer. The father of Steve Wozniak, who would cofound Apple Computer, worked there.

Terman witnessed this industrial explosion and wanted to tie it ever more closely to Stanford. In 1951, the Stanford Industrial Park opened—an experiment in corporate-academic marriages. At 8,180 acres, Stan-ford is one of the largest tracts in the Bay Area. While much of it was (and is) pasture, the bequest from Leland and Jane Stanford bars selling off any land. So the school, always looking to reap additional income from otherwise unused farmland, decided to lease 650 acres on the edge of campus to high-tech—"smokeless" industries, which would work in concert with university departments. When Charles de Gaulle toured California in 1960, according to Michael Malone, he wanted to see two places—Disneyland and the Stanford Industrial Park. Varian Associates took up residence the first year, followed by Eastman Kodak and 150 other companies. In 1956, the year before it went public, HP became the flagship tenant and its far-flung empire remains headquartered there today.

Shockley Semiconductor Lab was supposed to get space in the in-dustrial park, too. At the beginning, though, it leased a run-down cin-derblock shed—formerly a storage facility for apricots—down the road, at 391 San Antonio Road in Mountain View. Shockley had grand designs for his semiconductor start-up and intended to bring together a Dream Team of scientists—those with the highest "mental temperature," as he arrogantly put it. Initially, he invited former colleagues at Bell Labs to join him. While they professed allegiance to the lab and their fondness for the state of New Jersey, they probably also knew Shockley's failings too well. They declined. Shockley then had to expand his search to those who *didn't* know him.

In early January, he contacted a twenty-eight-year-old physicist at

Philco in Philadelphia. His name was Robert Noyce, another Ph.D. from MIT, who was working on the company's attempt to build a better high-frequency transistor. He was Shockley's key hire. Inventive and winning, Noyce had led a charmed life. In the small Iowa town where he grew up, where he mowed lawns and raked leaves, the townsfolk all knew him. His Iowa high-school yearbook tagged him the Quiz Kid, "the guy who had the answers to all the questions." At Grinnell College, Noyce won the Brown Derby Award, given to the student with "the best grades with the least effort." As a senior, he took physics from Grant Gale, who happened to be a university classmate of John Bardeen and was persistent enough to get Bardeen to send him two of Bell Labs' earliest transistors. Not at Stanford, not at MIT—Grinnell College in Iowa offered the first show-and-tell demonstration of solid-state electronics to college students, and it was this coincidence that nudged Noyce to semiconductor research.

Not even a college prank gone awry seemed to taint him. At the end of Noyce's junior year, his dorm friends planned a luau, still a popular fad in the wake of the war in the South Pacific. So, one night, Noyce and a friend stole a suckling pig from a local farmer. It was delicious, but by next morning a contrite Noyce was before a magistrate. On the scales of prairie justice, pig thievery weighed heavily. Noyce avoided prosecution only because of Gale's intervention and only by agreeing to a one-semester banishment. His exile was to the actuarial department of Equitable Life in New York City. Those who called this a trifling punishment never worked for a life insurance company. But since that pignapping could have destroyed his career before it began, the point was well taken. Noyce returned to Grinnell the following spring. Having piled up so many extra credits in prior years, he had no problem graduating on time. And he was the only senior who could carry on an intelligible conversation about term life insurance.

A month before Shockley called him in 1956, Noyce had delivered a technical paper at a Washington conference that impressed Shockley. He told Noyce of his plans to bolt Bell Labs and stake a new claim in California. "It was like picking up the phone and talking to God," Noyce recalled years afterward. "He was absolutely the most important person in semiconductor electronics. Getting that job meant you would definitely be playing in the big leagues." Noyce did and would, and apparently was pretty cocky beforehand. He arrived in San Francisco at 6 A.M. after a cross-country flight. By noon, as Tom Wolfe told it in a memorable *Esquire* profile in 1983, Noyce had found a house to buy and only then headed off to his afternoon audience with Shockley. The eccentricities of the great man notwithstanding, Noyce, the minister's son from Iowa, also was intrigued with the West. "All ions wind up in California, if they meet their dream," he told an interviewer.

In February 1957, Shockley set his sights on another emerging science star, Gordon Moore, a twenty-seven-year-old former pot scrubber and a chemist from Caltech. Though a native Californian, Moore was working on weapons propulsion research for the Applied Physics Lab at Johns Hopkins University in Maryland. He had gone east because he could find no technical jobs in California. If Noyce was ebullience, Moore was nonchalance. Noyce liked a party; Moore liked to fish, so much so that later he once would spend an afternoon on Ireland's most famous golf course . . . with a rod and reel in hand. Moore's biggest adventure as a child had been almost burning a neighbor's house down with a home-made rocket.

Shockley heard about Moore by asking around for the names of young, talented chemists. Moore was born in Pescadero, a coastal village over the mountains from Palo Alto; his father was deputy sheriff of the county and his mother's family ran the only general store. Moore wished to return home, just as Shockley had. Shockley offered him a job and Moore signed on, along with two dozen other ambitious renegades from a range of academia and industry. They included: Jean Hoerni, a multilingual Swiss chemist at Caltech; Gene Kleiner, an industrial engineer and toolmaker at Western Electric; Jay Last, an expert in photo optics from Corning Glass; and Sheldon Roberts, a metallurgist at Dow Chemical. These were science's best-and-brightest, all but two under thirty. Shockley called them "my Ph.D. production line." However, he was less adept at more mundane hiring tasks; David Packard couldn't believe it when Shockley asked him how to find a secretary and where to buy pencils.

Before they were even offered jobs, Moore and Noyce both should have sensed Shockley would be no ordinary boss. Shockley told Noyce (along with Kleiner and several of the others) that he'd have to pass some psychological exams. For a full day in New York City, Noyce was subjected to ink blots and IQ tests. (What would Fred Terman's father have thought?) Moore went through a bunch of questions personally administered by Shockley, stopwatch in hand. Kleiner realized how ridiculous the exercise was and kept telling Shockley how much he loved his mother. Shockley didn't get the joke, but at least he had the sense not to buy into the results too heavily: Noyce was adjudged a wallflower and an unlikely leader, but Shockley took him anyway.

If anyone needed psychological guidance, it was the boss. Shockley was a wizard at physics, but a bozo at management. One of his ideas was to post salaries on a bulletin board. This managed to infuriate both those who were being shortchanged as well as those doing well enough that they didn't want others to know. Between his flip-flop decision making and contempt for people, he had no idea about running a company. It was like Dilbert's boss on crystal meth. Worse, while a place like Bell Labs provided infrastructure and personnel to solve problems, Shockley Semiconductor had no margin for error. Kleiner, for example, was in charge of preparing silicon, but his shop had none of the equipment he needed. "Shockley knew nothing and was of no help," Kleiner remembers, "but he wasn't smart enough to admit it. Yet he would still tell me what to do, so everything I built was a monstrosity. We once had to raise a roof to fit in something he designed and it never worked. I was completely surprised that Shockley knew only about theory."

Others fared even less well. Shockley's managerial incompetence paled in comparison to his abiding paranoia and volcanic temper. He once told Noyce, without irony, that one in ten individuals was psychotic. The engineers were required to give each other report cards. When a secretary cut her hand on a metal object sticking out of a door, Shockley was convinced it was an act of malice. He ordered lie detector tests for staff members he suspected, which led employees to call the lab "391 Paranoid Place." When Sheldon Roberts was questioned, he was reduced to examining the offending metal object under a microscope and proving it was merely a defective thumbtack. When engineers brought in a promising report from the labs, Shockley would make them stand around— while he called somebody back at Bell Labs for a second opinion. Shockley was everything that Hewlett and Packard two miles away were not— untrusting and uninspiring. Noyce and Moore, among others, tried to maneuver Shockley out of operations and make him a consultant.

So it hardly helped when in late 1956 Shockley became an international luminary, gaining the imprimatur of science's greatest honor. On November 1, he learned he would share with Bardeen and Brattain the Nobel Prize, for their invention of the transistor. "We couldn't believe it when he won," Kleiner says. "We hoped it meant maybe he wasn't crazy." The following month, in the darkness of a Stockholm winter, as his wife and his mother looked on, Shockley received his medal from the King of Sweden. At a celebration at Rickey's in Palo Alto, Shockley treated his minions to a champagne breakfast. In a well-known photograph—about the only one documenting his company—Shockley, in open collar and sport jacket at the head of the table, is raising a glass and beaming with pride.

Stockholm may have been a moment of glory, but it was no substitute for revenue. Nearly a year in business, Shockley had yet to sell a single transistor. "He may have won the Nobel Prize for the transistor," Kleiner says, "but he never actually manufactured one." The mutineers' attempt to move him aside ultimately fizzled, mainly because his absentee financial backers failed to grasp how bad the situation was. Moreover, they were reluctant to depose a Nobel laureate. During the summer of 1957, Shockley went to Cape Cod for a vacation. When he returned, he discovered that his critics were following his entrepreneurial example and had decided to quit. On September 18, eight of them were out the door, including the key people of the lab: Noyce (his favorite) and Moore, along with Hoerni, Kleiner, Last, Roberts, Victor Grinich, and Julius Blank, who was known for sending memos folded into paper airplanes. Shockley had no magic left. He couldn't believe it. And who could blame him in 1957, at a time and place where job mobility typically meant nothing more than moving to another office? His disciples were youngish men who deserted the high priest of electronics in the name of ambition. Shockley branded them "the Traitorous Eight," though, in their crew cuts and white short-sleeve dress shirts, they hardly looked the part. Given the source of the remark, says Kleiner, "we took it as a compliment." Shockley never again spoke to most of them; his widow still rails at them.

Their future was to found Silicon Valley's greatest hardware company.

It wasn't just serendipity that these eight men jumped ship at the same time. Shockley's pathology was a unique motivator, but they all had been talking and plotting an escape together. "We lacked a Nobel laureate," Kleiner says, "but we liked each other—and not much anyone else." Initially, they looked for a company that would hire the entire group. During the summer, Kleiner sought advice from his father back east, who went to Hayden Stone & Co., the Wall Street investment house, where he had a brokerage account. The firm sent a senior partner out for a look, who took along with him a thirty-one-year-old Harvard MBA to carry his bags on the eleven-hour flight to the Coast. The MBA's name was Arthur Rock, son of a candy store owner from Rochester, New York. They all met in a hotel room over a weekend. Nobody in New York quite understood what could be so bad about working for a Nobel laureate, but Rock was curious, especially because he had already handled a few electronics deals. Rock was the one who ultimately concluded that the group was too large to be hired en masse; instead, he suggested, they should start a company themselves. "So we did," Kleiner says with as much of a smirk as he can allow, "especially since our last boss didn't turn out to be so good." Moore calls the group "accidental entrepreneurs," the first to merge the money of the East with the electronics frontier of the West.

The venture-capital industry largely didn't exist in 1957. There were a few outfits in Boston and New York—run by families like the Rockefellers and the Whitneys—but nothing on the West Coast. Hayden Stone thought the best way for the rebels to get funding was from a corporation, much as Shockley had obtained it from Beckman. Rock approached twenty-five to thirty potential suitors—"everybody on the New York Stock Exchange who might have heard of semiconductors," Moore says. But nobody seemed interested in funding a business that had nothing to do with theirs—which was the essence of venture capital. Finally, they found someone who could see beyond his own company when Rock talked to the industrialist Sherman Mills Fairchild.

Fairchild, who was IBM's largest individual stockholder and whose father had been an early investor in Big Blue, was a prominent CEO, as

well as a celebrated bachelor. The story goes that he met Jerry Sanders—
one of his marketing executives—while they were dating the same
women. (White-haired Sanders, now the CEO of Advanced Micro Devices and still looking a bit like Colonel Sanders, was immortalized in
the 1980s for being photographed in a terrycloth robe on his giant, ornate
four-poster bed. That at least was better than sporting a pink suit, which
supposedly got him canned from a job when he showed up wearing it at
humorless IBM.) Sherman Fairchild, who had invented such devices as
the aerial camera and hydraulic aircraft brakes, was looking for ways to
diversify his business. His Fairchild Camera and Instrument on Long
Island had gotten rich during the war and was now getting into missile
and satellite systems, which would rely heavily on transistorized devices.
At the time of Rock's first inquiry, the Traitorous Eight was only seven
because Noyce had yet to turn. Fairchild's concern that the group lacked
management experience was quelled when Noyce came aboard as captain. This was the critical last step to getting Fairchild's check.

The deal, struck in a handshake between Rock and Fairchild, was
both simple and unique: In return for a $1.5 million investment from
Fairchild, the Traitorous Eight would create a subsidiary to make semiconductors. They would own a piece of it—even though they each had
to put in only $500 and even though Fairchild agreed to cover any losses.
There was only one caveat: If they succeeded, Fairchild would have the
right to buy them out over a five-year time period at a predetermined
price—beginning at $3 million and building to $5 million. Hayden Stone
and several executives would get a third of the pie, with the remainder
split between the Eight. That meant at least $250,000 each, which was
pretty good considering that their annual salaries were about 5 percent
of that. This meant that the entrepreneurs got a piece of the action, but
only up to the point when the piece got too big. It also worked out to be
a pretty good deal for Sherman Fairchild, which was only fitting because
it was the first venture-capital investment in the Valley.

It was obvious from the start of Fairchild Semiconductor in late 1957
that Noyce—confident and charismatic, a "scientist-cum-charmer," as
one of his marketing people put it—would be the leader. Moore, low-
key and a technical virtuoso, would run research and development and
serve as his confidant. Writing in *Esquire* more than a quarter-century
later, Tom Wolfe himself was enchanted. "Bob had a certain way of

listening and staring," he wrote. "He would lower his head slightly and look up with a gaze that seemed to be about one hundred amperes. While he looked at you, he never blinked and never swallowed. He absorbed everything you said and then answered very levelly in a soft baritone voice and often with a smile that showed off his terrific set of teeth. The stare, the voice, the smile—it was all a bit like the movie persona of the most famous of all Grinnell College's alumni, Gary Cooper. With his strong voice, his athlete's build, and the Gary Cooper manner, Bob Noyce—all of twenty-nine when Fairchild began—projected what psychologists call the halo effect. People with the halo effect seem to know exactly what they're doing and, moreover, make you want to admire them for it. They make you see the halos over their heads."

Because of both personality and job history, Noyce ran a loose ship. Hierarchy was an enemy: There were no reserved parking spots, no mahogany-paneled dining room, no private offices (not even for Noyce), no multilayered array of middle managers, and certainly none of Dr. Shockley's quirks. At sales meetings, they served brownies and whiskey, which probably helped, since early on many on the sales force had no inkling what a transistor actually was. Late nights and early mornings were spent over a beer at Walker's Wagon Wheel. One engineer kept an ax in his cubicle, and it wasn't for transistor production. Management strategy was an oxymoron, because Noyce didn't do a lot of managing; he was never the prototype for William Whyte's Organization Man. But California casualness masked another part of the emerging tech lifestyle: Everybody worked like beavers. If there was palaver at the watercooler, it was more likely to be about breakdown voltage and P-N junctions than Willie Mays and Orlando Cepeda. More than HP—at heart, a company that didn't rely on semiconductor technology—Fairchild Semiconductor was the first of the nerd-driven companies that embodied the Silicon Valley culture. It was also fundamentally different from the formal East Coast way of doing business—a funny thing, since Fairchild Semiconductor was owned by a New York company.

Imagine what happened when the young CEO of Sherman Fairchild's company came for a visit to his Western outpost. His name was John Carter and he had arranged for a limousine and chauffeur to accompany him. Tom Wolfe had a field day with the contrasts. "So Carter arrived at the tilt-up concrete building in Mountain View in the back of a black

Cadillac limousine with a driver in the front wearing the complete chauffeur's uniform—the black suit, the white shirt, the black necktie, and the black visored cap. That in itself was enough to turn heads at Fairchild Semiconductor. Nobody had even seen a limousine and chauffeur out there before. But that wasn't what fixed the day in everybody's memory. It was the fact that the driver stayed out there for almost eight hours, *doing nothing.* . . . John Carter was inside having a terrific chief executive officer's time for himself. He took a tour of the plant, he held conferences, he looked at figures, he nodded with satisfaction, he beamed his urbane Fifty-seventh Street Biggie CEO charm. And the driver sat out there all day engaged in the task of supporting a visored cap with his head. People started leaving their workbenches and going to the front windows just to take a look at this phenomenon. It seemed that bizarre. Here was a serf who *did nothing all day* but wait outside a door in order to be at the service of the haunches of his master instantly, whenever those haunches and the paunch and the jowls might decide to reappear. It wasn't merely that the little peek at the New York–style corporate high life was unusual out here in the brown hills of the Santa Clara Valley. It was that it seemed *terribly wrong.*"

Noyce wanted his little company to have nothing that reeked of social or class division. But he was not without affectations. Noyce was the son of a Congregationalist minister, but he was no ascetic. He scuba-dived, hang-glided, para-glided, whitewater-rafted, and drove a Porsche. Folklore has it, according to Robert Cringely in *Accidental Empires*, that Noyce once waited on a long line at his savings bank to get a cashier's check for $1.3 million; when the teller inquired why, he explained he was off to buy a Learjet. His divorce in 1974 became a little scandal in Silicon Valley, though he seemed to have gotten the last laugh. When elders of the Los Altos Country Club turned their noses up at him and his new bride, Cringely wrote, Noyce simply duplicated the club's facilities on his own property nearby. Over time these sorts of things helped build the Noyce legend. Along with Hewlett and Packard, he was among the first of high-tech's celebrities. Engineers revered him; in time, Wall Street grew to love him; and the press, when it finally discovered the Valley, delighted in making him an American parable.

Chapter III
Belief

Within a month of leaving Shockley, the Traitorous Eight set up shop a mile away, in a Mountain View building that, ironically, lacked electricity at the outset. They had their work cut out for them. In 1957, a decade after the Bell Labs discovery, American production of transistors had grown to 30 million. A typical one cost only a few dollars, down from forty-five dollars several years earlier, and its price would drop by the 1990s to milli-fractions of a penny. Transistors were ubiquitous in portable radios and hearing aids, and were beginning to infiltrate car dashboards, watches, toys, and the space program, which had been jump-started by Russia's *Sputnik I* launch. Fairchild shipped its first transistor—packaged in a Brillo box—to IBM. The connection with Big Blue would be a long one for Bob Noyce and Gordon Moore.

By the end of 1958, Fairchild Semiconductor had half a million dollars in revenue—a lot given its youth—and was turning a profit. The next year, the company was hired to make transistors for the Pentagon's Minuteman nuclear missiles, which—like NASA space vehicles—needed onboard computers. That meant any transistor-maker had a ready, willing, and able market. Fairchild already had made significant advances in silicon production: Jean Hoerni created the "planar process," by which a layer of oxide was added to the top of the semiconductor, protecting the transistor "inside" the oxide frosting from dust, static, and other contaminants that could disable it.

Nonetheless, like any maker of transistors, Fairchild Semiconductor faced a manufacturing paradox. Its increasingly tiny product, precisely because of size, was impossible to work with. It was one thing to work with a single transistor in a research lab, but it was fantastically more complicated when you connected scores of them and then attached other electronics components like capacitors and resistors, all in areas that were measured in thousandths of an inch. Multiple transistors could be made on a single piece of silicon, but they had to be separated, picked up by tweezers, attached to leads, and then reconnected to something else. Each connection had to be done manually; it took forever and it was often unreliable, because the wires were relatively large compared with the other parts (assuming the semiconductor had been cooked and doped correctly in the first place). Women, who comprised most of the production line, were either hunched over microscopes or on their knees looking for lost pieces. The process was absurd—Albert Einstein meets Rube Goldberg. The scientists gave the problem a more formal description: the "tyranny of numbers." Overcoming it would be the next great step in the evolution of silicon.

The trick was to construct everything on the same thin wafer of semiconductor material—a "solid circuit," an "*integrated* circuit"—eliminating the wires and the tyranny. This was "the monolithic idea," as theorists described it. "I was lazy," Noyce said. "It just didn't make sense having people soldering together these individual components." The integrated circuit would reduce "size, weight, etc., as well as cost per active element," he wrote in his lab book. Jack Kilby, a thirty-five-year-old engineer at Texas Instruments in Dallas, thought similarly. Experimenting with germanium, he was fabricating components on the same slice and using gold wire to connect them. In September 1958, he produced the first integrated circuit, but kept his discovery a secret. Four months later, at Fairchild, Noyce did the same thing—and thought he was the first.

Noyce's device was better than Kilby's, because it eliminated wires altogether. The technique for making it, much refined thereafter, was complex and costly—and Noyce at first didn't realize how revolutionary it would be. It involved layering different materials on top of the silicon and using photolithography to engrave circuit patterns on the surface. Then, extremely fine lines of aluminum—whose widths now are measured in millionths of an inch—were deposited in place in order to make contacts *between* the separate layers and thereby connect the solid-state circuit. The electronics were now seamlessly rooted in the semiconductor itself, the circuit completed electrically rather than physically.

It was a kind of latter-day alchemy, converting worthless elements into something precious. As manufacturing technology improved, the parts got smaller and the etched metal lines got finer. This made the integrated circuit—a "chip" the size of a fingernail—more and more powerful, and ushered in the "minicomputers" of the 1960s that took computing power away from the technoids in white lab coats. Silicon real estate began to look like Tokyo: The piece of semiconductor that once housed a single transistor eventually held a thousand, then a million, of them, at a size so small that they were invisible and conjured up the image of angels dancing on the head of a pin. In its magnified state— which is how it starts out, as engineers design and plot circuitry—the chip resembles a labyrinthine city map, filled with avenues and alleyways where electrons run to their appointed tasks.

The respective inventions of Noyce and Kilby got tangled in patent litigation that lasted more than a decade. If it had been a boxing match, Fairchild would've been declared the winner on points. But it didn't much matter. Near the end, Fairchild and Texas Instruments called a truce and worked out a licensing agreement allowing each to flourish. Kilby had the notion about an integrated circuit first and built a crude one. But Noyce's creation was clearly superior to Kilby's. Not only did it get rid of all the wires, but it was made of silicon, which was more resilient in the clean room. Inventing the integrated circuit was marvelous—and Kilby never got the fame he deserved—but, as Henry Ford proved, mass production was the Holy Grail. The laws of economics proved just as powerful as the principles of physics. Noyce understood that: To win over skeptical customers, he sold his integrated circuits for less than it would take to buy the components and build the circuits themselves. Soon, semiconductors gained traction and turned into a multibillion-dollar industry—beginning a second Industrial Revolution. As early as 1960, *Business Week* put a photo of a transistor on its cover.

Noyce's integrated circuit had set his company up for glory, fleeting as it turned out to be. In 1959, less than two years after Fairchild Semiconductor was born, the moneymen back in New York called in their chips. Sherman Fairchild exercised his option and made the Traitorous

Eight rich. They each received $250,000, a five-hundred-fold return on their initial investment. The profits from Noyce's California subsidiary were needed to bolster the bottom line at headquarters. While this buy-back was part of the founding deal, it was also the death knell for Fairchild Semiconductor. The parent company back home kept up its meddling—"the tail wagging the dog," as Moore put it. The executives back east were what are now called "seagull managers"—they would fly in, eat some free pastries, make a lot of noise, shit over everyone, and then fly off. But for all its interference, Fairchild's Eastern establishment had never provided enough support for a manufacturing operation that was constantly in need of upgrades. The damn thing about integrated circuits was that they became obsolete in six months, when some other chipmaker made them smaller and faster. To compete, the lab and the production line had to keep reinventing themselves.

The Fairchild buyback also meant that Noyce had to shuttle to and from New York. Plus, Noyce was not able to provide stock ownership to his employees, which hindered both future recruitment and current morale. The story goes that another Fairchild director once asked Noyce incredulously: "Isn't the fact we're not going to fire them tomorrow incentive enough?" It wasn't enough, as Noyce had come to realize, along with the fact that greed could be a magnificent motivator. If your company's stock did well, he said, "you got several years' salary at once." Or more.

In 1961, not four years after creating Fairchild, the first group of founders resigned—Jean Hoerni and two others—to form another semiconductor start-up that became Teledyne. "I had become Americanized," Hoerni explained. In Europe, "you had to wait until someone died to get a good position. But, like Americans, I'm too impatient for that." Thus began the diaspora of Fairchild talent that would gut the company and, at the same time, pollinate a grove of other high-tech saplings in the neighborhood—"Fairchildren," as Adam Smith christened them. In many lobbies today around Silicon Valley, there's a souvenir poster showing how at least one hundred companies trace their genealogy back to Fairchild, giving it a patina of history that no other organization in the Valley ever attained.

But even before Hoerni left, Fairchild suffered a more humiliating, telling loss. While Noyce was the acknowledged leader, the company had brought in a more seasoned manager, Ed Baldwin, to run the place. In 1959, he and his deputies quit to found Rheem, another semiconductor start-up. In just its infancy, Fairchild was learning the "what goes around, comes around" reproof. The Traitorous Eight had left Shockley— now a few turncoats were leaving Fairchild. Bell Labs begat Shockley begat Fairchild begat Rheem. It wasn't exactly the corporate culture of IBM, which could just as well have included caskets among the benefits.

Those in commerce tend to have a poor sense of irony. So, when Fairchild suspected that Baldwin's group had walked out the door with the recipes for its transistors, it reacted in the good old-fashioned American way: It sued the pants off Baldwin. Electronics may have been new to the Valley, but lawyers were not. *Fairchild* v. *Baldwin* would be the first in a never-ending parade of trade-secrets lawsuits that were a necessary corollary to the culture of entrepreneurship. But wasn't it a bit hypocritical to go after employees who were only doing unto you what you had once done unto others? Noyce and Moore became famous for going after CEOs who went after their trove of talent; in the 1980s, T. J. Rodgers at Cypress Semiconductor made a point of framing each of the cease-and-desist "lawyer's letters" Moore sent him.

As it turned out, Fairchild's legal pursuit of Baldwin's team became moot and Baldwin's victory, if any, was Pyrrhic. Baldwin left at the same time Noyce was inventing the integrated circuit. Once this new device hit the market, the old Fairchild recipes were worthless. Rheem, the first spin-off from the spin-off, had learned its own lesson: There were no keys to the castle because the locks were changed too often. Though any electronics company might fairly object to disloyalty, it could rest assured that the very nature of the business would discipline most miscreants. There was little stopping the free-for-all that the Valley was becoming.

By early 1968, all of the Traitorous Eight had left Fairchild except Noyce and Moore. The company had thirty-two thousand employees around the world and annual sales of $130 million. The first chip back in 1959 contained a single electrical circuit, but the latest from Fairchild had almost a thousand. Counting transistors was on its way to becoming absurd, much like trying to add the number of sand grains on the planet—silicon by any other name. Each Apollo moonshot would use a full million chips. In time, Moore would say there were "more transistors made in the world every year than raindrops that fell on California during the same time period"—50 quadrillion of them, give or take.

But despite Fairchild's pivotal role in fulfilling Shockley's manufacturing dream, the realities of business had settled in. Mismanagement from the home office—Sherman Fairchild had died—set the place up to fail, in both production and spirit. Charlie Sporck, who was running Fairchild day-to-day, with Noyce in a more senior role within the Fairchild corporate hierarchy, took a large group to go to National Semiconductor. Sporck, an operations wizard, would make National a powerhouse. Rivals like Motorola and Texas Instruments (old home of Jack Kilby) sailed by Fairchild in revenues. When Sporck left, Noyce was incensed that he wasn't given the title of CEO of the Fairchild parent. So, in the summer of 1968, Noyce and Moore became the last of the Fairchild founders to move on. "There is a tremendous advantage in being able to get rid of everything you've done in the past and start over with a completely new slate," Moore says now. "It's awfully hard to do that in an established company. There's too much invested, too much to protect. It's hard to find any companies that have been successful in one area who can do a sharp right or left turn and do something else. Companies are like supertankers."

Size and inertia allowed Fairchild to linger during the 1970s, ultimately selling out to Schlumberger Limited. After Noyce and Moore departed in 1968, Fairchild began hiring away employees from its archenemy, Motorola Semiconductor of Phoenix, including its CEO, Lester Hogan. So much for Fairchild's outrage over Ed Baldwin's treachery nine years earlier. Now it was doing the same thing and using the most obvious lure. Hogan got so much money that for a few years, according to Dirk Hanson's *The New Alchemists*, wealth in Silicon Valley was measured in "hogan" units. Hogan's team was known as "Hogan's Heroes," but hardly deserved it because Fairchild never regained its former status. In the 1990s, its once-legendary headquarters in Mountain View were bulldozed in favor of another celebrated start-up, a company called Netscape. Entrepreneurs love that kind of rising-from-the-ashes symbolism.

They believe they are the driving force of capitalism—the "gales of creative destruction" that sweep away the old technologies and usher in the new. (Hey, all that money is just a byproduct.)

As they left Fairchild, much as eleven years earlier at Shockley's lab, Noyce and Moore wanted to stick together but had only a vague idea what to do next. Though this time they had the luxury of some wealth, they again sought out Arthur Rock, who by then, on a Pacific whim, had left Manhattan for San Francisco, where he had become the preeminent financier of high-tech start-ups. Rock was known as a "venture capitalist," which suited him fine because he made the term part of the lingo. He provided Noyce and Moore not only with access to cash—they needed a few million—but with credibility. If Art backed it, it must be good.

Rock was more than happy to see them. Fairchild had achieved a certain resonance in the corporate and financial communities, even with its defections and slow burnout. Moreover, Noyce and Moore were willing to take money out of their own pockets, which was always a pleasing sight to a banker. Noyce and Moore each would pony up $245,000—a sizable chunk, roughly 10 percent, of their net worths. That left $2.5 million or so for Rock to raise, plus the $300,000 he would invest himself. It was easy—Noyce and Moore had become Rock's stars. Without even a business plan to show prospective investors—the days of derrick-delivered legal disclaimers had yet to arrive—it took him all of two afternoons to get backing for what they called NM Electronics. (Moore, Noyce Electronics was rejected because it sounded like "more noise.") According to Rock, it took "only as long as it took me to get hold of the twenty-five individuals I called."

The rich got richer: Among the twenty-five were the other six members of the Traitorous Eight, who got to buy the early stock for five dollars a share (compared with Noyce's and Moore's one-dollar shares). At that time, institutions like universities weren't allowed to invest in ventures perceived as too risky, but some months later, Grinnell College, Noyce's old stomping grounds, where he now was chairman of the trustees, got to invest its own $300,000—a sum that would multiply over and over and make Grinnell the envy of every development office in America.

Noyce and Moore quickly agreed that their company needed a better name. Modern branding specialists they weren't. Moore suggested Integrated Electronics, whose first syllables were a play on "intelligent," which they knew, because they were. Noyce shortened it to one word. Intel had been born—the new chip on the block. In July 1968, Noyce and Moore hung out a shingle in Mountain View, near Fairchild. The logo was the now-omnipresent lower-case blue "intel," with the *e* dropped down halfway, emphasizing the "integrated electronics" wordplay and perhaps suggesting the exponential nature of transistor growth.

Rock was chairman of the board, Noyce the CEO, and Moore second in command. A local paper ran a nice story about them, even including both the founders' home addresses. But high-tech was fast losing its small-town flavor. Jerry Sanders, the flamboyant Fairchild marketing whiz, remarked that any gearhead in the Valley could now switch jobs without changing parking lots. Lack of fealty to one's employer became a hallmark of the Valley.

Chiefly because of Fairchild and its Fairchildren, the Valley was pushing its boundaries. Unlike transistors on chips, there were only so many buildings you could pack into a parcel of real estate. The sun still shone, but nourished fewer and fewer fruit trees. The orchards were giving way to paved progress, in the form of office parks and a spiffy new interstate running along the Santa Cruz Mountains, from San Jose to San Francisco. Upon his retirement as Stanford provost, Fred Terman complained that "the price that is paid for all these blessings is annoying traffic congestion around 8 A.M. and 5 P.M. in the ten-to-twenty-minute drive between home and work." Today, of course, the traffic seems part of a Möbius strip, and that drive can take an hour.

Intel would become the greatest industrial engine ever in the Bay Area, another accidental start-up. Without Shockley's incompetence, Intel—like Fairchild—never would have happened. Along with Apple, Intel was the first of the mass-market companies; their rates of growth, along with their profits, just exploded upward and changed the expectations of both entrepreneurs and investors. The rise of Intel would also be critical down the road for another company—Microsoft, which was founded in 1975.

Joining Noyce and Moore atop Intel management was a thirty-two-year-old Hungarian émigré named Andy Grove, whose personality would shape the new company as much as theirs. Grove regarded himself as

one of the cofounders and bristled at accounts that did not so note—
even though the Intel prospectus didn't. It also did not help his ego that
his fortune was only a pittance compared with Moore's—less than a half
a billion in early 1999. Grove lives sanely—one of the least extravagant
lifestyles in the Valley—so disposable income was beside the point. He
skis, he bikes, he listens to opera with his wife of more than four decades.
He has none of the expensive toys of his Valley brethren, he still lists
his number in the phone book, and he flies coach except when he's saved
up enough United upgrades. But he loves the limelight: For a time, he
wrote a weekly column on management for the local paper; then he was
on the cover of *Fortune* magazine, over and over as one of the "Ten
Toughest Bosses in America," an honor he apparently liked.

Grove's path to California was difficult. He was one of the early immigrants who are now integral to the high-tech culture. Born András
Gróf in Budapest, an only child, he survived both the Nazis and the
Soviet takeover (though his Intel bio strangely omits any mention of his
heritage) before escaping to Austria and then to the United States, where
he took his new name. Grove lived in the Bronx with his uncle and
attended the City College of New York. His graduation in 1960 made
The New York Times: "A Hungarian refugee who three years ago didn't
know horizontal from vertical—in English," the paper noted, "today [is]
at the head of the class of engineering students." After graduation, he
left for Berkeley—largely because of the weather—to get his chemical
engineering Ph.D. in 1963.

At twenty-seven, Grove went to work for Fairchild Semiconductor in
R&D, where he began to earn the reputation that would follow him for
the next thirty-five years: aggressive, disciplined, tactless, unyielding,
combustible—and gifted at getting things done. Some called him "the
Mad Hungarian," and that's when he was in one of his better moods.
"There's a saying about Hungarians," Grove says. "They go in a revolving
door behind you and they come out ahead." Today, having appeared on
the cover of so many business publications and resplendent on the jacket
of his very own book, the five-foot-nine Grove is trimmed, tanned, and
turtlenecked in black—almost dashing by the standards of the Valley.
But back at the beginning of Intel, he was known for the odd mixture of
thick black glasses, gold chains, mustache, and muttonchops. Tom Wolfe

described him as an exemplar of "California Groovy," which wasn't to be confused with his management style, of which he didn't have one. What you saw is what you got; he was nothing if not blunt.

In a laissez-faire corporate culture, somebody had to crack the whip, to assume command of the chip. If Noyce was the spirit of Intel and Moore the heart, Grove was the fist. As the head of operations and, from 1987 to 1998 the CEO, he was responsible for executing strategy, budgets, and, if necessary, employees. "He'd fire his own mother" was a refrain from both Grove's admirers and detractors. Grove permitted no mirth around the corporate campus; about the most fun Intel workers had was looking out the window at the Great America Amusement Park just across the way. His annual "Scrooge Memo" advised employees to work the full day before Christmas. His "Mr. Clean" inspections were something out of a bad TV commercial. His practice of "constructive confrontation" became an official management technique: Discuss problems before they became crises, the higher the decibel level the better. Grove could yell better than anyone, and the Hungarian accent only made him more intimidating. This was not Hewlett-Packard.

Best personifying Grove was the Late List: Start time every morning was eight o'clock. It didn't matter that you might have worked late the night before. (Hours were still flexible: Show up anytime before eight.) If more than 15 percent of any division's employees showed up late over the course of a month, then any employees who showed up after 8:15 had to sign in with the guard, and their names were then forwarded to management. In theory, violators suffered more from the corporate opprobrium they felt than from any specific punishment. But the Late List, which on occasion included Gordon Moore and even Andy Grove, was more an embarrassment than anything else. Some malcontents signed in as Mickey Mouse or as Grove himself; others just made sure to come in so late they were mistaken for customers. And predictably, the press ridiculed it. But Grove's Late List wasn't retired until the late 1980s.

Lacking a clear business plan, Intel's founders knew enough not to re-create Fairchild. The way to make money in electronics wasn't to be the best competitor in an existing market, but to create and corner an entirely new one. Noyce and Moore saw that mainframe computers—the devices their chips had overtaken—were becoming part of the corporate mainstream. But as wonderful as computers were at logic functions like adding and subtracting, they were lousy places to store information. To be more useful, a computer's "memory" needed to be faster and more accessible. Integrated circuits controlled the brain of a computer, but nobody had figured out how to put them to work in the memory banks, where nonchip, nonsilicon technology still ruled (using "core" magnetic fields for storage). Noyce and Moore intended to come up with the first mass-market memory chips. Tilling a field no one had yet sown, they would do to memory storage what the transistor did for electrical current.

In the spring of 1969, less than a year after its founding, Intel came out with its inaugural memory chip. Crude and simple, its memory was more like that of a flea than an elephant. But it served notice to established semiconductor makers that Noyce and Moore were doing it again. Within a few years, Intel introduced a far more powerful chip, the random-access memory "1103," which became the first commercially viable device in semiconductor memory. When compared with the next invention, though, the 1103 became just a memory. Intel had hardly forgotten that there was also a market for logic chips.

Employee No. 12 was a young engineering Ph.D., from Stanford naturally, named Marcian E. "Ted" Hoff, who had gotten his start in electronics in the Westinghouse Science Talent Search. In the summer of 1969, Intel had been asked by Busicom, a Japanese manufacturer of business machines, to help make complex logic circuits for a desktop calculator. The legend of Noyce extended to Asia by now and no Japanese engineer could design what Busicom wanted. Each calculator would have up to a dozen components, to handle such different functions as computation and printing. The thirty-two-year-old Hoff was in charge of the Busicom project and quickly recognized that the proposed calculator would be too complicated to build at a reasonable price. Instead, he theorized that the various chip functions be condensed into, and onto, one. By 1971, he had developed the first prototype.

Miniaturizing the brains of a machine—its central processing unit (CPU)—had uses way beyond a calculator, not that Hoff or anybody else at Intel in 1969 had even conjectured what would come. A "microprocessor," as it was called, could execute logic and arithmetic functions akin to what computers did, but on a different scale. The old ENIAC and then the mainframes of the 1950s were bigger than a bakery; the current minicomputers, devoid of vacuum tubes, were down to the size of a closet and then of a breakfront; Hoff's sensationally tiny invention presaged machines that could sit on your lap. Custom features would one day be delivered by "software"—the programs that ran the microprocessor. With the proper set of commands, a computer could do math, balance ledgers, display words on a screen, play games, send E-mail, and draw dinosaurs. Since programs could be altered (from a keyboard), the computer could perform many different tasks, making it unique among human inventions. Indeed, sufficiently reduced in size (and price), general-purpose microprocessors could be put in virtually any appliance or device, and given capabilities never imagined. And there was a world beyond computers. Dishwashers, stereos, bombs, satellites, cars, pets, pagers, jewelry, sneakers, ski bindings—all would someday be able to "think" because of this new way to harness a sliver of silicon. Combined with sensors in a car engine, a microprocessor could adjust the air-fuel mix in a carburetor; in a toaster, it could determine just the right crispness. Wherever a lever or gear or other mechanical device operated before, electronics could now do the job.

Intel's Japanese customers gave up on Hoff's idea, but Noyce and Moore allowed him to continue. By early 1971, along with ex-Fairchild engineers Federico Faggin and Stan Mazor, Hoff had created the first programmable "computer on a chip," as Intel would market the microprocessor. (In succeeding years, other engineers would come forward to claim that, no, their custom chips were really the first microprocessors; but so far, the scientific community has dismissed those claims.) When you see those annoying TV commercials of Intel technicians in spaceage bunny suits, it is the profits from that microprocessor they're dancing about. You're also seeing a brilliant marketing scheme, which succeeded in turning a commodity into a renowned brand. In time, memory chips—the first Intel product—became a minor part of the business.

The first microprocessor, containing 2,300 transistors, was called the 4004—a digital pun, as well as roughly the number of components that the new device replaced. Its processing power was limited to four "bits" of digital data (a "0" or "1" in the parlance of the binary numeric code that a computer could understand)—so primitive that it could perform only 60,000 calculations per second. (Today's Pentium II can do 600 million.) Less than a year later, Intel came out with the 8008, a chip that incorporated all the essential parts of a computer other than the plastic casing. It held not only the CPU, but the other components like input and output circuits as well. Trouble was, apart from a fee from Busicom for the 4004 and another company for the 8008, Intel simply wasn't making a lot of money from the chips.

But then it got lucky. Shortly after Busicom received its first 4004 microprocessor—and faced with new competition in the calculator market, including a new Texas Instruments model designed by Jack Kilby—Busicom asked Intel to renegotiate the price. Noyce could easily and justifiably have laughed at the request. Given Busicom's exclusive contract for the 4004, Hoff had another idea. "For God's sake," he told Noyce during the negotiation, "get us the right to sell these chips to other people." Busicom got a better price and Intel got its creation back. Not that the sales and marketing force much cared at the time—they regarded the 4004 as a white, if miniaturized, elephant. If only twenty thousand mainframe computers were sold a year and even if Intel could commandeer that market with its new microprocessor, what was the point? Only if there truly was demand for a small computer—something called, say, a "personal computer." One of Intel's engineers liked to say he'd never own a computer he couldn't lift.

Undaunted, Intel launched its campaign to sell the 4004 to whoever might want it. "Announcing a New Era in Integrated Electronics," proclaimed an ad in a trade magazine in the fall of 1971, "A Microprogrammable Computer on a Chip." The strategy was sort of like the *Field of Dreams* declaration that "if you build it, he will come"—precisely the opposite instinct of products driven by market research. But just who would be coming?

Wall Street certainly had confidence. Around this same time, not surprisingly, Intel offered stock to the public (as well as its employees). This was something Shockley Semiconductor and Fairchild never had done. This was now real money, a taste of what the Valley of the Dollars had to offer. As with the innards of a chip, counting zeroes was part of the game. Noyce and Moore were the largest shareholders, between them holding more than a third of the company. Even at the initial offering price, they would each achieve seven-zero wealth—at least $10 million. By today's standards of Internet affluence, that's chump change, but it was a lot of money in 1971 for someone whose name didn't end in Getty or Du Pont. Playboy happened to go public the same day as Intel, at about the same price, but within a year, Intel's shares were trading at more than twice Playboy's offering price. "Wall Street has spoken," declared *Financial World* magazine. "It's memories over mammaries." *Time* anointed Noyce a "financial genie." Today, Moore is the richest person alive in California, with a fortune topping $10 billion, enough to buy out Steven Spielberg, George Lucas, and Michael Eisner combined. (Noyce, a heavy smoker, died in 1990 at age sixty-two and his fortune, like David Packard's, was diluted by inheritance and gifts.)

When Intel came out with the 4004, Moore—not typically given to hyperbole—declared his company's device "one of the most revolutionary products in the history of mankind." "We were the revolutionaries of the time," he said, tweaking the noisy longhairs on the Berkeley campus and elsewhere. Maybe so, but it would take some time for the revolution-in-miniature to begin. The early chips didn't sell and there weren't any computer programmers around to think up specific uses for them. It was the "hackers"—young, rebellious, mischievous software artists—as much as the corporate suits who launched Moore's revolution. And it was two teenagers in Seattle who together bought one of the first 8008s and built a machine to measure traffic flow on city streets. They started a company called Traf-O-Data that went nowhere, but a few years later, they founded something else which became the blood brother of Intel. The teenagers with the Intel chip back in 1972 were William Henry Gates and Paul Allen—and their second little company was Microsoft. Even today, Gates keeps a poster of an Intel chip in his office.

The problem with the early chips was part metaphysical, part physics. In the suites of corporate America, the essence of a computer's appeal wasn't merely utilitarian: It also signified power, status, mystery. Unless one seriously believed that everybody at the company needed to have one of these electronics wonders, there was no reason to think smaller or cheaper, and every reason not to. Image aside, the more concrete issue was simply that the chips weren't powerful enough. That was a function of design and manufacturing. The progression from Shockley's transistor to Noyce's integrated circuit to Hoff's microprocessor had been formidable. But the fact was, chips weren't yet legitimate substitutes for main-frame computers.

Intel's far more sophisticated 8080 chip, introduced in early 1974, was the runaway breakthrough. It was ten times faster than the 4004 and was introduced at the same price of $360. It was so successful that for a time "8080" was the latter part of the company's phone number. As the business got started and efforts concentrated on R&D, Intel's sales during its first year were under $3,000. Just six years later, in 1974, the figure had grown to $135 million (on its way to the current $25 billion), making Intel the No. 5 chipmaker in the world at the time. The payroll ballooned from 42 to 3,100. In just a five-month period, it recovered its R&D costs on the 8800. The company moved out of its cramped Mountain View space and built huge new offices six miles to the east, in Santa Clara. The company is still there, its huge INTEL INSIDE logo emblazoned on the roof of the main building. Only a few walnut trees down the road serve as a reminder of what was once the Valley. The logo on the rooftop was Moore's idea, a dig at a former colleague and current competitor. "So Jerry Sanders would have to see it every time he took off from the San Jose Airport," he says. More than anything else, Ted Hoff's microchip had let Silicon Valley surpass the Boston area when it came to supremacy in high-tech.

The 8080 was the latest embodiment of a prophecy made by Gordon Moore back in 1965 that propelled the Valley into the modern era. Writing in the trade publication *Electronics*, reflecting on a graph he had sketched out, Moore observed first that the number of components on a chip had doubled every year for the prior six (resulting in increasing processing speed, since electrons had less distance to travel to complete their circuitous rounds). Then, almost casually, he suggested that this astonishing doubling of speed could continue every year for a generation, with a corresponding, and equally extraordinary, decrease in the cost of each chip. "If these economics applied to cars," venture capitalist John Doerr preached years later, "they would cost ten dollars, go thousands of miles on a tank of gas, and when they got old, you'd just throw them away." (They'd also crash a lot more.)

Moore would later pull back a little and say it would take eighteen to twenty-four months to squeeze twice as much on a piece of silicon. Because Moore's suggestion proved so accurate, it took on the mythical status of a scientific axiom, just a few rungs down from Newton's. It was called "Moore's Law," coined by—depending on whom you ask—either Carver Mead, Caltech's famed engineering theorist (and Woodside week-ender), or Arthur Rock, the financier who might also have noted that his own wealth seemed to double every few years. Even the humble Moore would come to refer to his own law by name, though it took him until the 1990s to sheepishly take credit. (Moore's Law also led to the prolif-eration of other "laws," increasingly less profound, including Andrees-sen's Law of Internet stocks: "The number of millionaires will double every eighteen to twenty-four months, unless there are company profits, in which case there will be even more.")

Moore was wise enough to understand that miniaturization was inev-itable, but not so shrewd to see it all the way through to its logical conclusion. Mainframe companies like IBM and General Electric and Honeywell would continue to dominate the computer market, but who could envision demand for "personal computers," and, more to the point, who would want to make such a device? The big machines could cost a million dollars, so why manufacture a small one for a few thousand bucks? In 1965, Moore didn't see a market for personal computers any more than anyone else; at best, he imagined home computer terminals hooked up to a mainframe. Chips would do fine in Chevy carburetors and G.E. toaster ovens, but personal computers? When an engineer in the mid-1970s proposed that the company build a computer suitable for the home, Moore asked what use it might have. The only idea he got back was a housewife sorting her recipes. At least the engineer gave a woman that much credit. "I could just picture my own wife, Betty, at the stove with a computer beside it," Moore recalls now. He rejected the proposal. Bob Noyce saw the future of microprocessors . . . in smart wristwatches: Ted Hoff meets Dick Tracy. You can't be a genius at every-thing.

Yet even without marketing prescience, Moore's Law reflected an en-gineering reality that, as a practical matter, was unparalleled. Do the math: Double anything every eighteen months and, in a thirty-five-year period (say, 1965 to 2000), the ratio of the thing goes from 1 to 8.5 million. Intel's Pentium II chip (circa 1997) has 7.5 million transistors on it, each one smaller than a bacterium. Only the barriers of the atomic universe—there's only so small you can go—will repeal Moore's Law.

And that may not happen for another ten or twenty years. In manufacturing, the ability of an industry to double performance and halve cost was unheard of and led every semiconductor cheerleader to say as much over and over.

In other industries, Moore's Law might have been a fiscal death sentence. Unless they dramatically raised demand, relentlessly lower prices never helped a balance sheet. But the beauty of hardware is that when power and speed increase, so do the tricks a chip can perform. Today's $2.50 singing Hallmark card contains more power than the ENIAC. It may be that the computer you buy now will cost half as much in eighteen months; but more to the point, there'll be a new piece of hardware out by then that's twice as fast. Nobody wants to be working in Slowsville; nobody's interested in an old machine. Intel didn't die, instead becoming the most stable of companies in Silicon Valley. In early 1999, Intel was worth $225 billion, placing it behind only Microsoft and General Electric in the pantheon of publicly traded companies. Its stock has made thousands of millionaires.

Intel's success raises the question of whether Moore's Law derives from the laboratory—or from the marketing department, where it becomes a self-fulfilling prophecy of sorts. Why can't microprocessors, say, quadruple in speed over a given time period—isn't it just a matter of manufacturing mastery, etching a few more lines and cramming in a few more components? Sure, there's a learning curve in the factory. But perhaps it's also because Twice As Fast is enough to get a lot of folks to go out and buy a new model, even if the current one works just fine. We might call this Barnum's Law, a corollary to Moore's: There may not be enough high-tech suckers born every minute to justify a new fabrication plant, but in eighteen months, you can bank on it. Otherwise, as Moore himself acknowledged to me, sitting in his eight-by-eight cubicle at Intel, "our business model falls apart. If we get to the point where the world says, 'I don't need any more performance—these computers give me everything I need,' then we're in trouble. You're seeing some of that now—people buying PCs for $1,000 rather than a higher-performance device for $2,000. I worry less about being able to build more complex chips than people saying 'Who needs it?' " Fortunately, the easiest answer to that question is teenagers. Games, still one of the most compelling uses for computers, need speed.

The business culture of the Valley doesn't fear instant obsolescence—
it thrives on it. *Only the Paranoid Survive,* Andy Grove called his 1996
"how-to" book (a title that amused more than a few Intel staffers and led
some to christen it Grove's law). Author Tim Jackson, in *Inside Intel,*
reported that the "house joke" is that the company's photocopiers run
on paper already marked "Intel Confidential" at the top of every page.
When I walked out of the main building with Gordon Moore one afternoon, the security guards asked to search his briefcase. Well before
Noyce officially stepped down in the late 1970s, it was Grove's persona
running the company. Instability, velocity, turbulence—these are the
quicksands on which the Valley's technological foundations rest. Here
today, gone tomorrow. Everyone keeps running, reaching for destiny or
trying to avoid it.

In 1971, the same year Intel came out with the first microprocessor and went public, the Valley finally got its name. What
previously was the Santa Clara Valley was now associated with the semiconductor material driving electronics. It was increasingly difficult not
to notice that the orchards were disappearing and that there were now
all these engineers with their strange talk. Don Hoefler was in the midst
of writing a series of articles for *Electronic News,* a weekly trade. One of
the Fairchildren, Hoefler got his start as a publicist for the Traitorous
Eight. As he finished writing his series, he couldn't come up with a title.
A local CEO suggested "Silicon Valley USA" and Hoefler used it,
and in the great creative spirit of American journalism, others began to
copy it.

To this day, "Silicon Valley" exists on no map, no census report, no
freeway sign. Yet it has become part of the global lexicon—not just as
the capital of high-tech, but a great new marketing tool. Today, there's
Silicon Valley Towing, Silicon Valley Samurai, Silicon Valley Book-
keeping, Silicon Valley Bank (with a branch in Beverly Hills), Silicon
Valley Power Wash, Silicon Valley Toxics, Silicon Valley Engine &
Hose, Silicon Valley Pest Management, Silicon Valley Psychological,
a Presbyterian church in Menlo Park that bills itself as "The Church
of Silicon Valley," and Moshe Mendelsohn, O.D., Silicon Valley Eye

Physician. Some people will do anything to avoid saying they're from San Jose. Even the San Jose *Mercury News* has taken to eliminating the first two words from many of its promotional materials. Now it's "The Mercury News: The Newspaper of Silicon Valley." Three decades ago, Dionne Warwick sang about a woman who pined for a pastoral place down the road from San Francisco, asking "Do You Know the Way to San Jose?" Today, she'd just as soon stay on the freeway until she reached Big Sur.

The Silicon Valley name is not without its drawbacks. A few years back, Ralph Kiner, the longtime broadcaster for the New York Mets and master of malapropisms, was doing a game from Candlestick Park, up near San Francisco. It was a pleasant evening, rare for the windy stadium. "They're coming in from everywhere in the Bay Area tonight," Kiner told his listeners, "from Napa Valley to the Silicone Village."

Meanwhile, William Shockley's corporate supernova had faded out. Renamed Shockley Transistor in 1958, the lab was still able to attract intellectual capital, trading on the name of the Nobel laureate in residence. Shockley Transistor even got around to manufacturing semiconductors, though Shockley resisted development of an integrated circuit. But profits were another matter. In essence, Shockley never understood that he was running a business and not a research institute devoted to his whims. And for all his skills, he never figured out how to get along with anybody. At one point, he gave up recruiting those crybaby Americans and journeyed to Munich to find employees. He thought German engineers might prove more amenable to a tyrant. That didn't work either.

In the spring of 1960, Shockley Transistor was sold to a Massachusetts competitor, which kept Shockley in charge. The following summer, he finally got to move the company into comfortable quarters in the Stanford Industrial Park. But even under new ownership, the company never made money. In four years, the company was sold again, and faded into the oblivion that eventually claims most start-ups. Shockley himself, at fifty-five, retreated to the university. Old Fred Terman, at this point vice-president of Stanford, gave him an endowed chair in engineering. Coming full circle, Shockley even did some consulting back at Bell Labs. But nothing he did in science could equal his embarrassing detour into eugenics.

Amid the social turmoil of the late 1960s, Shockley took up the cause of white supremacy, specifically what he saw as the genetic inferiority of blacks. "Dysgenics," he called the problem, defining it as "retrogressive evolution" caused by "excessive reproduction of the genetically disadvantaged." The problem, he argued, couldn't be remedied by education or other public policy, presaging the debate caused twenty-five years later by the 1994 publication of a book titled *The Bell Curve*. "My research," he declared, "leads me inescapably to the opinion that the major cause of American Negroes' intellectual and social deficits is hereditary and is racially genetic in origin." His solution: cash bonuses to those who agreed to be sterilized, at a rate of $1,000 for each IQ point below 100. As with any other endeavor in his life, Shockley crusaded. In repeated, unsuccessful pleas for funding to the National Academy of Sciences, in lectures, at cocktail parties, on TV talk shows, and in a *Playboy* interview, he pressed his case, seeming to exult in the controversy it caused. It was, as he once said in another context, like striking a match to a bale of hay tied to a mule.

For some reason, unsympathetic audiences suggested that his scientifically couched theory might, lo and behold, be racist—which probably had something to do with the dearth of scientific evidence for his argument. How, for example, could he fail to factor in cultural and social influences? When the *Atlanta Constitution* compared his theories to Nazism, Shockley screamed libel and sued for millions. A jury ruled in his favor—and awarded a dollar. Shockley was so discredited he was "libel-proof"—it was impossible to tarnish his reputation further. Yet, with absolutely no self-doubt, he continued to tell his wife that his research on intelligence was the most important of his life; appropriately enough, she was a psychiatric nurse and by all accounts a good listener. Due in part to the unsuccessful lawsuit, he began to tape-record all his conversations and take Polaroids of his blackboard before he erased anything, the better to disprove naysayers next time. One engineering colleague, William Spicer, remembers that Shockley's home answering machine announced that all calls were recorded, beeping merrily as it took down every word. Shockley's home was filled with cassette tapes, neatly arranged and cataloged.

Even Spicer, one of his few remaining friends, couldn't figure him out. "I know he had no relationship with at least one of his children," Spicer recalls. "But one time his cat was sick and he called the vet every few hours from the lab. When he found out the cat had died, he broke down and cried." In personal dealings, Shockley had learned little. Before he would grant an audience to visitors, they had to undergo screenings by an assistant—shades of recruiting the likes of Noyce and Moore. Students seeking internships weren't even considered unless they had scored above 700 (out of 800) on each of the verbal and math parts of the SAT.

T. J. Rodgers, the CEO of Cypress Semiconductor and Woodside provocateur, had Shockley for a teacher when he was a twenty-six-year-old Stanford graduate student in the mid-1970s. "I took 'Physics of Semiconductors' and I'll never forget the first class," Rodgers says. "It was a fifty-student room and it was packed. Shockley assigned a month's worth of reading to be done in a week. Next class, only six of us showed up. Shockley said, 'Now I know who's serious.' Then he tossed a piece of chalk at me and told me to write down Schrödinger's Wave Equation on the blackboard. I only got part of it right. So he made us memorize it by heart." Rodgers admired Shockley's intellect and became as close to him as the teacher would permit any student. "I would have dinner with him," Rodgers recalls, "and he'd be almost emotionless, like Mr. Data on *Star Trek*. The biggest reaction you got out of him was ever so slight a smile and those glinting eyes. He didn't want to concern himself with mere mortals. He only wanted to talk about things in terms of reason and information. When he wanted to prove a point about something he'd said, he'd find one of his tapes and play it back for us."

One final examination was disrupted by demonstrators, dressed in white robes, barging in and chanting, "Shockley is a motherfucker! Shockley is a motherfucker!" The professor responded with bizarre equanimity. "Why," he asked his guests, "do you say that?" He wrote down their accusations on the board and then offered up a defense. They, in turn, started to throw things at him. Shockley picked up his Polaroid, which only incensed the demonstrators more. Only the physical intervention of Rodgers and another student, formerly of the Navy, prevented Shockley from getting pummeled. "The entire scene was absurd," Rodgers says, with part lament and part bemusement. "And Shockley didn't get it."

In the tinderbox of campus unrest and political turbulence nationwide during the late 1960s and early '70s, Shockley's racial theories won more notoriety than anything he had done in a physics laboratory. Harvard and Yale banned him from speaking. At Stanford, he was reviled. His Lincoln Continental was spray-painted, he was burned in effigy, and editorials called for *his* sterilization. Outside his office window in the sandstone engineering quad, demonstrators routinely gathered with transistorized loudspeakers to shout "Off Pig Shockley!" As the journalist T. R. Reid wryly observed, this gave Shockley "the experience, probably unique in engineering history, of watching his own invention used to provide hundredfold amplification of demands for his death." At one rally, a microphone malfunctioned and Shockley, who had come to watch, was kind enough to repair it, apparently without irony. He simply didn't want the show to end—better ignominy than anonymity. He would have a burger at the student union just to incite a debate, as well as to get his digestive juices flowing.

Shockley's musings on genetics would continue for almost twenty years, as he dabbled in a bid for the U.S. Senate in California (his platform was dysgenics), and boasted that he had personally donated to a sperm bank for breeding geniuses. The Repository for Germinal Choice outside San Diego reportedly still has some of the sample on ice. When Shockley died of prostate cancer in 1989 at age seventy-nine— forgotten—his chief activity was growing corn.

Unlike the men he brought to California, and so many who followed in his wake, Shockley never reaped the millions he so coveted. While others had turned silicon into gold, as Michael Riordan and Lillian Hoddeson wrote in their history of the transistor, "due to fate and his own obstinacy, Shockley never got a chance to enter this Promised Land himself." He led the way and parted the waters, but his followers passed him by. Andy Grove at Intel might have personality quirks, but he got his product out and—more important—had the respect, if not the affection, of those he tormented. Grove, still the chairman of Intel, was smart, efficient, funny, and on occasion could even laugh at himself—he was human.

If Fred Terman is the "Father of Silicon Valley," and Bob Noyce the "Mayor," then William Shockley might deserve to be remembered as the Moses of the Valley. But in fact, he's not remembered at all, his name more associated with the actor who played the saloonkeeper on TV's *Dr. Quinn, Medicine Woman.* There are no physics libraries named for William Shockley, who should be the silicon icon. His lab at 391 San Antonio, once the portal of Silicon Valley, is now a "1-stop ergonomic superstore!" home to "bionomics solutions" and some great desk chairs. Though they did wonder why there were so many old electrical fuse panels in the closet, the folks who own the store only know about Shockley because an old colleague or historian occasionally pops in. The city council of Mountain View has yet to put up a historical plaque for the place. Who else could be a more tragic figure on the journey to Canaan?

Shockley had the dream first, yet it escaped him. And rightly so. He did himself in.

Chapter IV
Prophets

Up in the madrone-filled hills of Los Gatos, on the south-western edge of Silicon Valley, lives one of the personages of the computer revolution, the soul of the old machine. Some would say Steve Wozniak was key to launching the revolution, as much as or more so than his childhood friend Steve Jobs. Now approaching fifty, the Woz (rhymes with Oz), as he sometimes signs his name and is usually called, doesn't spend his time inventing the future. Instead, he teaches those hackers who might. And, of course, he does it in a garage.

Los Gatos, a less horse-ified version of Woodside, is where Olivia de Havilland and John Steinbeck used to live, but now it's Steve Wozniak there. Every summer on Blackberry Hill, in one of the spare houses he owns, Woz spends hundreds of hours teaching a carpeted garage full of local Lexington Elementary kids—half boys, half girls—how to use a computer. The brand is Apple, which makes sense enough because he cofounded the company in 1976. On one Friday afternoon, Wozniak, still bearded and a little more bellied than he was back then, is dressed in a striped cotton sweater, white jeans, and white sneakers. If it weren't for all the cables and hardware running along the floor and across the ceiling, he could pass for a dentist.

But the students, and a few parents watching off to the side, know who that is, sitting up front by the whiteboard with his three laptop computers and his laser pointer. "I'm the Woz," he tells fifteen fifth-graders, including the boy with blue-green hair, his son Gary. There's no affectation, no boast about it—he's just giving them his name. His voice almost chirps, all the more so when he's speaking fast. "What's the best computer?" he asks the students. "Mac!" they intone back, using the other name, short for Macintosh, that Apple products go by. One of the young assistants that Woz employs turns to me and insists the kids know instinctively to give this rousing show to the teacher. Who wouldn't? This is the inventor teaching them how to use *his* invention. It's like taking Sunday school with Jesus or getting batting tips from the Babe. But Wozniak is quite blasé about it, much as you'd expect from any fifth-grade teacher. This Everyman quality adds to his charm—his fortune of $50 million or so notwithstanding—and makes more remarkable what he did when he was twenty-five. Wozniak is uninhibited, comfortable with the quieter profile he's assumed in middle age. By contrast, Jobs seems to be gnashing his teeth by the mere act of conversing. (When I first talked to him, Jobs—in black T-shirt and sandals—asked me why I was wearing a tie. I told him, with a smile, that I was trying to show him some respect. He just rolled his eyes.)

Each student in Wozniak's garage gets a Powerbook laptop loaded with software, and an account with America Online for E-mail and Internet access. Wozniak pays for it all, just as he has every year since 1991. If they want to keep the machines, he'll sell them at cost, though there's a rumor that he gives a lot of them away. This is the "Woz class," to the students. Most of Lexington Elementary takes it at some point. The instruction is great. But the other good reason for being in the Woz class seems to be the perks. Even with his teaching mission, Wozniak seems to understand his audience. They have limited attention spans over the course of five hours, which may be why they have unlimited access to Jolt and Surge to wash down the unlimited pizza.

What are the other fringe benefits? Woz's place is the best-equipped fun house this side of a Barb Ellison party. In addition to the $5.5 million indoor-outdoor pool complex overlooking San Jose—with built-in underground speakers—Woz has arcade games everywhere. There are games to simulate driving, skiing, hockey, and bowling; the venerable Pac-Man; even paleolithic pinball machines that say "Tilt!" None require quarters. In all, in the upstairs and downstairs, there are twenty-nine different machines, which means that fifteen fifth-graders have it pretty good—sort of like a drug addict with free run in a pharmacy. For kicks more than surveillance, the toys are monitored by minicameras, which occasionally are hooked up to the Internet, which has the strange effect of letting parents at home watch what their kids aren't learning about computers at Wozniak's house up the road.

The little cameras are of much amusement to Woz. In his office, he's got one pointed at his desk, so that any Internet wanderer can see what he's doing. His favorite camera is the one at the Blackberry house. Remember that in early 1998 the president of the United States took a room at Steve Jobs's place in Woodside? "Clinton could've stayed here," Woz says. "We would've turned off the Bathroom Cam." It's typical Wozniak prankstering, but, sadly, it's also a dig at his former partner and former friend. The two of them, born and raised in the Valley, were of the place, nurtured by it and then remaking it in their own image. Once upon a time, the two owned the keys to the kingdom. Their story remains the decisive turning point in the history of Silicon Valley.

Intel's wildly successful third-generation microprocessor, the 8080, led to the first widely used personal computer. The primitive Altair 8800 was offered as a mail-order kit in the January 1975 issue of *Popular Electronics*. It could do almost nothing useful—the software industry didn't yet exist—and nobody much thought about spreadsheets or word processing. (In time, the joke became that the PC had four applications: spreadsheets and word processing . . . and spreadsheets and word processing.) The Altair had but 250 bytes of memory, less than a millionth of a 1990s machine. In short, it did not compute. Fully assembled, it was a rectangular teal box about the size of a microwave oven, with no keyboard or screen or even a joystick. It could be programmed only by flipping a row of eight toggle switches in the correct sequence, each one representing a digital yes or no; its only program made a bunch of lights blink on the front of the box—a parody of a computer by modern standards.

But those deficiencies didn't deter the magazine from announcing a new era, especially since the Altair cost only about $400 and required little more than a soldering gun and a hobbyist's dedication to put together. "PROJECT BREAKTHROUGH!" declared the cover. "World's First

Minicomputer Kit to Rival Commercial Models." The manufacturer, out of Albuquerque, was Micro Instrumentation and Telemetry Systems, whose acronym, MITS, was only slightly more catchy. (At least "Altair" was an in-joke, based on a faraway land in a *Star Trek* episode, the TV show of choice for hackers.) The chance to build your own machine— to create a little sector of the emerging digital universe—was irresistible (and it sure beat dealing with illogical, unpredictable humans). Several thousand readers sent in checks for the Altair and a generation of weenies was born; in the time of Bill Hewlett and Dave Packard, these same people were radio "hams" and backyard munitions experts. The Altairists weren't out to demystify computers or to show the public that the machines weren't electronic portents of evil. But their hobbyist fascination with computers filled a void left by the complete lack of interest in the consumer market shown by IBM and Hewlett-Packard and other established companies, even Intel.

Popular Electronics fudged the issue of the Altair's practicality. The computer, said the magazine, promised "manifold uses we cannot even think of at this time." The task of giving the primordial PC something to do would be up to a new breed of tekkies. In Menlo Park, just north of Palo Alto, a group of these computer hobbyists formed just after the Altair came out. They tinkered, they talked technique, they had electronic show-and-tell with their latest devices—it could just as well have been guys' poker night out, except that the cards had been replaced by gadgets. These "hackers," as many called themselves, were both industry pros and obsessed amateurs, including a few teenagers—so different from the scientists-in-white-coats who years earlier lorded over machines like the ENIAC. The mavericks called themselves the Homebrew Computer Club and met every other Wednesday in the auditorium at the Stanford Linear Accelerator.

Some Homebrewers preached politics, about how technology might foster community and democracy. Others talked of physics. Steve Wozniak just liked the gadgets. The son of a Lockheed missile designer, Woz grew up in and with Silicon Valley, in the town of Sunnyvale, just another of the flatland suburban pockets sprouting up after World War II. His neighborhood of Eichler-designed homes was surrounded by orchards, so much so that he remembers riding through them on his bicycle to get to school. "I was lucky to be in a comfortable and beautiful environ-

ment," he says. It was during Woz's childhood that Shockley and Fairchild started semiconductor companies, and Noyce and Hoff invented
the miniaturized wonders that made a personal computer possible.

While other boys welcomed the New York Giants to the Bay Area and
rooted them on against the Los Angeles Dodgers, Wozniak followed the
local technology companies, attending electronics shows with his father.
Other boys collected baseball cards, but he traded computer manuals.
Wozniak lived in his own enchanted electronics world and in it he was
Merlin. After school, he and a friend would sneak into computer labs at
nearby companies to conduct experiments. At thirteen, he won a science
fair by building his own addition-subtraction machine from transistors
that a Fairchild engineer gave him. When he daydreamed in class, his
doodles consisted of schematics for rudimentary computers. He still remembers the first time his father showed him an early integrated circuit
from Fairchild. "I got interested in electronics because of him," Wozniak
says. "He showed me things on a blackboard, started teaching me equa-
tions. I read little books about Tom Swift, this engineer who would design
anything to solve any problem, whatever the conflict of the story was.
For an alien from outer space, he'd make some electromagnetic thing.
And I just admired that when you can design things, you can make neat
things happen. I wanted to be like Tom Swift."

But the Woz of early lore wasn't so much about technical ingenuity.
He was about pranks. If his father gave him an interest in electronics,
his mother taught him how to laugh. Pranks were a way to show intellectual prowess and rebellion at the same time. They were also just fun.
Some pranks were simple; others required great effort, which he was all
too happy to put in. Once he put a metronome in a friend's locker at
Homestead High and half the school thought it was a bomb. Later on,
he ran Dial-A-Joke, a free phone service that specialized in the Polish
variety—as told by Woz in his best Warsaw dialect. Wozniak, of course,
was of Polish descent. His number supposedly was the most frequently
dialed in the entire 415 exchange. When the Polish American Congress
complained, Woz—showing rare diplomacy—switched to Italian jokes.
The best part of his antics was the delight he took in them. There wasn't
much point in the gag if you couldn't howl about it afterward, over and
over. (Even today, he can't resist. He likes to accumulate phone numbers
in his 408 area code with consecutive matching digits—all seven, if

possible. One of them, with two sets of repeating digits, happened to be the phone number for Pan Am reservations—minus the 800 prefix. Woz got frequent calls by mistake—from people forgetting to dial 1-800 first—and he routinely promised free trips to exotic lands or special discounts to New York City for passengers willing to endure twenty hours on an old propeller plane routed through Billings, Montana. This, as one writer noted, was the "rare case of the prank phone call *initiated by the recipient.*")

His most celebrated prank came when he and Jobs heard about "blue boxes," which let users illegally make free long-distance calls anywhere. *Esquire* had run a piece about a subversive character named Captain Crunch who traveled the country showing young renegades how to beat Ma Bell. Captain Crunch got his name from the breakfast cereal boxes that included free whistles that just happened to work like a blue box: Toot into the phone and the crack AT&T network opened up a long-distance line. Inspired, and having found an AT&T technical manual, Woz built his own box. Not content merely to call Fresno or Fargo, or to make a reservation at the Ritz in London, or even make a call that goes around the globe so many times it begins to echo, Woz dialed up Rome, trying to reach the pope in the middle of the night. Using his best Eastern European accent—Polish or otherwise—Wozniak claimed to be Henry Kissinger. Maybe he'd have been better off imitating Andy Grove. The Vatican switchboard put him through to the bishop who'd do the translating, but that guy didn't fall for it—no doubt because he already knew about Dial-A-Joke.

In 1968, Wozniak, at eighteen, began a wayward detour through academe. With his folder full of 800s on SATs, he could have chosen any school. Instead, in his first journey outside California, he went off to college in Colorado ("I'd never seen snow—that was it for me"). From there, he returned to the Valley to attend community college, and finally enrolled at the University of California at Berkeley to study engineering and computer science. But more significant than whatever courses he wandered through, Wozniak hooked up with Steve Jobs, five years younger than him. Jobs had grown up in Mountain View of all places, not far from Shockley and Fairchild, raised by adoptive parents who knew little of high-tech.

After the Jobs family moved to Los Altos, Woz met him through a mutual school friend, just before Jobs turned fourteen. While they shared an interest in computers, they were driven by different compulsions. Woz, like Walter Brattain, was a tinkerer. He liked the intellectual challenge of creating something and of understanding the way things worked. Jobs, by contrast, seemed to see electronics as a means to an end, much as Fred Terman did. For Woz, the fun was in the chase; he once told an interviewer that in playing tennis, "the winning isn't as important as the running after the ball." Jobs just wanted to win and, better yet, to sell all the tickets to the stadium. Woz had no ambition, Jobs had nothing but. That desire, combined with his freight-train intensity and golden tongue, made Jobs formidable, creating the "reality distortion field" so often used to describe his charismatic effect.

Woz had built those mischievous blue boxes, but Jobs brought them to market, such as it was. Woz was attending Berkeley and wasn't looking to make money. Jobs, forty-five miles down the peninsula at Homestead High and an occasional visitor to Woz, saw a business. Between them, they sold plenty on campus. When Jobs arrived at Reed College up in Oregon in 1972, he found a whole new source of customers.

Like Wozniak, Jobs had little passion for college. Like others of the time and age, he had no taste for scholastic discipline, and looked the part with his long hair and wiry beard. He had decided against going to Stanford, he subsequently explained to historians Paul Freiberger and Michael Swaine, because "everyone there knew what they wanted to do with their lives. And I didn't know what I wanted to do with my life at all." Reed was known as a haven for the counterculture. Jobs's floundering might have seemed at odds with the preternatural ambitions he soon exhibited, but he was still only a teenager and the success with blue boxes was essentially a lark. It would be three years before Woz came up with the ultimate machine.

Jobs lasted a year at Reed, never quite fitting in. He discovered vegetarianism, meditation, and Eastern religion; then picked apples on a commune, went home and worked for a Sunnyvale start-up named Atari; and then headed off to India, barefoot, to find spiritualism and his inner geek, dysentery notwithstanding. It was the kind of flaky thing that would only set him further apart, generationally and symbolically, from the Hewletts and Packards, Moores and Groves of Silicon Valley. With their engineering mentality and strait-laced ways, those men were hardly typical. But Jobs was from another dimension, the product of an entirely peculiar time.

When his funds ran out after three months, Jobs reappeared in California, moving back in with his parents and back to a job at Atari. By then, Wozniak had taken a leave of absence from Berkeley to go work on calculator design at Hewlett-Packard. He may not have been the prototype for the HP Way, but Woz was barely twenty-two and the company's tent was big enough to accommodate a talented junior engineer, even one who lived for pranks and games. Jobs, too, had once worked at HP—at age thirteen, winning that summer job in classic fashion. He had been building an electronic counting device and needed extra parts. So he just picked up the phone and called Bill Hewlett, the cofounder of the company. Hewlett offered him the parts—and a job.

Now, seven years later, Jobs was working again in the Valley. He and Woz were reunited on the same continent, within several miles of each other. Atari was a Sunnyvale start-up making video games for arcades. Conceived by Nolan Bushnell, Atari had developed the revolutionary Pong, the mother of all computer games and the bane of bowling alleys coast to coast; "Pong" consisted of two players hitting a round object, but the paddles and ball were electronic apparitions on a screen.

Jobs was hired at Atari simply because he showed up one day. Al Alcorn, an Atari cofounder, says: "The personnel director comes to me and says he's got this hippie who looks like he just came out from the mountains. The personnel guy asks, 'Shall I call the cops?'" Alcorn was intrigued and hired him instead, even though Jobs gave no indication he knew anything about technology. The engineers grumbled to Alcorn, who recalls: "One of them asked me, 'What did I do to deserve this? The kid smells and has no skills.'" When Jobs announced he was leaving to go to India, Alcorn asked him to run an Atari errand in Germany. "To this day," says Alcorn, "I cannot imagine Steve Jobs in Bavaria—there's not much of a vegetarian diet there." Somehow, Jobs survived the mission.

The critical thing Jobs brought to Atari was Woz, who himself had earlier been offered a job at Atari, after demonstrating his own smartass variation of Pong to Bushnell and Alcorn. After a player missed, Wozniak's said, "Oh Shit!" on-screen. Woz stayed at HP, but had the best of both worlds. After hours, he went over to Jobs's Atari offices and played arcade games till dawn—and for free. "The best thing about hiring Jobs," Alcorn says, "is that he brought along Woz to visit a lot. Creativity often means concealing your sources." Woz wound up being responsible for a single-player version of Pong that Bushnell had wanted. The game was called "Breakout." Jobs said he could make it for the company. Sort of—just as long as Woz visited. During four all-nighters, Wozniak designed "Breakout" and Jobs assembled it. He and Jobs were so exhausted they each got mononucleosis. But the game worked.

Atari was started in 1972 by Nolan Bushnell, who himself was a cross between the impish Wozniak and ever-peddling Jobs. Bushnell's Atari ushered in the world of adrenalized, addictive video games that to this day represent the "killer application" for the home computer owner. A generation ago, the now quaint Pong was no less intoxicating than today's Doom. While Steve Jobs was still growing a beard in high school, Bushnell was the prince of consumer electronics. And far more than Hewlett-Packard or Intel, Atari was the first Silicon Valley company that had an electronics product with mass appeal. Its disappearance simply highlights the ephemeral nature of success in the Valley.

The twenty-nine-year-old Bushnell looked larger than life. A bear of a man at six foot four, with dark curly hair, full beard, and a beaming, round face, Bushnell could fill a room even before he started talking up an idea. His smile hid the heart of a huckster and his pipe gave him an air of dignity. Bushnell's idea for Atari could not have come along at a better time, given the revolution in the Valley. Raised outside Salt Lake City in a Mormon family, he didn't have Wozniak's or Jobs's advantage of being around a community of engineers and the new electronics. His father was a cement contractor. But one day in third grade, his teacher Mrs. Cook put him in charge of the "science box" for the unit on electricity. With all the batteries and wire and old gadgets he could find at home, Bushnell began experimenting. By the time he was ten, he had a ham-radio license. His mother didn't go near his room out of fear she'd be electrocuted. One time, he set the family garage ablaze with a liquid-fuel rocket mounted on a roller skate.

During engineering school, at the University of Utah, Bushnell learned two important things. First, the university's giant mainframe had a great shoot-'em-up video game called Spacewar; like so many other useful creations, video games first came out of the Pentagon as a way, the generals hoped, to teach battle strategy. Second, Bushnell became the best carnival barker in the county. Working on commission at the knock-down-the-milk-bottles concession, Bushnell made a bundle. Merging these interests—games and huckstering—is what begat Atari. Bushnell also learned that he wanted out of Utah. After graduation, he escaped to the excitement of Silicon Valley, going to work for Ampex, the old company that pioneered sound on tape. "It was the only place I saw Nolan wearing a tie and a lab coat," says Alcorn, a former Ampex colleague who left with him to found Atari (who was succeeded as head of Atari R&D by Ted Hoff, coinventor of the microprocessor, proving yet again how nobody quite quits in Silicon Valley, just recycles to another company). "Nolan was a mediocre-to-bad engineer, more interested in the stock-investment group he formed. But he was full of wonderful ideas and knew everything there was to know about the coin-operated amusement business."

Spacewar inspired Bushnell over the course of several years to create a similar game on a smaller, cheaper computer. The result was Computer Space, the first retail arcade game using an integrated circuit. Bushnell's dentist hooked him up with another patient, the only arcade-machine manufacturer west of the Mississippi. But the game barely sold. Not because of its simplistic lines and pixels, Bushnell realized, but because the instructions were too complicated. "To be successful," he said, "I had to come up with a game people already knew—something so simple that any drunk in any bar could play."

The game was Pong, and it was designed in three months by Alcorn, who was then only twenty-four years old. Were Pong any simpler, it would've bored Koko the Gorilla. It required no instructions. It was table tennis on an arcade screen (later available as a plug-in for home TV). *Pong* was also the hollow sound the "ball" made when hitting a player's "paddle." Bushnell and Alcorn started Atari with a mere $500 in funding and a $2,000-a-month royalty stream from Computer Space and other games. Their company car was a turquoise Oldsmobile wagon. Bushnell tried to hire creative types like himself, asking engineers what games they played, as well as how to wire a light switch at a house; it was amazing how few could answer either question. Venture capitalists had no interest in investing. "It may sound ludicrous," Bushnell subsequently observed, "but people thought the idea of playing games on a television set was the stupidest idea they'd ever heard of." One venture capitalist asked him, "So who's going to pay a quarter to play a silly game?"

"Atari" roughly meant "checkmate" in the ancient Japanese pebble game of "go," which Bushnell adored. As Bushnell liked to explain it, "atari" was a "polite warning to your opponent that you were about to take his stones." It was also Bushnell's second choice for a name. Initially, he called the company "Syzygy," which meant a lining up of the planets and was the last *s* entry in the dictionary; it was cute, but a roofing contractor already had the name and Bushnell's lawyers quivered.

Bushnell decided to test the Pong prototype in a corner of a Sunnyvale bar named Andy Capp's—right next to the Computer Space machine. At first, the regulars weren't sure what to make of the paddleball bouncing back and forth on the screen, awaiting somebody with a few quarters. Then two people stepped forward, put in a coin, and watched as the ball bounced and the score changed. Maybe this was too complicated, too. After all, if each player didn't bother to turn his knob, there wasn't much of a game. But at 3-3, one of them figured it out, then the other. The game was under way, and another game and another. The "pong" noise drew others in the bar. By closing time, the story goes, everybody in the house had played.

The next morning, there was a line out front—and it wasn't for beer. That night, the game stopped working—not because of any defect, but because the coin bucket had spilled over into the guts of the machine. Bushnell knew he had a hit, a once-in-a-lifetime consumer creation that defined a genre. Coney Island meets Silicon Valley. Within several years, Pong and games like it became an item of furniture at saloons and malls and the kind of amusement parks that Bushnell first worked at. More than ten thousand Pong machines were sold, for several thousand dollars apiece, adding up to tens of millions for Atari. By 1975, Sears was retailing the home version. Christmas morning in America—*boing! boing!*

boing! Sears sold three million of them. Bushnell was giddy: When his Sears partners visited the Atari packaging factory, Bushnell got in a box and circled round the floor on a conveyor belt.

The following year, Bushnell cashed out, selling Atari to Warner Communications for $28 million. (Disney and MCA passed on the bidding.) Bushnell pocketed roughly half. His fiscal timing was good. His young company still hadn't peaked—sales would hit more than $400 million in 1980, when Atari licensed the home version of Pac-Man—but it was closer than anyone dared realize. In 1983, Atari lost $536 million, at the time the largest puddle of red ink in American corporate history. Japanese competitors, with a better sense of the games market and of how to harness the power of the microprocessor, developed superior products. That's why Nintendo and Sega are now household names and Atari is a relic; the Japanese companies gave the industry a go and took Atari's "stones." The demise of Atari became as legendary in the Valley as its sensational rise.

But Bushnell didn't take his millions and act like any self-respecting Valley tycoon of the mid-1970s and just find a comfortable California contemporary in Palo Alto or Atherton. Sure, he had the new wife, the new Rolls-Royce, a yacht named *Pong*, a Learjet, a condo in Aspen, a spread in Paris, and the swagger to appear in the *San Francisco Chronicle* in a hot tub with a hot number. No, he did it even better: He discovered Woodside, which previously had been the domain of horsies and wealthy San Franciscans desirous of peace and quiet in the hills. Flamboyant to a fault, Bushnell bought the old sixteen-acre Folger Estate, a megalomansion from a different era. This was a real estate—a thirty-seven-room, sixteen-thousand-square-foot Edwardian manor house nestled in the redwoods and built at the turn of the century by James Folger III, heir to the Folger's Coffee fortune (those tiny little crystals and that really big bank account). The original Mr. Folger arrived in California during the Gold Rush and was shrewd enough to run a coffee business in addition to his digging. The Folgers got rid of the place because their granddaughter, Abigail "Gibby" Folger, was one of Charles Manson's murder victims in 1969 in the Hollywood Hills, and the family was haunted by the memories left by her childhood in the four-story Woodside house.

Bushnell filled the great house with eight children, talking stuffed animals, all manner of puzzles and toys, an ice-cream parlor, a theater, and furniture that ranged in taste from velour to raw silk. The rocker Neil Young and his wife—fellow Woodsiders—were frequent guests. In time, though, like the man who owned it, the apple-green estate became run-down, badly in need of renovation; Bushnell's wife joked that she was on a first-name basis with the plumber. In part to solve his own serious financial problems, Bushnell put the property on the market for $8.9 million, settling for barely half that in 1997. He was off to London for a year and then on to L.A.

In between, ever the entrepreneur, Bushnell founded a range of businesses. Some did well: Chuck E. Cheese, the chain of hyperkinetic entertainment restaurants combining pizza and video games, mascotted by a rat; and Lion & Compass, the upscalish Sunnyvale hot spot that replaced the Wagon Wheel in the high-tech pecking order. (To this day, Lion & Compass is operated by Bushnell's brother-in-law, and still runs a stock ticker near the bar and has phone jacks at the tables, a tribute to the days before cellular.) Other businesses, like online gaming and an "interactive dining room" for adults, got bursts of publicity because of Bushnell, but never went anywhere. Still others of his ventures got mucked up in litigation, where all good ideas go to die. He even got himself sued for $50 million by a former employee who alleged that Bushnell infected her with herpes; the San Jose *Mercury News* ran the story on the front page, though it didn't bother to note his later exoneration as prominently. And then there were the personal robots. Somewhere out there in the ether of the Valley, they still talk about his Androbot venture, though in August 1983 the proposed stock offering for the company had to be pulled just before its scheduled issue. That Wall Street debacle cost Bushnell millions and got him enmeshed in a long-running litigation with Merrill Lynch, the primary backer—a legal battle that continued into 1999.

Androbot Inc. was the 1980s version of today's typical Internet start-up. No earnings, no profits—no problem! All you needed was a little hype, which Bushnell was perfect at. *Playboy* paid a visit, along with an NBC camera crew. The robots were the hot gizmo of the 1983 Consumers Electronics Show in Las Vegas. Bushnell intended $500 personal robots to be the biggest thing to hit since Pong. Bob (for "Brains on Board") would vacuum the house and fetch you a beer; Fred (for "Friendly Robotic Educational Device") would teach the kids math. There was even AndroMan, who could plug into the home version of Atari—Pac-Man unleashed! Bushnell explained to Herb Caen, the San Francisco columnist, "Wouldn't you like someone to come into your room in the morning and say, 'O great and omniscient one, are you ready for your coffee?' "

The only problem was that despite the on-board microprocessors, sonar, and infrared sensors, Bushnell's robots couldn't do any of their appointed tasks. They were merely engineering dreams; at best, waist-high Bob could navigate a room without bumping into the bookcase, not much better than a dachshund with good eyesight. For that kind of companionship, you could buy some tropical fish. And for the kind of money Bob cost, consumers wanted something more useful—for which the engineering challenge was insurmountable. Nonetheless, in the underwriting dreams of Merrill Lynch, Androbot, incredibly, was supposed to be worth $90 million. "More perfume than substance," said one financier, presaging the Internet-driven high-tech market of 1995 to 1999.

Bushnell seemed to be an entrepreneurial dilettante, bored the minute a good idea required execution. At Atari, Alcorn designed a security system in Atari's lab to warn the engineers when Bushnell approached; otherwise, Bushnell would jitterbug around with the prototypes and order something entirely new. "Golden retrievers have a longer attention span," Alcorn says now. Reflecting on his life a few years ago, Bushnell acknowledged as much. "I've always got to force myself into completion," he explained to an interviewer. "I like being the guy with the machete hacking his way through the jungle, never to go that way again. That means that I have to get good and talented people to take care of the other things."

It usually didn't happen that way, even at Atari. In many ways, Nolan Bushnell might have felt more at home if he'd been part of the Homebrewers.

The Homebrew Computer Club of Steve Wozniak was like organized religion. Individually, the members did their own thing and thought their own thoughts. Together, they were a congregation. "It was a unique time in the Valley," recalls Jim Warren, one of the early Homebrew members. "The main characteristic of people was a willingness to share with each other—standing on each other's shoulders rather than on each other's toes. We were part of the late 1960s—hippies, antiwar, members of the Free Speech Movement—that was going to fix all the problems of the world, even if it took a year or two." Warren was a little more unorthodox than the rest. He once had taught math at a local Catholic women's college. But when the nuns found out he was throwing nude parties at his house, they suggested he might not be the ideal role model, and he resigned. The subculture that Homebrew embodied could not have been more different from the worlds of Intel or Hewlett-Packard. Whereas the captains of hardware by and large were conservative, nine-to-five Republicans, this new breed of rebels got up at noon and worked till dawn, taking a shower when it suited them. It made political sense that Intel could trace its roots to the 1950s; it made just as much sense that Apple could not.

Despite the success of the blue boxes—Ike Turner bought one—Wozniak was not looking to commercialize a product. But hanging out with other hackers every two weeks not only exposed him to new ideas, it spurred him to outdo them. Building a better computer wouldn't so much mean glory as a chance to impress his friends—the "personal" in "PC." He'd own his own dream machine, designed just the way he wanted it. After all, he didn't have the money to buy a fully equipped Altair. The other club members quickly recognized Woz as a technical wizard, and that only made him more eager to create.

Attending a computer show in San Francisco, Wozniak found a small semiconductor firm that was selling microprocessors for twenty dollars. It wasn't an Intel chip, but MOS Technology's 6502 was more than enough for what he had in mind. He bought a bagful, not forgetting his Homebrew friends. Around one 6502, using parts he "borrowed" from HP, he built a computer. It wasn't really a finished offering like the Altair, but a board of circuits that could then be hooked up to a keyboard and monitor. Despite its unfinished look, the machine was an improvement on the Altair because it used fewer components. "I'm into it for esthetic purposes and I like to consider myself clever," Wozniak told Steven Levy in *Hackers*. "That's my puzzle, and I do designs that use one less chip than the last guy. . . . Every problem has a better solution when you start thinking it differently than the normal way." That's what drove him. "I would have something to show off and I hoped that other people would see [my inventions] and say, 'Thank God, that's how I want to do it,' and that's what I got from the Homebrew Club."

And it might have been enough. His friends were duly dazzled and gladly accepted copies of the digital blueprints he handed out. But Steve Jobs, who had started coming to Homebrew meetings and was more kibitzer than inventor, viewed this as a trifle. If a few hobbyists were this captivated, what would happen in the larger marketplace? Jobs intuitively understood the significance of what Gordon Moore had recognized a decade earlier: Because of miniaturization, chips would inevitably get cheaper (and faster) and would necessarily become available to just about anybody. Jobs thought: What if Wozniak could be persuaded to make his computer in quantity? The trick was to convince him that the thrill he got in an auditorium could be multiplied out by a business. Jobs was relentless and manipulative, and Woz went along. He would continue to be the engineer and Jobs would be the master salesman. Together, to the other Homebrewers, they were "the two Steves." Jobs got local stores to lend them parts on twenty-nine days' credit. Their negotiating routine, at times, was like Abbott and Costello. At one supplier, Jobs was going through his well-rehearsed script to get the price down, while Woz kept saying he just wanted the parts. Jobs was across the table and tried to signal Woz by giving him a kick underneath. Instead, Jobs missed and slipped right under the table.

It was now early 1976. Jobs was twenty-one, and Wozniak going on twenty-six. They began to take the idea of a business seriously: Woz sold his HP-65 calculator and Jobs, his VW bus—though they had every intention of making enough money to buy back their treasured possessions. But before they advertised in a trade publication, their embryonic product needed a name. How they settled on "Apple Computer" depends on which version of history you believe. It could have been a reminder of Jobs's fond feeling for his orchard days in Oregon, or his occasional all-fruit diet, or a play on "bytes," or that the name would appear on the first page of any phone book, or a reference to the Apple record label of the Beatles. Or "Apple" could just have been a way to create a sweet-sounding, all-American brand name in a high-tech industry not known for radiating warmth. Nobody conjured up Johnny Appleseed when they thought of "Fairchild Semiconductor" or "International Business Machines." The original Apple logo was unimaginative, showing somebody sitting under an apple tree. But the company soon came up with the friendly rainbow apple—one bite missing—that remains the best-known trademark in the computer universe (even with the latter-day "lemon" jokes). Jobs was selling an image as much as a maze of circuits. (The bite was supposedly added so nobody thought the apple was a cherry.) Hewlett-Packard was run by engineers and their customers were engineers. Intel didn't market directly to end users. Apple was different, selling to everyday consumers.

In person, with his glower seemingly too strong for his rail-thin body, Jobs radiated zeal. But even behind the scenes, he was an instinctive marketer with a flair for showmanship. At one National Computer Conference in Anaheim, then the leading digital gathering in the world, Jobs wanted some attention for his company. The conference's organizers had barely consented to include personal computers in the same category as industrial-strength mainframes; but there was no way Apple would be allowed in the main exhibition hall. Relegated to the Disneyland Hotel, Jobs rented out the entire theme park and invited all ninety thousand conferees. Few had ever seen a stunt like Jobs's.

The first Apple machine—the Apple I, as it came to be known—sold for $666.66. It was available at computer stores around Silicon Valley and by mail order. Working out of Jobs's garage at 2066 Crist Drive in Los Altos—would he have preferred an *h* be added?—Woz and Jobs sold about 175, a number that said more about supply than demand. In order to continue, they needed capital. While Hewlett-Packard and Atari had indulged their respective employees' tinkering—on the job and off— neither company was now interested in bankrolling it. Woz and Jobs were told the Apple wasn't practical, wasn't marketable, and hadn't even been designed by real engineers. If HP had a blind spot, it was its inability to nurture the start-up mentality, though the company in time did enter the personal computer market. Neither was Intel interested in the Apple boys. When Jobs came by, Gordon Moore dismissed his machine as just another device in which to put a microprocessor. Even Bob Noyce, by this point the éminence grise of Silicon Valley, passed up the chance to back the two Steves personally. His second wife, Ann Bowers, had heard of their start-up, but he laughed it off. Later on, she did fine in her own right—as Apple's vice-president of human resources and a recipient of jumbo stock options.

Throughout the latter half of 1976, Wozniak worked on a new, more powerful computer, the Apple II. This one would be a more elegant design—like a typewriter—and be more functional, too. At twelve pounds, it had a regular keyboard, a power supply, and eventually, a floppy disk drive to store data. The machine came fully assembled, in a tapered, beige plastic casing with the Apple II logo on top. It was easy to program and could generate color graphics and motion. All you had to do was hook it up to a TV set or other monitor. Some of the Homebrew crowd were disappointed that the machine didn't come as a kit, but acknowledged its technical panache. Whatever its foibles to a hobbyist, the Apple II represented the mass-market product Jobs was looking for— the machine responsible for an entire species of small computers to follow. It was at Jobs's direction that an industrial designer was consulted about the computer's packaging. The Apple II *looked* like an appliance that a normal person could use, without having a fourteen-year-old tech-noid hovering nearby. The Apple I, raw and intimidating, its silicon innards always showing as if part of an uncompleted experiment, could never have that broad appeal. Jobs set the Apple II's price at $1,298, though Wozniak still protested at the crass economics of it all.

Even when the Apple II was on the drawing board, Jobs set out to find backers. As self-assured as he was, he knew that his lack of resources and experience stood in the way of creating a real company. It was great that two of the first employees were local high-school students, but nei-ther was likely to be running the accounting department. Nolan Bushnell was interested in providing capital, but his dough was tied up in Chuck E. Cheese. In his first critical hiring move, Jobs first persuaded Regis McKenna, who represented Intel and was the Valley's best-known flack, to handle Apple. Initially, McKenna told him to "go away." But Jobs, a cross between a ferocious tiger and a nudge, kept coming back. Finally, McKenna gave in and came up with the multicolored logo, as well as the notion to put an ad in *Playboy* to get attention.

McKenna and Atari's Al Alcorn introduced Jobs to Don Valentine, an early venture capitalist and the founder of Sequoia Capital. Witnessing Jobs in full grunge regalia, Valentine uttered to McKenna one of the memorable lines in Jobs lore: "Why did you send me this renegade from the human race?" Missing out on a fabulous opportunity, Valentine passed Jobs off to a recently retired, amiable thirty-four-year-old who had started at Bob Noyce's Fairchild and went on to run marketing at Intel. His name was A. C. Markkula, Jr., whom everyone called "Mike," since the "A. C." stood for Armas Clifford. When Intel went public and its stock kept going up, Markkula made millions. Now, in the sunset years before he hit forty, he planned to spend time with his family, ski the runs at Tahoe, and play some guitar. Then he met Jobs and Wozniak.

Though an engineer by vocation and gadgeteer by avocation, Mike Markkula's skill was political: He understood how the Valley worked and he knew its players. That was the way he had thrived amid the internal battles at Intel. While different in style from Jobs—who wasn't?—Markkula saw the same potential for personal computers as Jobs did. Markkula also could identify with Woz's hacker instinct. It was Markkula, for example, who urged that a floppy disk drive be part of the Apple II (rather than using cassette tape storage) to make it useful for businesses storing accounting data; under the name "Johnny Appleseed," Markkula also wrote some early software for the machine. But it was the evangelism of Jobs that finally won Markkula over. In return for a third of the company, Markkula put up $91,000 of his own money and helped the Steves write a business plan. Just as important, he agreed to oversee the business side and help raise several hundred thousand dollars more from the established venture capital community.

To do that, Markkula called Arthur Rock, of Fairchild and Intel renown. Rock was skeptical, signing on only after Markkula cajoled him to attend a few computer shows. "Nobody was at any of the booths other than Apple's," Rock recalls. "I couldn't even get close enough to touch the machine. It was like Willie Mays–plus." Or like Steve Jobs, whom Rock calls the "most Svengali-like person I've ever met." Apple would be yet another gold strike for Rock, who became a member of the board of directors. Illustrating the ever-growing network of the Valley, his decision to invest was based on a presentation Jobs and Wozniak made at an Intel staff meeting. Venrock, the venture arm of the Rockefeller family, ponied up close to half a million dollars as well.

With a grown-up and third cofounder on board, and some cash in the bank, Apple went legit at the beginning of 1977. It formally became a corporation and moved out of the garage and into small offices in Cupertino, where its sprawling campus would one day be built. The lobby displayed the prototype of Woz's Apple I circuit board, with a little plaque, "Our Founder," underneath—proving that even Jobs could show some humility. Jobs quit Atari and Woz finally left HP, resigned to the vulgar fact that he'd now actually have to make a living from his hacking. When the company got around to ID badges, somehow Jobs was only No. 2. The sport of Homebrew was giving way to the lucre of the marketplace. In the span of just a few years, the PC industry mutated from carefree teenager into avaricious capitalist.

The Apple II would be the core of the Jobs-Wozniak-Markkula business. It made its debut at the inaugural West Coast Computer Faire in April 1977, organized by Jim Warren, the pot-smoking, hot-tub hippie who made Woz seem establishment. Warren edited a magazine for programmers and was now becoming entrepreneurial. Big computer shows were popping up around the country, but not in the Bay Area. He decided to put one together in downtown San Francisco. It was an odd sight at the Civic Center—a futuristic gathering in one of the city's grand, old halls, with the feel of a carnival and the fervor of a revival. It was also a colossal success and Warren made a killing. More than thirteen thousand paying customers showed up—beyond anybody's wildest dreams, including those of the poor people standing in lines for hours to get in. Watching it all, Markkula remarked to Wozniak in wonder, "This shows the revolution is really happening."

Steve Jobs managed to get the biggest, coolest booth right at the entrance to the exhibition. Serving as sentinels at the door, the magnificent Apple IIs were the hit of the show. Woz had the best time of all. Unable to resist a prank, he distributed a brochure for the "Zaltair," which supposedly ran circles not only around the Altair but the Apple II and every other personal computer in the building. "Imagine a dream machine," the brochure read, according to historians Freiberger and Swaine. "Imagine the computer surprise of the century, here today." The joke was aimed at Jobs, whose salesmanship Woz had come to know so well.

At the beginning of 1978, just a year after incorporating, Apple was already turning a profit. The company's take was more than $2 million. Markkula was always finding new dealers to carry the Apple II. Sales more than tripled the next year and grew fivefold the next, in part because of the educational market. Phenomenally, three years later, the total stood at $335 million, more than a third of the entire PC business. Apple entered the Fortune 500 faster than any company, and Steve Jobs was the youngest person ever to make the Forbes 400 list of tycoonery. The little electronic lark Steve Wozniak thought up had sold more than 300,000 machines by late 1981 (on its way to a total of five million over a seventeen-year period). The horsepower of the ENIAC was now available to Everyman, to be used by just one person rather than an army hunkered down in a climate-controlled citadel. With memory devices and other accessories to connect to printers and phone lines, the Apple II sold from $1,298 on up.

From nothingness, the personal computer had become the fastest-growing industry in American history, a billion-dollar triumph spurred by the dream of one college dropout and the engineering virtuosity of another. During one decade, Apple alone reached $1 billion in sales, a capitalist feat accomplished by only one other company—Xerox, after its 1959 launch of the copying machine. Apple was not only a commercial success—the beginnings of the Information Era—but the societal one that Jobs dreamed of just as much. In part because of shrewd advertising and in part because he and Woz were lucky, Apple Computer became a symbol and an image—a revolution in a box. T-shirts and bumper stickers proliferated in a way that no Fairchild chip could match. Hackers started a club called The Apple Core, with a newsletter titled *The Cider Press.*

As the Apple II became more popular, independent programmers wrote more software to give the machine things to do. The box represented nothing more than potential, but it needed instructions. A piano is hardware; the sheet music that makes it sing is software. The interplay between hardware and software was a chicken-and-egg cycle that would replicate itself in the PC industry later on as IBM and other competitors entered the field. As a computer manufacturer—say, Apple in the 1990s—lost market share, it would also see the range of available software diminish, which would lead to a further reduction in market share. But that was a long way off.

In 1978 and 1979, there were roughly a hundred programs available for the Apple II, offering chess, bridge, war games, graphics, and rudimentary word processing. The key item of software—the "killer app," as it would be called now—was VisiCalc, an electronic spreadsheet introduced in 1979 by two Harvard MBAs. VisiCalc wasn't merely a way to balance a ledger. It allowed the user to do complex financial forecasting. If you changed one row or column of numbers in your chart, the program would adjust all other dependent numbers. Doing that kind of arithmetic manually would be absurd and not even a minicomputer had the capability. VisiCalc could do it in a blink. Backed by Arthur Rock, it sold 150,000 copies in 1981 and its practicality helped make the Apple machine a business tool. And why not? VisiCalc wasn't about flashing lights or performing other tricks—it was about profits and losses, debits and credits. It was about money.

Even for the boys. (Not that Wozniak could resist monkeying around with it and coming up with "VisiCrook," which undid some of VisiCalc's antifraud features.) Twelve days before Christmas Eve, 1980, the founders of Apple deliciously cashed in by going public. For a long time, Apple's was the most spectacular high-tech IPO since God took Earth public; the state of Massachusetts thought the price so speculative it banned trading in Apple shares. The IPO created the first of many price explosions in Valley real estate, especially in the more prestigious addresses like Woodside and Los Gatos. Wall Street loved the stock and so did the press. "Not since Eve has an Apple posed such temptation," *The Wall Street Journal* crowed. Even before the stock went up more, Jobs and Markkula were each worth $155 million on paper. A few years later, *Time* magazine featured Jobs sitting on the floor of his living room, legs crossed and barefoot. The point was to show the new tycoon—vegetarian and unaffected—living a simple life. The truth was, he hadn't gotten around to buying any furniture: He could afford to buy *any* couch, so the choosing was hard.

Woz would have been worth $155 million as well, but earlier he had given away some of his stock options—not just to family members, but to five Apple engineers who he thought had been gypped when the equity pie was divided up. He also sold a chunk of shares at a steep discount to forty other employees. "The Woz Plan," he called it. So Woz had to settle for $90 million, still enough to pay for his own long-distance phone

service. Scores of Apple's employees became millionaires. Apple's early
financial backers also were rewarded. Arthur Rock watched with delight
as Wall Street converted his $57,600 investment into $14.1 million. The
Rockefellers made almost six times that.

All this in a few years. The PC industry hadn't evolved from the
established computer companies—IBM and the so-called Seven Dwarfs
making mainframes (RCA, G.E., NCR, Honeywell, Burroughs, Control
Data, and Sperry Univac). This nascent business was an example of
spontaneous generation, from Homebrew hackers and an underground
of amateurs. But Apple soon bred competition. Radio Shack and Com-
modore and even Atari, among others, started selling their own personal
computers. Then, in late 1981, IBM finally took notice. Thirty-eight years
earlier, Big Blue's CEO had sneered: "I think there is a world market
for maybe five computers." Now, the multinational Sasquatch of mainframes that did an annual business of more than $25 *billion*, and for
years ignored the geeks-in-garages, wanted a piece of the PC industry.
It was like General Motors deciding that Matchbox needed a run for its
money. International Business Machines—its very name said big.

Meanwhile, Apple was going through its first rough patch. This was
unavoidable for a company that had grown so quickly but given so little
thought to the competition. In the spring of 1980, it announced the Apple
III, which was faster than the Apple II, had a built-in disk drive, and
was intended to appeal to the higher-end business crowd. It looked good,
too. But the new machine was a disaster technically and had to be with-
drawn from the market until 1981.

IBM's discovery of personal computers that year surprised no one, but
it nonetheless represented another high-tech watershed. The "IBM Personal Computer," matter-of-factly named, was the result of a year-long
crash effort by a maverick IBM development team based in Boca Raton,
Florida, which called itself, with characteristic immodesty, the "Manhattan Project." In uncharacteristic fashion, the slow-moving IBM managed to assemble, in a year, a machine that in days gone by might have
taken that long just to be funded. The machine was bigger than the Apple
II and looked more like something you'd find in the offices of a company
like, well, IBM. The point was not to try to be mom-and-pop like Apple.

IBM embodied American computers, and because of that hegemony
rather than any technical superiority, the IBM instantly became a player

in PCs. "IBM's entry erased any lingering doubts that personal computers are serious business," stated an article in *The New York Times*. Within two years, it dominated the market—like some celestial object hovering above a planet. Nevertheless, in character, Steve Jobs poked fun at it. "Welcome IBM. Seriously," read a sly ad that Apple placed in *The Wall Street Journal*. The taunt, as a subsequent Apple executive observed, was like Little Red Riding Hood going after the wolf.

Apple had its own celebrated ad—during the 1984 Super Bowl. Sixty seconds long, it ran just once on national TV and has been called the most effective corporate commercial ever broadcast—an event unto itself. The ad—for Apple's newest computer, the Macintosh—relied on dark Orwellian imagery more than Madison Avenue words. The setting was a great hall full of gray-clothed, bald-headed drones listening indifferently to their leader, who was projected on a huge screen. A brightly clothed woman bursts in, wielding an anti-establishment hammer and then flinging it at Big Brother. The screen blows up, the workers are liberated, and the ad concludes, "On January 24th, Apple Computer will introduce Macintosh. And you'll see why 1984 won't be like *1984*." IBM, of course, was Big Brother; Apple represented individuality. Its specific product aside, the Super Bowl ad perfectly captured Apple's counterculture view of itself—a company as a cause, not merely a cash register. Not Intel before, not Netscape or Yahoo after, could pull that off. Apple has a gift shop on its main campus to which tourist buses still pull up. But that cult status wouldn't be enough to keep Apple flying high.

The $2,495 Macintosh was the emotional high point of Apple's history, and the model name that become synonymous with the company. Sleek, cute, friendly—a triumph of taste—the Mac had more pizzazz than the Apple I or II. You turned it on, and as the machine came to life, the first thing you saw was a smiley face. (Apples still do that.) The Mac's breakthrough was a "GUI"—a graphical-user interface with such innovations as icons, pull-down menus, and a movable, handheld "mouse" that allowed your finger to control the action on the screen. You only had to "point and click," rather than type in a list of commands. The GUI on the screen meant WYSIWYG (pronounce it Whiz-ee-wig)—What You See Is What You Get. It transformed the computer screen into a visual metaphor—a desktop.

Even so, the GUI wasn't an Apple creation. It had its roots in the prior decade at Xerox's estimable Palo Alto Research Center. PARC, as everybody called it, had the mundane mission of R&D aimed at maintaining Xerox's stranglehold on the copier market. Its team of engineers was unmatched, as skilled as any roster of inventors and theorists that, say, Stanford could put up. Xerox was interested in possible deals with Apple and contacted Jobs, who asked for a tour. Three weeks before Christmas, 1979, Xerox opened the doors to him. He left with a fine holiday gift—the realization that a GUI was possible and that PARC management didn't seem to understand the commercial implications. Apple's Macintosh represented the GUI's ultimate fruition—a decade before any other computer did. It was, Jobs proclaimed, truly the "insanely great" machine "for the rest of us."

But while the Macintosh could legitimately be considered the progenitor of all PCs to follow, the Mac itself was hardly a raging sales success at the beginning. It was slow, it lacked digital storage space, and few software applications existed for it. Those problems could and would be solved. Yet however popular the Macintosh might turn out to be with individuals, it needed to be attractive to big corporations and that didn't happen. Corporate America was more comfortable with the IBM product. Not only was it IBM, the PC looked and acted the way a computer was supposed to act—it was ugly, unintuitive, "a man's computer designed by men for men," as one trade columnist put it. WYSIWYG it wasn't, but who had the bright idea that computers were supposed to be easy?

Jobs's standing at Apple was hardly helped by the Mac's commercial failure. It was no secret that he was prone to tantrums and that the most charitable description of his behavior was "idiosyncratic." The stories about him became a combination of Silicon Valley gossip and in-house clinical file. Since everybody in the Valley at one time or another seems to have worked at Apple, they all could repeat the stories. The best one had nothing to do with Apple at all.

Remember "Breakout," the game that Wozniak and Jobs had done together at Atari? Jobs told Wozniak that they were getting $700 for the project and he gave Wozniak his $350 share. But Nolan Bushnell actually had paid $5,000—$1,000 plus a huge bonus because "Breakout" had been done so economically, using few chips. Always the innocent, Wozniak didn't find out about it until Bushnell years later mentioned something about the fee. According to Bushnell, Wozniak was in tears, and muttered, "Oh, *he's* done it to me again." When another version of the story became public in 1998, Jobs called Woz and professed not to remember anything about any extra money. With characteristic grace, Wozniak told me that he doesn't like talking about it. As for Jobs, in a moment of rare self-awareness, he once asked journalist Michael Malone, "People think I'm an asshole, don't they?"

At Apple, Jobs's worst performance was the way he handled the Macintosh development effort. He handpicked an engineering team, then commandeered a separate building and flew a skull-and-crossbones outside (with the Apple logo replacing an eye on the flag); Jobs liked to say, "It's better to be a pirate than to join the Navy." The anointed Mac programmers got freshly squeezed juice every morning, first-class plane travel, and a Bösendorfer grand piano and video arcade in the lobby. All other Apple engineers were treated like inferior court subjects. When the Macintosh didn't rejuvenate the company in 1984, the swashbuckling Jobs was primed to walk the plank.

In early 1983, Markkula and Jobs had brought in an established marketing pro to run the company—John Sculley, a senior executive at Pepsi. Markkula wasn't interested in the position and even Jobs, at twenty-eight, understood he had neither the temperament nor the Wall Street standing to lead Apple into the corporate markets it needed. Sculley was in line to run Pepsi but came within Jobs's magnetic field. As Sculley recalled it, Jobs turned to him during one meeting of the courtship and asked, "Do you want to spend the rest of your life selling sugared water—or do you want a chance to change the world?" Sculley decided against sugared water. Apple had a new CEO and Jobs thought he had a mentor. For his part, Sculley believed he had a prize pupil, someone he could help "become the Henry Ford of the computer age."

It didn't turn out that way. In a power struggle that enthralled the

Valley—the son who creates the father who destroys him—Jobs and Sculley vied for the support of Markkula and the rest of the board of directors. Jobs lost. It wasn't so much personal. Now almost a decade old, Apple could no longer survive on its past. Its stock price had fallen from the sixty-dollar range pre-Macintosh to below twenty dollars in the spring of 1985. The company needed a new business plan for the Mac and other products, and Jobs, though a Valley celebrity, didn't have one. Steve Wozniak was already long gone. In 1981, he had crashed his small plane practicing landings at a local airport and taken five weeks to recover from amnesia. Afterward, Wozniak realized he had a life outside Apple: He returned to Berkeley to finally get his undergraduate degree as a matter of pride. And while he retained nominal employee status—including his Employee No. 1 badge—he no longer was an integral part of any Apple engineering team. (The Woz always loved Apple, but eventually he came to own more stock in Microsoft.)

Now Jobs was out, fired by Sculley in late May 1985. Jobs temporarily carried the title of chairman, but at age thirty he was seemingly through at Apple. He dumped all his stock—less a single share—and, a few months later, started a company called NeXT.

For the next dozen years, with Jobs in exile, Apple soldiered on, making an adequate living in an era of PCs dominated by IBM and IBM "clones." Those machines were hardly special. Unlike Apple's hardware, the IBM PC had a design anybody could copy—an "open architecture" in digital parlance. How so? IBM didn't own the computer's "operating system." That privilege belonged to Microsoft. Just as a microprocessor was the computer's brain, an operating system was the built-in software that served as a central nervous system, controlling the machine's basic functions—turning it on and off, organizing files, displaying commands, and communicating between the chip and the disk drive. Applications like word processing and a spreadsheet ran "on top" of the operating system. All IBM really provided was a casing. Because the PC's design was open, anybody could make one—and lots of hardware companies such as Compaq Computer did. IBM was correct that this allowed its machine to set an industry standard, which had advantages, the most important of which was assuring IBM a central role in determining market perceptions and next-generation features.

Apple didn't figure that out. Not only did it squander market share by refusing to lower prices of the Macintosh and other Apple computers, it didn't license the Mac operating system to outsiders and reap the royalties. If you wanted an Apple machine, you had to buy it from Apple. For a company that believed—correctly—it had the superior product, that kind of arrogance came naturally and might have made sense. Corporate isolation wasn't a bad thing if you believed you were in paradise. But as a cold business proposition, Apple's failure to adjust doomed it to an insignificant market share. From 20 percent in 1983 to half that in 1985, the trend was unmistakable.

In 1993, Sculley himself was forced out by plunging profits, to be replaced by Michael "the Diesel" Spindler, who lasted until 1996, when Gil Amelio became CEO. By late that year, Apple Computer was nearing a death rattle—billion-dollar losses, failing morale, and its share of the personal computer market down to a pitiful 3 to 4 percent. "The Fall of an American Icon," declared a *Business Week* cover. It was a long way from the days of the Jobs garage.

In the face of Apple's crisis, Amelio took steps to shore up the company's financial position to even permit it a future. And in late 1996, he had the remarkable notion of returning to the past—and bringing back someone named Steve Jobs, now all of forty-one. Jobs's NeXT Software Inc. was hardly the successor to Apple that Jobs had envisioned. His other company—Pixar—had done much better. Established by George Lucas as a supercomputing division of his film company, Pixar was bought by Jobs in 1986 for $10 million. Jobs turned it into a digital animation studio that went on, with Disney, to make the hit *Toy Story* in 1995 and then immediately to launch a stock offering that made Jobs a billionaire for the first time. But NeXT was the company that Amelio wanted because it owned an advanced operating system that might replace Apple's. In December 1996, Jobs agreed to sell NeXT for $430 million and to come to Apple as an adviser. It was the return of the prodigal son—if not a reconciliation, at least some measure of personal vindication. At a Mac conference the following month, Steve Wozniak even consented to appear onstage with Jobs.

NeXT Software proved of little use to Apple, but Jobs's return was far more than symbolic. In the summer of 1997, in the ever-revolving door at Apple headquarters, Amelio was given the boot. The new "interim CEO"? Steve Jobs, returning to his first love. In the time since, he has willed a rebirth of sorts for the company, including profitability over the course of a full year. (For Apple, this was a big step.) The heart of the new Apple was the iMac, a futuristic-looking computer-and-monitor-in-one, as brilliantly marketed and packaged as anything Jobs had done in his youth. The iMac—something George Jetson might like—came in translucent teal and its curviness made even the Macintosh of 1984 seem passé. At $1,299, it also was a bargain for an Apple and a strategic entry into the low end of the consumer market.

By early 1999, because of strong iMac sales (and five new colors), Apple's market share approached 10 percent and its stock price rose to its highest level in years. Bit player that Apple remained, Jobs could be forgiven for gloating. "We still have a soul," he told an auditorium of employees gathered to hear about Apple's newfound profits and new-found passion. F. Scott Fitzgerald once wrote that "there are no second acts in American lives." Steve Jobs was out to prove him wrong. Unfortunately, his once-and-future company had lost the war long ago and no moral victory was going to change that. His dream company amounted to little more than that anymore. Jobs had indeed "put a dent in the universe," as he long ago had promised. But others had moved in. And Apple bore much of the blame. Given Apple's refusal to license its operating system, IBM's decision to allow clones of the PC made that machine the industry standard.

But IBM also wound up ceding control of the marketplace, much as Apple had done. With Big Blue hubris, it never imagined anybody could take it over. The usurpers were two companies: Intel, which would make the chip inside, and Microsoft, which would write the operating system. Thanks to IBM, which made one of the all-time blunders in American business history, both Intel and Microsoft were able to do business with a range of hardware manufacturers. IBM, foolishly, had failed to demand exclusive licenses. This would be key to the monumental success of both Intel and Microsoft; for IBM, as Larry Ellison later said, it amounted to "a $100-billion mistake." Intel was already an established Silicon Valley company and doing fine. But up near Seattle, Microsoft was barely a blip on the Valley's radar screen. If IBM was the Spanish Armada, Microsoft soon appeared as a more nimble fleet.

Microsoft might have remained a blip, except for Gary Kildall. William Shockley was the tragic, prophetic figure of the early Valley, but

Kildall was his counterpart a generation later. In every life, there are moments that can go either way—the road not taken, lips not kissed—and Kildall's cost him the future. His failure is as compelling as any story of success.

Gary Kildall, a Ph.D. in computer languages, was one of the first great software programmers. Like a fellow named Gates, who was thirteen years younger, he was raised in Seattle and a math whiz. He had whimsy, too, like Steve Wozniak: He liked to rewire the neighborhood phones to eavesdrop, especially when the conversations involved his sister's boyfriends. Kildall's family ran a school of navigation known as the Kildall College of Nautical Knowledge. He was big and handsome—six foot one, slender, with a soft voice and a full reddish beard. (Everybody in high-tech seemed to have a beard at some point.) In his jeans and boots, he looked more like a cowboy than a geek.

In the early 1970s, Kildall was teaching computer science at the U.S. Naval Postgraduate School in Monterey, California. Earlier, he had joined the naval reserve, and to avoid going to Vietnam, he used his engineering background to get the Navy teaching job. Down the coast from Silicon Valley, across the Santa Cruz Range, the Monterey peninsula was a haven for him. It was more countrified than the Valley, and the neighboring village of Pacific Grove, where he lived, gave him a seaside taste of the Northwest (even with the palm trees). The village is bounded by water on three sides and the mountains on the fourth. Kildall's specialty was translating computer programs from one language to another, and from one kind of device to another; these "languages," written, not spoken, were the tools used to create the programs that gave PCs something to do. It was the golden lexicon for engineers who could barely speak a coherent sentence.

As early as 1972, Kildall had bought an Intel 4004 microprocessor to experiment with. Intel then hired him as a day-a-week consultant to write code for its 8008 and 8080 chips. During that work, in 1975, Kildall figured out a way to short-circuit the tedium of translation and interdevice communication. His personal solution: an entire operating system to control the microprocessor's basic functions. He called it CP/M (for Control Program/Monitor or Control Program for Microcomputers). Even before the Altair appeared, it was the first "disk operating system" for a PC—"DOS" for short. His friends say he wrote it by himself, effortlessly, which said less about the difficulty of the task than about his aptitude for code. They also wondered why anybody would possibly want a single-user operating system. For Kildall the professor—like Wozniak the hacker—the joy was in the invention, not the commerce. Kildall initially toyed with Intel chips in hopes of presenting his father with something of use at the navigation school.

Apple had its own system, based on a non-Intel chip. But with all the other PCs that followed the Altair, CP/M became a standard. By the late 1970s, Gary Kildall's software ran 500,000 machines, and programmers were writing applications designed to run with CP/M. Intel could've bought CP/M outright for $20,000, but turned down the chance, believing that microprocessors were already far afield from its core memory business. Along with his wife, Dorothy, Kildall formed Intergalactic Digital Research Inc. (He subsequently dropped the cosmic reference.) She managed the numbers and he wrote the code. When the company was hiring, he was known to interview applicants while wearing roller skates and a toga. Rejecting the model of mainframe programmers, who charged a fortune for operating systems, Kildall anticipated the volume business of PCs and typically charged only seventy-five dollars—even though, remarkably, at the moment of creation there was hardly any market at all. The company, begun in an old Victorian house in Pacific Grove, sold thousands and the Kildalls made millions.

Gary Kildall liked the money and loved the toys he could buy. He owned planes, speedboats, Jet Skis, motorcycles, a stretch limo, a Corvette, a Rolls-Royce, Formula One race cars, two Lamborghini Countachs (among the first in California)—and a Ford pickup. But for Kildall, like Woz, money wasn't the point. He was just an academicized version of Wozniak, delighting in discovery and telling everybody—including possible competitors—about it. "What a rush!" he'd exclaim upon getting a program to work—sometimes calling a colleague in the middle of the night to share his glee. Cashing in was just a by-product. "I'm not a competitive person," Kildall told people. However, later in his life, in his attempt at an autobiography, he wrote: "I learned that computers were built to make money, not minds."

As IBM's Boca Raton team was secretly developing its personal computer in 1980, it lacked the time and ability to invent an operating system. So it set out to buy one. CP/M, the market standard, was the obvious product. IBM mistakenly believed it was owned by a small company outside Seattle called Microsoft, run by twenty-four-year-old Bill Gates. Formerly Micro-Soft (which beat out Allen & Gates Inc.), Microsoft was founded in 1975 by Gates and Paul Allen in Albuquerque. Its first project was to write a basic computing language (not surprisingly called BASIC, for Beginner's All-purpose Symbolic Instruction Code) for the MITS Altair and then other machines. Gates was so sure of the business that he had dropped out of Harvard to pursue it. By now, Microsoft was the biggest provider of computer languages for PCs. But it had little to do with operating systems (except for reselling CP/M as part of a hardware accessory that allowed it to run on Apple IIs) and was a bit player to Kildall's Digital Research Inc. (DRI). IBM didn't know this. When IBM representatives asked about CP/M, Gates told them it wasn't his and sent them to Kildall (proving that even a tyrant can do a good deed). The IBM suits arrived in Pacific Grove the very next day for a meeting.

But Gary Kildall wasn't there.

It became the stuff of lore—so much so that the facts barely matter in the histories of both the Valley and Microsoft. The legend goes that "Gary went flying"—a famous three-word summation uttered by Gates, derisively referring to the free-spirited thirty-eight-year-old Kildall's love of planes and indifference to boardroom protocol. "Gary went flying"— too busy to be troubled by the most powerful technology company on earth, leaving the meeting for his wife to run. That's the Microsoft, and popular, version—and since winners tend to write history, it's the prevailing one. It's not the only one. Gary Kildall is the *Rashomon* of Silicon Valley.

Almost two decades later, the story of that August 1980 day still reverberates around Pacific Grove in the way that doormen at the Watergate still talk about the break-in. Pacific Grove is a scenic coastal town off Highway 1, known largely for the squadrons of Monarch butterflies that arrive from Mexico every spring. The Kildall tale is now just as much a part of the place as the butterflies: This is the place that really gave the world Microsoft because "Gary went flying." Tom Rolander just wishes the story was told with a little nuance.

Rolander—today the president of PGSoft, a boutique programming firm with seven DRI alumni at the corner of Fourteenth and Light-house—was then DRI's chief tekkie, Kildall's best man and constant flying companion. He had met Kildall in the computer lab at the University of Washington, then worked at Intel for three years before joining Kildall. As Rolander tells it, "Gary went flying" is myth—and he should know because he was with Kildall that August day.

Rolander has a condition for telling me the tale. Since Kildall was so fond of taking customers and friends up in his planes, Rolander wants to re-create the ambience. I'll have to go with him in his immaculately restored 1948 four-seater Navion. Rolander, fifty-one, has been flying since he was seventeen and also is a competitive marathon runner. He lives for tomorrow's race, not to analyze yesterday's performance. But as we fly down the Monterey Coast, over the spot where John Denver crashed and died, and then along Big Sur, it is a bittersweet chance to recollect.

They were in a plane that day two decades ago, Rolander says, but that had nothing to do with the ultimate demise of DRI and ascendancy of Microsoft. Bill Gates indeed called Gary Kildall one day in August 1980 about a "big company," whose identity Gates wouldn't reveal. IBM's people called shortly thereafter, saying they would arrive from Seattle the next day. "Gary and I had already scheduled a morning appointment up at the Oakland Airport with Bill Godbout, a CP/M distributor," says Rolander. So they flew up in Kildall's single-engine Turbo Arrow, allowing the IBM meeting to begin with Dorothy Kildall. In retrospect, it was a stupid decision by Gary Kildall, if for no other reason than it didn't show IBM enough respect. But it wasn't his style to worry about style. He had created CP/M, it was making money, so why should he worry about corporate high jinks? Like Nolan Bushnell, he could've cared less about the details. But the devil was in the details and not just up near Seattle.

Before discussing its secret PC project, IBM presented Dorothy with its standard, overreaching nondisclosure agreement, to wit: The meeting taking place . . . never took place. And if somebody one day decided it did, anything IBM told DRI was confidential, whereas anything DRI told IBM was not. The absurd one-sidedness of the IBM form underscored its Big Blueness and made Dorothy Kildall wary. She refused to sign and telephoned DRI's lawyer, Gerry Davis, to come over from Monterey. IBM began to boil. Here the big dog had come all the way to Pacific Grove, and now they were being jerked around by a flea of a company doing business in a Victorian.

Kildall and Rolander arrived at the meeting at 801 Lighthouse in the afternoon. "Gary just didn't see it as a big deal—so what if a big, plodding company like IBM wanted to get into microcomputers?" Rolander remembers. "We thought we'd get a few hundred thousand dollars of business and that would be it. So Gary signed the IBM form." This, Microsoftians and others continue to say, did not happen. Instead, they insist, "Gary went flying."

The bigger stumbling block, lost in the legend, was IBM's insistence on licensing a new version of CP/M (called CP/M-86) for a flat fee of several hundred thousand dollars rather than for ongoing royalties of ten dollars a copy. IBM also wanted to rename the Kildall operating system PC-DOS. Kildall acceded to neither condition. "He was earning millions from royalties," says Rolander, "so why would he give that up? And back then the product had name recognition. Gary even thought of renaming the company after CP/M. He was headstrong—and why not? Almost any PC at that time that wasn't an Apple was using Kildall's operating system."

Kildall didn't budge and IBM left. The next day, he departed with his family for a cruise off Miami. On the same commercial flight to Florida, Kildall ran into some members of the IBM team who were returning to Boca Raton. They talked cordially but still could strike no deal. Kildall didn't think much of it. Foolishly and unluckily, as it turned out, he saw IBM as just another potential entrant in the PC market. Plus, he simply didn't like them. Perhaps IBM sensed as much, for it went back to Gates, hoping he would intervene with DRI. But Gates's game plan then shifted. He may not have had Kildall's talent for programming, but his instincts for business were a whole lot better. Microsoft was a tiny operation, and

PROPHETS
113

while Gates recognized that hopping in bed with the planet's biggest computer maker had risks, it also gave his forty-employee start-up the strategic opportunity of a lifetime. To Kildall, programming was religion; to Gates, it was just a means to build a secular kingdom.

All Kildall got out of it was a lifetime of regret. He trusted human nature and believed that technology would advance quickest if different companies worked together. He couldn't imagine being knifed in the back—certainly not by Gates. The two had known each other since Gates was a thirteen-year-old hacker in Seattle and Kildall was getting his doctorate. In the late 1970s, as Gates and Allen were preparing to move from Albuquerque either to Silicon Valley or back to Washington, they discussed merging their young companies. It didn't happen, but there seemed to be a gentleman's agreement that neither would get involved in the other's business: DRI would stay away from languages, and Microsoft would leave operating systems alone. This might explain why Gates originally sent Big Blue to Pacific Grove. It does not explain what Gates did next.

In the wake of DRI's failure to make a deal, even though Microsoft had no operating system of its own, Gates set out to satisfy IBM himself. It was that kind of single-minded drive that Kildall never had. Among Gates's transcendent gifts was the good sense to give folks what they wanted, rather than indulging in some pure intellectual exercise. Gates also had the luck of a man who would someday be worth close to $100 billion. In one of the great, terrible serendipities of the Kildall story, Gates's cofounder, Paul Allen, just happened to know of a PC operating system just across town. Tim Paterson, of Seattle Computer Products, had written Q-DOS—for Quick and Dirty Operating System—a close imitation of CP/M, which DRI had never tried to keep proprietary. Seattle Computer knew Kildall and did business with DRI.

In late September 1980, Allen first contacted Seattle Computer. For a mere $75,000—and never mentioning the magical letters of its *own* customer, "IBM"—Microsoft bought Paterson's operating system and renamed it MS-DOS, for "Microsoft Disk Operating System." (IBM called its version PC-DOS.) And that was just the first of the two best software deals of all time. Microsoft then arranged for getting royalty payments from IBM—*and* continued ownership of the operating system. That meant Microsoft could license MS-DOS to anyone else manufacturing

"IBM-compatible" machines—and on pretty much its own terms. (The result is pure profit—at, say, up to fifty dollars per PC—since the cost of each additional copy of MS-DOS is negligible.) Compaq Computer sold $300 million worth of clones in its first year alone.

IBM did not envision the breadth of those hardware competitors, but Microsoft did just fine by them, as they became utterly dependent on Microsoft. IBM enabled Microsoft to establish a beachhead for its operating system—from which Microsoft could turn around and stick it to IBM. By its own hand, IBM created the monster that would take its place as the symbol of high-tech supremacy. (IBM compounded its mistake by pricing its PC too high, creating the opening for clone manufacturers.) David became Goliath. In time, for its investment of $75,000, Microsoft made a quarter-billion dollars from its pre-Windows product, establishing itself as the dominant operating system. Its easier-to-use Windows operating systems—modeled after the GUI of Macintosh—would generate billions more.

Understandably, Gary Kildall was less than pleased when he heard about IBM's deal with Microsoft—all the more so when he learned how suspiciously similar Gates's MS-DOS seemed to his own CP/M. Kildall thought it was thievery, though it was not his nature to sue and the copyright law hardly would have made his case easy. (It was one thing if Seattle Computer had outright copied Kildall's source code, quite another if CP/M had merely been imitated.) However, the threat of litigation was enough to bring IBM back to Pacific Grove and, this time, it seemed more accommodating. It would offer CP/M on its PC—paying royalties and agreeing to call it CP/M-86—just as it would offer Microsoft's operating system. The market would decide which was the better product. All DRI had to do was promise not to sue. "Gary said sure," recalls Rolander. "Why not compete? We thought it would be a fair basis to do so, especially since we had a newer, better version about to come out."

But the competition was rigged and IBM's deal with Kildall was a ruse. When the IBM PC hit the market, consumers could choose between operating systems all right—trouble was, Microsoft's DOS sold for forty dollars, CP/M for *six times* that. IBM never told Kildall about the pricing structure. "To this day," says Rolander, "I have no idea if Bill Gates had anything to do with this. But, regardless, it was a conscious choice to kill CP/M. Why would anybody buy it when the other cost $200 less?"

It's not clear whether DRI could have sued IBM successfully—after all, IBM was within its rights to buy CP/M just to kill it, and DRI simply got outfoxed. Whether the fox was IBM or a nerdy twenty-four-year-old remains a mystery. But given Gates's predilection for trying to annihilate competition—in spreadsheets, in personal-finance software, in Internet browsers—it doesn't take Oliver Stone to wonder about Gates's role in IBM's pricing. Then again, maybe IBM just liked Bill: Upon hearing that his company was considering doing business with Microsoft, the head of IBM supposedly remarked, "Oh, that's run by Bill Gates, Mary Gates's son?" The two had served together on the national board of the United Way. That certainly didn't hurt.

Kildall never did sue IBM or Microsoft or Seattle Computer. His naïveté was part of his appeal. It was also his fatal flaw. Kildall figured all along that the PC industry would have room for two operating systems. Two colas, three carmakers, dozens of computer manufacturers—why not MS-DOS and CP/M? At one industry forum, Gates and Kildall appeared on the same panel. Ever the idealist, Kildall maintained the PC market was vast enough for the both of them. Gates was more clear-eyed and prescient about the inevitability of a software monopoly. "There's room for just one," he replied. PCs needed certain industry standards that would promote the efficient development of compatible software and accessories. It was no different than typewriters having the same keyboards or houses having uniform electrical outlets.

As a matter of principle, Kildall also refused to enter the market for such software as spreadsheets and word processing. He thought that one company doing operating systems and applications presented a conflict of interest, ethically if not legally. Maybe so—the antitrust police have yet to resolve the theoretical issue—but Microsoft built an empire out of integrating MS-DOS and applications.

Kildall took a long time to recognize how Gates did business. In the mid-1980s, while still at DRI, Kildall became interested in the new technology of CD-ROMs—the first seeds of electronic "multimedia"—which could store huge amounts of digital data. With Rolander, he founded KnowledgeSet, which produced the first interactive encyclopedia on a CD-ROM. To share his ideas and hear others, Kildall planned a conference near Pacific Grove at the Asilomar resort. On a visit to see family in Seattle, Kildall mentioned the conference to Gates. Three months later—lo and behold—Gates called Kildall about his *own* Microsoft CD-ROM conference to be held in Seattle and wouldn't-Gary-just-be-honored-to-be-a-keynoter? Kildall agreed and gave up on Asilomar. Unlike Kildall, Microsoft didn't even have a CD-ROM on the market; yet Gates-the-imitator would now be seen as a leader in the field. Gates may not have come up with a good idea, but once again he pounced on it—Kildall be damned.

According to Rolander, it wasn't until after Kildall spoke at the Microsoft conference—as the two of them sat and talked to industry gadfly Esther Dyson during a plane flight—that "Gary realized he had been had by Bill yet again." It might have been better if he hadn't finally realized it.

Kildall stayed with DRI, but became increasingly distracted. He and Dorothy divorced, and he spent much of his time gallivanting between California and Monaco and places in between. DRI eventually was sold to Novell and basically euthanized; Kildall still reaped millions more. He remarried, but it, too, hardly gave him peace. The IBM episode, and the canonization of Bill Gates, haunted him. Gates was on his way to becoming the richest man in the galaxy, a billionaire times ten and then one hundred, leaving Kildall in the dust. As in Gold Rush days, the discoverer hadn't reaped the prize—the exploiters of the discovery had. Kildall conceived the idea for the operating system; but not only was Gates getting the credit, Kildall was left to answer for his screwup, over and over and over.

"When he met with the son of Sony's Akio Morita in late 1987," Rolander remembers with a grimace, "the first question Masao Morita asked Gary was, 'Were you really flying when IBM came calling?' " When all you've become to a ranking Sony executive is a historical asterisk—the answer to a trivia question—who wouldn't be embittered? The only reason Gary Kildall's name appeared anymore in the same sentence as Bill Gates's was to note What Might've Been. He had become the Tucker of the software business. In earlier times at DRI, several programmers working late in the Victorian at 801 Lighthouse spoke of footsteps on the stairs and other mysterious incidents; now, it was just the ghost of Gates. Under other circumstances, Kildall might have been content to reap the intellectual and pecuniary benefits of being one of the PC industry's pioneers. He even got his own TV show for six years, on PBS about computers. (Bill Gates may own a TV network someday, but he'll never get a TV show.) But the unrelenting comparisons to someone he once considered a friend proved too much. Kildall couldn't help himself. At one point, he was keeping a list of his own inventions that he believed Gates ripped off. The fact that it had the ring of truth was beside the point.

Kildall was hardly consigned to a lifetime of resentment and, ultimately, self-destruction. Wozniak demonstrated that there was a different way to react to an overbearing rival and Kildall was surely aware of the parallel. In 1983, according to a trade journal, he told a friend, "Steve Jobs is nothing. Steve Wozniak did it all, the hardware and the software. All Jobs did was hang around and take the credit." Yet Wozniak flourished in his postgarage days—while he blew $25 million on legendarily unsuccessful rock concerts, he also finished college, funded countless worthy causes, and still teaches computers to schoolchildren. "I had two goals in life—to be an engineer and to teach fifth grade," Woz liked to say. Kildall, by contrast, drifted into depression and alcohol. "I saw less and less of him," Rolander says. The new toys, funded by the $120 million sale of DRI to Novell—a $3 million Learjet, an oceanfront estate along the 17 Mile Drive in Pebble Beach, another home in Texas—weren't distractions enough. In the early 1990s, he set out to tell his story in a memoir that he acknowledged could be construed as "sour grapes." The 250-page manuscript took shots at Bill Gates, calling him "divisive," "manipulative," a man who "has taken much from me and the industry." But the book was never published during his lifetime and still hasn't been. "His son is terrified about publishing and then being sued by Gates," says Rolander. "On his list of concerns, getting sued is numbers one through nine." The son would like his father to be remembered for more than the last, ignoble episode of his life.

Shortly before midnight, on Friday, July 8, 1994, Kildall walked into the crowded Franklin Street Bar & Grill in Monterey. A few minutes later, in circumstances that neither the police nor coroner could ever pin down, Kildall was on the floor beside a video arcade. He had struck his head on something, but it wasn't clear how: Had he gotten into a barroom brawl or had he just fallen dead drunk? Kildall himself couldn't remember. Refusing treatment from paramedics, he walked out, still wearing his Harley-Davidson vest. He made it home, but with a mortal injury just as insidious as the psychic wound that IBM's visit inflicted fourteen years earlier. In two separate visits to the Monterey hospital that weekend, no one discovered the blood clot between his skull and brain. Three days after hitting the floor, Gary Kildall was dead at fifty-two. His obituary didn't even make all the newspapers around Silicon Valley.

His career mistake was not coming in first place. The notion that a plane flight changed his destiny—that the wheel of fortune might have spun in his direction rather than Microsoft's—overlooks Kildall's failings and ignores Gates's prodigious abilities. No matter what the details of that first IBM meeting, Kildall could never have been Bill Gates, and to his friends, that's a good thing. "I miss him still, very much," says Tom Rolander. "I miss him especially whenever I'm flying alone."

Kildall's ashes were returned to his native Seattle. They were buried not far from where Gates had begun construction on his $60-million lakeside home. In California, three hundred mourners attended the memorial service at the Naval Postgraduate School. Bill Gates was not among them. Through his public-relations minions, he issued pabulum about death-at-a-young-age and "a loss to the industry," but that was all he could muster. Tone-deaf and devoid of heart, Gates couldn't have cared less about what had become of the man he beat. "The morning after Gary died," Rolander says, "I E-mailed Bill, but got no response. I also told him later about there being a memorial service. I didn't hear from him on that either." Bill Gates may have all the instincts of a great white shark, but he has its warmth as well.

Wozniak, Jobs, Bushnell, Kildall—they

were part of a time in Silicon Valley that was vanishing. Tinkerers, hackers, dreamers, idealists would be overwhelmed by the moneymen and hucksters who viewed technology as nothing more than commerce. Yesterday's empires were accidental; tomorrow's would be premeditated. The archetype of the new breed was someone named Larry Ellison.

Chapter V
Oz

Bill Gates may be the undisputed monarch of the modern software kingdom, but its clown prince—and the man who showed Silicon Valley how the game is played—is fifty-four-year-old Larry Ellison. His company, Oracle, is the second-largest software enterprise in the world after Microsoft—started in 1977 with just $2,000 and worth $50 billion at the beginning of 1999—but that's not really the point.

Both Gates and Ellison got their breaks at the expense of IBM, but the similarities stop there. If Gates is high-tech's stereotypic techno-weenie, Ellison is the alpha-male playboy from central casting—tall, thrice-divorced connoisseur of long-legged blondes (preferably employees), defendant in a sexual harassment suit, and the life of any party, as long as he's the center of attention. "As long as Stanford keeps turning out beautiful twenty-three-year-old women, Larry will keep getting married," a friend and colleague once quipped. Sure, he can be wayward—like the time he left his two young children alone in the house one night while he went out rollicking—but that's just part of a rascal's charm (except to one of the kids, who refused to go back for years).

Gates is the nation's richest citizen; Ellison is only in the Top Ten. Gates is rough-hewn, Ellison is smooth. Gates radiates cool, Ellison is all charisma, wit, urbanity. Gates is dull, even when he's singing a lullaby to Barbara Walters; Ellison is the funniest character in the land of bores, with a theatricality matched only by Steve Jobs. Gates's $60-million house in the suburbs of Seattle looks like a Marriott Conference

Center; Ellison's $40-million estate under construction in Woodside is so refined it'll be built not with nails but with wooden pegs. The name of Gates's company sounds like something you use at a Laundromat; Ellison's suggests a sixteenth-century shrine to a god, even if it's one who likes to say "fuck you" a lot. Scheduled for completion in the new millennium, the Ellison house was designed by a Zen priest and "should be the most authentic Japanese structure outside Japan," Ellison says. "It's a 'Balance of Elements,'" explains the site plan. "Air, Earth, Time, Water, and Wood." How could he have left out the Archery Range? Even more than the ginkgoes and bonsai pines, the centerpiece of his twenty-three acres is a huge man-made pond of purified drinking water. One day, Ellison will be able to jump in from the deck of the master suite and swim across to the spa. "Smart" sensors will inform the gym who's coming, what music to play, and the proper lighting. Give Ellison credit: If you're not going to give the money all away, spend it in style.

Gates owns a plane and a boat; Ellison owns a fleet, a flotilla, and a raft of other toys that won't fit in that house. Gates knows code, Ellison preaches samurai culture and forty-seven topics he read about last week. Gates epitomizes dweeb; Ellison looks preternaturally youthful, which has something to do with obsessive bodybuilding, a great tan, and perhaps a few trips to the plastic surgeon. Ellison even has his very own stalker—who goes by "Jill" and manages to sit in the front row whenever he has a public speaking appearance; she likes to lunge for him, but Ellison's armed bodyguards act as intermediaries.

Just as Gates is just "Bill," Ellison is known around the Valley as "Larry"—who else could "Larry" be? As in, "Did you see that model Larry was with?" Or: "Looks like Larry had another face-lift—maybe it's because of that model Larry's with. Is that why he has the beard?" Multibillionaire Larry may only be the second-richest Californian—next to Gordon Moore, chairman emeritus of Intel—but Gordon's mating and dating habits never, ever made it into *Vanity Fair*. While Moore once was the exemplar of the Valley, Ellison now is its poster boy—the emblem of the Valley's transformation into a celebrity town. Unlike the modest Moore or the pizza-eating, cola-chugging, culture-starved techno-nerds of the movies, Larry *acts* rich. Others use him as a benchmark. Call it the Ellisonification of Silicon Valley.

"Someone once accused me of being unnecessarily interesting," Ellison said a few years ago. If most of Ellison is an act, he is a master showman. Microsoft makes software that you and I use at home or work. Oracle software, on the other hand, is invisible; you're more likely to know the name of the governor of Minnesota than to have a clue about what Ellison's company makes. But it's important stuff: Oracle software organizes hundreds of critical databases for the military, for government agencies, for traditional businesses and online retailers. When the CIA needs to access how many pastry chefs in Europe it has on retainer, and which of them went to college with Benjamin Netanyahu, it uses Oracle software. So, too, does the KGB and England's MI-5. When American Airlines needs to reconfigure routes based on passenger capacity on its 767s, it uses Oracle software. But the company has little profile—no dancing bunnies, no brightly colored logo, few ad campaigns. It does have Larry Ellison. Among so many contenders, no citizen of the Valley is more willfully full of himself.

He owns, in no particular order, three Bay Area homes; a giant $50,000 plasma-screen TV (just three inches thick, it hangs flat on the wall like a picture); a koi pond full of personally named specimens (all eat better than you and me); a brown Bentley convertible; a Porsche Boxster; two customized Mercedes; a Jeep; an $875,000, 627-horsepower silver McLaren roadster (0 to 60 mph in 3.4 seconds); a Gulfstream 5 (0 to 600 mph in 3.4 minutes, plus *two* bathrooms); a Marchetti turbojet fighter from Italy; a Cessna Citation jet (which he sometimes uses to shuttle from the local airport near Oracle to Oakland, twenty-five miles away, where he keeps the Gulfstream); and a bunch of acrobatic planes, including a purple one, no bigger than a VW, for his teenage son. And he's trying to convince the U.S. Bureau of Alcohol, Tobacco and Firearms to let him import a used Russian MiG-29 warbird with all the trimmings (the better to fly by Gates's house at supersonic speed). Ellison has his very own hangar—complete with staff, lounge, and the big photo on the wall of Larry in military garb for his F-16 flight of fancy, courtesy of the Air Force. All those planes can be a maintenance nightmare. In 1997, after his annual cherry blossom party at home, Ellison and neighbor Joe Montana went up joyriding in the Gulfstream despite the fog; when they came back, according to one of the partygoers, Ellison was muttering over and over: "Fuckin' radar never fucking works, fuckin' radar never fucking works." He loves that F-word.

The two Ferraris, alas, are gone. Ellison says they "blew up" on him when he tried to go too fast. "Too fast" in a Ferrari? Ellison liked to pull right in front of semis on the freeway and then zip away from the blaring air horns. When Ellison's pulled over for speeding, his M.O. is to refuse to cooperate with the police—particularly when it's the same officer who ticketed him the day before—and call his lawyer on his cell phone to come negotiate. Sometimes, he'd plea-bargain "up": Ellison wouldn't show up for court hearings until an arrest warrant was issued; then he'd avoid a speeding violation by admitting a "failure to appear." It cost more, but he didn't get points on his license.

Actually, Ellison only blew up one of the Ferraris. A coworker had ordered a 348 model months before they came on the market. Ellison decided he needed one, too. Within weeks of getting it, though, he encountered a small problem while racing along the freeway. A dashboard indicator warned, "Slow Down." This meant he should "Slow Down," because an engine malfunction was causing raw gas to pass over a catalytic converter and make it red-hot; this, in turn, ignited the tail lamps and wiring. But Ellison hadn't bothered to read the owners' manual. If the highway patrol couldn't get him to slow down, some dashboard light wasn't going to. The other Ferrari? It was a company car that the company took away during a financial crisis. Looked bad. Not a problem these days. Whichever car Ellison drives to work gets the reserved We'll-Tow-You spot right by the front door of Building 500.

Ellison's best friend—his only friend, the joke goes—is Steve Jobs; his dearest enemy is Bill Gates; a business partner is Michael Milken ("I didn't know him at Drexel—if I had, I'd be a rich man!"). His biographer called his book *The Difference Between God and Larry Ellison: God Doesn't Think He's Larry Ellison.* More than a few authors, employees, customers, and stockholders have marveled at Ellison's ability to fudge the facts—whether about the kind of neighborhood he grew up in, if he would join you for dinner, or what products Oracle offered. "Does Larry lie?" asks Ed Oates, an Oracle cofounder who's known him for twenty-five years. "We prefer to say Larry has a problem with tenses. For example, 'our product is available now' might mean it'll be available in a few months or that Larry was thinking about maybe one day developing the product." This had the added benefit of confusing competitors

who were trying to figure out what Oracle was up to. "Temporally challenged," another colleague calls it. "Not a lie, just a different version of
the truth." If Ellison wasn't in Fantasyland, at the very least he lived in
Futureland, where anything was possible. It was precisely that kind of
delusion that allowed him to start Oracle. One man's prevaricator is
another man's visionary.

Most of the stories about Ellison are actually real, not that it matters
for the legend. *The Wall Street Journal* published a story saying that he
signed his personal checks in green ink—the color of money. Ellison
says that's ridiculous. Probably so. Same for the time the plumber supposedly shut off the water in the house, only to hear a startled Ellison
scream from the shower: "I've got two women in here!" Or his oft-quoted
counsel to young entrepreneurs: "I have one piece of advice—Acura
NSX. Before it's too late." He tells me it is indeed a great line. "I wish
I'd said it. Feel free to attribute it to me." An adolescent going on fifty-
five—but driving a lot faster—runs the top software company in Silicon
Valley. Ellison bought four NSXs. If the truth doesn't matter to him,
should it matter if the tales *about* him are all true? "Does anybody tell
the truth all the time?" Ellison asked a *Fortune* correspondent in 1997.
That was right before he told her how he once convinced his daughter
that pepperoni pizza was made from snakes.

Then there's the story of what he did to his vanquished opponents—
including Roy Disney, nephew of Walt—in the 1996 Miami-to-Montego
Bay sailboat race. Roy supposedly made such fun of Ellison's boat that
he even suggested Ellison not bother to set sail on the 522-mile course.
Ellison did anyway—and won by several *hours*. As the story goes, he
then boarded his twin-engine Cessna Citation in Jamaica and swung
back over the course to buzz the losers, especially Disney. I ask Larry
about that incident one afternoon.

"You have it all wrong," he replies with mock anger. "After the race,
I shaved, showered, ate dinner, did a TV interview, and *then* took my
plane out over the Caribbean to find the other boats. I dove the plane
down to about 50 feet over the water, throttled up to 250 knots and went
right between Disney's boat and *Boomerang* owned by George Coumantaros, the powerful shipping magnate. I was below the tops of their masts
and then I pulled up as hard as I could so I turned right in front of them.

Perfectly legal—I was in international waters. It was one of those won-derfully immature acts for which"—he pauses for effect, then grins—"you can't pass up the opportunity."

"Andy Grove would never do that," I offer back.

"Well, Andy Grove doesn't know how to fly."

One-liner Larry. The year before Montego Bay, he invited Rupert Murdoch to crew on the famous Sydney-to-Hobart blue-water race off the coast of Australia (the same event that later, in December 1998, claimed the lives of a half-dozen sailors during a monstrous gale). Near the end of a prep race, Murdoch grabbed a piece of rigging at the wrong moment and it sliced off one of his fingertips. Murdoch sucked on his finger, somebody else picked up the digit and bagged it in ice, and off Murdoch went for microsurgery. Cracked Ellison, "He can still write checks."

Ellison seems to relish these moments of glibness. "So you like the CEO bad-boy, Randy Andy image?" I ask him.

"Wrong word," he says. "It was just good fun."

"Which you seem to have a lot of."

"I think I have a terrible habit of answering questions that I'm asked, but I just try to be as honest as I can."

Okay, but what about sex by the roadside in Woodside? That seemed pretty randy—and was featured in a Valentine Day's column in the *San Francisco Examiner* about "unusual places to do 'it.' "

"I was with this girl that I had lived with for four years. We were looking for land in Woodside. It was late in the afternoon. I just went along with it. We were off the road, but we were near a bridle path and were discovered." Lucky it wasn't a bridal path.

Ellison embodies all that is wretched and all that is hilarious about Silicon Valley.

In June 1998—almost three thousand miles from the Valley—Larry Ellison has *both* his big boats in the harbor off Newport, Rhode Island. There's *Sakura*, the 192-foot wonder yacht. (The guy who paid $125,000 at the Woodside charity auction for a cruise never got it—Ellison wound up selling the boat first.) And there's *Sayonara*, the carbon-fiber sailing rocket that Ellison raced in Montego Bay and Australia. He's here to race her in the 144th annual regatta of the New York Yacht Club and then head out to sea for the famous biennial three-day, 635-nautical-mile race to Bermuda. *Sakura* is anchored a half-mile off-shore for Ellison to bunk on. Wherever he sails *Sayonara* away from home—the West Indies, Australia, the Mediterranean—he takes *Sakura* along. Why waste your money at the Ritz?

Originally built by the takeover specialist Kirk Kerkorian, *Sakura* is the sixth-largest private yacht in America, a floating testament to capitalistic hubris. She's big enough for a whale hunt: If you took the boat out fishing, your guide would be sporting a wooden leg and spouting off about revenge. *Sakura* is *wider* (twenty-nine feet) than a lot of pleasure craft are long. Her name is Japanese for "cherry blossom," and she's a 3,100-horsepowered peach. Her decks are all gorgeous teak, and Ellison won't let you wear shoes on them, which is a problem if you didn't happen to bring along alternative footwear. *Sakura*'s got a full-time staff of eleven to tend the acreage, a cavernous VIP quarters nicknamed "Steve's Room" (for all the time Jobs has used it), four other staterooms, four living levels, an entertainment salon (with six audiovisual systems and Surroundsound), a pool, a satellite-linked communications center, and a collection of smaller boats in tow for waterskiing, fishing, and going to shore. Impressed? Ellison's putting *Sakura* on the market because he's buying an "interim" boat—270 feet long that'll do 39 knots up on its hydrofoils—while a 350-footer is built. That specimen will be the No. 1 jumbo craft in America. For Larry, size always matters.

When I saw him in California a few weeks earlier, he told me to come visit him in Newport, maybe go for a spin. So I did. When I got there, though, he E-mailed me from the boat to visit him a few days later. Apparently, this is a cat-and-mouse game he likes to play. It's born more of coyness than shyness. This is the CEO who boasted of public sex and whose face once adorned a billboard on Highway 101 near Oracle headquarters. (Maybe it's just a thing about grabbing attention by the roadside.)

Ellison is a bafflement—an introvert who craves attention. The game in Newport extends to the very people who invited him here for the week-long regatta before the big race to Bermuda. The New York Yacht Club, which used Newport for a hundred years to conduct the America's Cup, still regards the place as holy ground, where the Brahmin of the sea gather every summer to sniff brandy and each other. Its annual regatta and Onion Patch Series is a big deal, a coming-out party for summer— especially the more so for the five skippers sailing in *Sayonara*'s class, the vaunted "Maxi Yachts," known for their speed and exclusivity. The annual dinner at the club is de rigueur, a festival of ego. But Ellison doesn't bother to attend.

Even Ted Turner's here. "Captain Outrageous," the "Mouth of the South," the media mogul—he's nouveau yacht club, but the place has long accepted him since he successfully defended the America's Cup for them back in 1977. I ask Ted in the dessert line what he makes of Larry, another rogue-turned-mariner. "Meet my wife, Jane," he offers. We meet. I ask about Larry again.

"Oh, Larry," Ted says with a smile. "Larry likes to have fun. . . ."

"Like you, Ted?" I ask with a smile.

He volunteers nothing more about Larry, but notes that "I've won a lot of races, you know."

"Doesn't Larry have a pretty fast boat?" I ask.

"He's got Chris Dickson, too."

Chris Dickson, a New Zealander and America's Cup veteran, happens to sail with Ellison on *Sayonara*. "I'm the driver," Ellison loves to say. Who's giving steering directions is another matter. Dickson, who's been described as a tactician "with the eyes of a U-boat commander," is regarded as one of the four or five finest sailors in the world.

Unlike *Sakura*, the $3-million *Sayonara* is moored right at the dock in downtown Newport. But Ellison plays cat-and-mouse here, too. Rather than accompany his crew of twenty-two into Narragansett Bay for the race, Larry the seafaring dilettante keeps them guessing when he'll show. He doesn't participate in practices, load gear, test the electronic instruments, review water currents and wind forecasts, or simply hang out with those lugs who grind the winches and man the windward rail. None of those tasks can get him any glory. Ellison pays the crew okay—Dickson might pull in about $10,000 a week, the other hired help gets a few hundred a day and an occasional Rolex watch—but he shouldn't have to fraternize with them, should he? Typically, Ellison gets a lift to the starting line just ten to thirty minutes before the gun.

"What's Larry like?" I ask one of the deckhands.

"Hard to know," he tells me. "I've never had a conversation with him."

Why does Ellison get to drive the boat at all? They didn't let the white-haired matron ride Secretariat, they don't let George Steinbrenner bat for the Yankees, so why does Ellison get to stand behind those mammoth port-and-starboard wheels? The answer lies in yacht racing's Corinthian traditions. The notion is that true sportsmen should take the helm, not just write the checks. It's a noble sentiment, except when Chris Dickson is there beside you giving *a few suggestions* and most of the crew have biceps like Popeye.

Sayonara is a remarkable machine, designed by New Zealand's Bruce Farr and one of the swiftest mono-hulled sailboats ever. Downwind, she's been clocked approaching twenty-five knots. So stiff is *Sayonara* that when she crashes through waves, the water, rather than the boat, seems to give way. When Ellison took sailing buff Walter Cronkite out for a ride and let him drive, Cronkite exclaimed: "This is better than sex!" Or at least that's what Ellison says he said.

Sayonara's seventy-eight-foot, narrow, ultralight hull is essentially foam surrounded by a carbon-epoxy skin—as basic as *Sakura* is luxurious. "So thin," says one of the crew, that "it'll crumble if it collides with dirt." Her black carbon mast, eleven stories tall, is a toothpick strung tight under 58,000 pounds of tension. The sails are space-age Kevlar and Mylar. Two-thirds of the entire boat's weight is ballast, deep beneath its white hull, on which *Sayonara* is stylistically emblazoned, the *o* in solid bright-red like the Japanese flag. Weight is so critical, they don't carry lunch for day races. The boat is its own little corporation—a few round-the-year employees, an E-mail address, color-coordinated foul-weather gear, and baseball caps with the *Sayonara* logo. Given her racing schedule across the globe, she travels in her own special container, equipped with machine shop and supply lockers. Ellison sails her several weeks a year. He had a modest sloop twenty years earlier that he could barely afford; when tycoonery happened and he learned about the new class of Grand Prix sailboats, Larry decided he had to have one. The name "Sayonara" appealed to his fondness for things Japanese and happened to mean, roughly translated, "See you later"—not bad if you're in first place.

For the first Newport race, Ellison arrives in town only six hours ahead of time. His Gulfstream 5 flew in at four in the morning and he's barely awake when he boards *Sayonara*. The racing weather is miserable—a summer storm with gusts over fifty. But he wins easily. "It never ceases to amaze me how Larry can step right on the boat and get right into it," Dickson says. The other boats break rigging and rip spinnakers; Coumantaros's *Boomerang* doesn't finish; Turner's *Courageous* doesn't even start. By contrast, *Sayonara*, with Ellison and Dickson taking turns at the wheel, knifes through the swells and wins by several minutes. The crew is soaked, sick, and exhausted, and still has an hour's roller-coaster ride into port. Minutes after crossing the finish line, though, Ellison's tender from nearby *Sakura* picks him up and brings him back to the mother ship for a nice hot shower. What Corinthian spirit! Fortunately for the losers, Ellison stayed aboard *Sakura* and didn't fuel up the Gulfstream for a flyby.

Larry Ellison was born on Manhattan's Lower East Side, but grew up in Chicago. His nineteen-year-old mother was unmarried and his father long gone. When he contracted pneumonia at nine months, she sent him to live in Chicago with his great-aunt and -uncle. Lillian and Louis Ellison formally adopted him and gave him his name. Larry was raised in a middle-class Jewish home—first on the North Side; then, as a teenager, on the South. He didn't learn he was adopted until he was twelve and the fact of the adoption beset him into adulthood. Why was he abandoned? What had he done wrong?

His adoptive father didn't help Ellison's psychological makeup. Louis Ellison was a bitter man, having lost a fortune during the Depression. He also wasn't thrilled with raising another child. They argued constantly. The father told the son he wouldn't "amount to anything." It doesn't take a shrink to figure out wherefrom Larry's adult insecurities and ambition came. "It was a powerful motivation," Ellison told his biographer, Mike Wilson. "I think my dad had a wonderful effect on me. If fire doesn't destroy you, you're tempered by it." Maybe even emboldened by it. Ellison always had a vivid imagination. His adoptive parents were Russian immigrants; perhaps to tweak them, he'd answer the phone: "Russian embassy—Boronov here." Another time, during college, he avoided a speeding ticket by telling the police officer he was a medical resident and was racing to the hospital to assist in a "craniotomy"; the cop apparently was not suspicious of the woman in the car, who happened to be Ellison's girlfriend, or the couple in the backseat. Such inventiveness with the truth served him well when it came to making a name for himself.

Arriving on a Harley, Ellison attended the University of Illinois at Champaign-Urbana, but dropped out after two years. He hadn't kept up his grades, and during finals of his sophomore year, his adoptive mother suddenly died. Restless and intrigued about the emerging counterculture in the Bay Area, he drove west to Berkeley. He was captivated by how different the place was and planned to come back someday.

Returning home in the fall of 1964, Ellison enrolled at the University of Chicago, but lasted just one semester. Like Bill Gates and Steve Jobs, he never got his college degree; the new breed of self-made technologist, so different from a Hewlett or Packard or Noyce or Moore, didn't rely on formal education. Unlike Gates and Jobs, however, Ellison never seemed to get over this lack of credentials and always covered his academic failures with more colorful stories. But by chance he was exposed to computers at the university. He had no great high-tech aspirations, but learning how to program mainframes gave him some income.

At twenty-two, with no personal reasons to stay in the town where he grew up, Ellison left for California again in 1966—not to join the new world of electronics but because he didn't have anything better to do. Unlike with Shockley or Bushnell or even Kildall, ambition didn't drive Ellison to California. Only his T-Bird did. He arrived in Berkeley, took a few courses, and began a series of computer-related jobs over the next decade. In between guitar, Yosemite, and his first failed marriage, Ellison supervised technicians at Wells Fargo and debugged the IBM mainframes at Fireman's Fund Insurance. He and the jobs were largely forgettable. About the only memorable thing about Ellison was that he knew how to spend money, even if he didn't have a lot of it—buying that first sailboat, a nose job in Beverly Hills, and a $1,000 bicycle. "Champagne tastes on a beer budget," his first wife called it. The tale goes that Ellison and a friend would sit on Telegraph Hill in San Francisco and point to buildings they'd someday own if they struck it rich;

their empire, Ellison said, would be called Universal Titanic Octopus (an ironic name given how often Ellison later accused Microsoft of being just that).

In the early 1970s, Ellison found himself working at Amdahl, a company that made mainframes to compete with IBM's. There, he met a droll, unassuming twenty-six-year-old programmer named Stuart Feigin, who came to understand and appreciate Ellison perfectly. "It was my first day at Amdahl and I met Larry right away," Feigin recalls. "He worked across the hall. I was very timid and a hard worker. Ellison was neither. He talked about the computer industry, he talked about basketball, he talked about Israel, books, stocks, and religion. But mostly he talked about himself—how wonderful he was, how stupid everybody he worked for was, and how much money he was going to make someday. He had to become rich . . . because he always lived better than he could afford. He had to have a Mercedes, even if it was so rusty it was held together by paint. He barely could make the payments on his house: Whenever I heard him on the phone, he seemed to be talking to builders and bankers at the same time. I always felt I needed three years' salary in the bank. Larry didn't have three days'." That probably explains why Ellison is still known among contractors around the Valley as the last billionaire to pay his bills. He once gave his executive secretary $2,000 on her birthday—$2,000 he owed her for dozens of errands and lunch runs. Old habits die hard.

Feigin would later be Oracle employee No. 5 and a centi-millionaire, before retiring in the late 1980s to Lake Tahoe—content to hike, invest, do charity, and lead a strikingly modest life, given his material alternatives. (Now fifty-three, he measures his net worth by the size of the national park he says he'll bequeath someday.) Feigin, in short, was the levelheaded sort. Yet the effect Ellison had on him was mesmerizing. "Actually, you know, I was really impressed," Feigin says, not able to pinpoint exactly why. "He was irresistible."

Harnessing that appeal—Steve Jobs's charisma without the edge—was the challenge facing Ellison. But he wouldn't do it at Amdahl, which fired him in 1973 in a major round of layoffs. His emerging charms notwithstanding, he'd have to find another paycheck. He and Feigin rode bikes together on weekends and frequently had lunch. You might have called them friends, even if Ellison was always late—unless he didn't show at all—and never picked up the tab. Feigin refers to him affectionately as "the late Larry Ellison." Years afterward, Ellison became infamous for his tardiness and the fact he couldn't care less about it. He once kept a group of U.S. senators waiting while he befriended a woman he had met en route to Washington. During the Clinton Administration, he often accepted invitations to state dinners, then didn't show.

After Ellison left Amdahl, he told Feigin he was starting his own software company and asked if Feigin wanted to come along. "I'll make us all rich," Ellison promised him. Content at Amdahl and cautious by nature, Feigin initially resisted. "Here was a guy who couldn't show up for lunch on time," he says. "How could he run a company?" But within a year, the lure of Larry proved too great, even if the left side of Feigin's brain told him he was making a big mistake. His only error was not taking up Ellison's offer at the outset; not being a founder cost him, by his own estimate, close to a billion dollars.

From Amdahl, Ellison drifted just down the road to Ampex, Nolan Bushnell's old stomping grounds. Ellison began working on database software that was a crude forerunner of Oracle's systems. More important, he met Bob Miner. Miner was Ellison's boss at Ampex, but that was just a technicality—Ellison never considered himself subordinate to anybody. Miner, as Mike Wilson put it, was "the anti-Larry"—warm, direct, unaffected, a man who agonized over spending $14.95 on a CD. Miner was the one who originated the line about Ellison and pretty Stanford women.

The best Bob Miner story, true or otherwise in Ellison's telling of it, is when he brought a $2-million check to the bank to deposit on a Friday afternoon. (It was right after Oracle's public stock offering and Miner got to sell a tiny part of his stash.) As was his habit, Miner also wanted $200 in cash for the weekend. So he filled out a deposit slip for $1,999,800, along with a withdrawal request of $200, and proceeded to wait on line with the rest of the proles. When his turn at the teller came, she was understandably bewildered and summoned the bank manager, who took care of the transaction, informing Miner that he wouldn't have to stand in the teller lines ever again. Miner was appalled by the notion that his station had changed. By contrast, Feigin says, years earlier Ellison told him he couldn't understand why *all* lines in society weren't organized by wealth. "Why," Ellison asked, "should a rich person ever have to stand in line behind a poor person?"

Ampex provided Ellison and Miner and a third programmer, Ed Oates, a place to germinate the idea that became Oracle. (Oates eventually retired from the company; Miner died of lung cancer in 1994, at age fifty-two, beloved at the company yet all but invisible in the last few years as he lost interest in cleaning up after Ellison.) Ampex was where Ellison got his act together. "When I met Larry," Oates remembers, "he expressed a belief that he was 'the best there is, the best there was, the best there ever will be.' He just didn't have a stage on which to act: He needed to write the script and direct the action. As long as he was working for someone else, he was unable or unwilling to give his best or be dedicated enough to sacrifice himself. Since his motivation was recognition, money, and power—not necessarily in that order—he did indeed find himself once Oracle was founded. Failure would've meant admitting that he wasn't the best—so failure was not an option."

In the 1970s, the Valley was going through a transformation—one of several in its history. First had come the garage pioneers like Hewlett and Packard, then the silicon cowboys like Shockley and Noyce and Moore. Giant mainframe computers of the 1950s led to minicomputers of the 1960s and the birth of the PC. HP and Intel were revving their corporate engines. Jobs and Wozniak came together, just as Gates and Allen did in Albuquerque. Apple was formed in 1976, Microsoft a year earlier. Great companies are the flower of inventive minds, but they need the right growing conditions. Now, Ellison envisioned a new market—for software—and opportunistically pounced on it, but he was in the ideal place at the ideal time. Like Apple and Microsoft, Oracle's good fortune was the by-product of luck—filling in a gap in the high-tech landscape. "Despite every effort on his part, Larry managed to not screw it up," says Tom Siebel, a former Oracle vice-president who went on to start his own high-flying software company.

The very notion that software programs could lay the foundation of a company was new. Intel and HP were hardware manufacturers—the bricks and mortar of high-tech. So was Apple (even though what would distinguish its computers from others was its software operating system for the hardware). Oracle would be the Valley's first genuine triumph in pure software.

Until the late 1960s, what software existed was usually free, coming "bundled" with the computer—no different from the little prize included in a Cracker Jacks box. The term "software" itself had only been around for a decade. Stores were not filled with pretty boxes of business applications, word-processing programs, video games, and such; Bill Gates was still in high school and Marc Andreessen—the cofounder of Netscape—wasn't even born. Instead, in what Feigin describes as "the good old days of software," programmers by the thousands gathered in hotel ballrooms in places like San Francisco and New York, where they traded code. They were working at universities and corporations that used IBM mainframes, but software was considered little more than plumbing. "We had no idea you could make any money from software," Feigin says, "because it had always been given away. When IBM started to sell it, we were aghast at their violation of principles."

Software represented an entirely different universe from the hardware that had powered the Valley for a generation. Hardware was tangible—metal, plastic, flashing lights, whirring drives, and miles of wire. Someone who made hardware, in an Intel clean room or an Apple assembly plant, could place the fruits of his labor in your hands. Software, by contrast, was as much a concept as code on a piece of paper or holes in a punchcard. But without software, a machine was just a machine.

Far more than the expensive infrastructure of the semiconductor industries, software relied on the imagination of an individual. It needed neither factory to build nor natural resources to mine—just brain matter. The fiscal beauty of software was that once you wrote it, the cost of each copy was negligible and the profit margin was extraordinary. Creating a mass-market piece of software was like mining a vein that just kept producing.

The motive to sell software, of course, arose from marketplace demand. As Moore's Law predicted, computer processing power kept going up and prices kept going down. Computers and computer networks were becoming essential to global commerce—for inventory control, manufacturing analysis, record storage, and payroll preparation. Ironically, the exponential growth in digital databases—some of them consisting of billions of data—rendered keeping track of all the resulting behind-the-scenes information overwhelming. Magnetic tape was long on capacity, but notoriously short on organization; disk drives were faster, but lacked space. The biggest, most complete database was of little use to General Motors or Bank of America or the CIA if the company couldn't easily and quickly search, sort, and retrieve the data. Why go digital if it was just as unwieldy as a giant file cabinet?

Organization wasn't the problem: Even the most primitive database management systems—"hierarchical" and "network" systems—could find and update a specific account. Hierarchical systems were rigid and top-down—if you wanted a piece of information in the middle, you still had to start at the beginning. Network systems were an improvement, because they cross-indexed information—but only if you understood the mind of the programmer who organized the index.

The challenge for database management was how to manipulate the database to answer questions not anticipated by the original programmers and therefore not specifically defined in the database. It was these kinds of random inquiries—how many seats on the Miami-to-Dallas flight were sold in November or which platoon had the youngest infantrymen by average—that could be answered only with the assistance of a computer. Hierarchical and network database structures didn't help. In theory, something called a "relational database" would. In a Fortune 500 company, one big, central computer would maintain the database and be connected to scores of terminals, which allowed many employees access to the system at once.

At Ampex, Ellison, along with Miner, Oates, and others, was trying to create new database management software. But the CIA-funded product, code-named Oracle, never quite worked. Ellison moved on to a marketing post, then left for another database company in the Valley, Precision Instrument. Ellison was supervising software development, which meant contracting work out to freelance programmers. But he came up with a better idea: He'd start a new company with Miner and Oates, which would put in a bid of its own for CIA work. Miner and Oates would do the real programming work, while Ellison temporarily stayed on at Precision to coordinate the project.

Ellison's motivation for starting a company was not couched in life-affirming, world-changing, curiosity-satisfying drivel. He was honest enough to say he wanted wealth and power. Stuart Feigin figured that out the first day he met Ellison at Amdahl Corp. "Gene Amdahl always had wanted to build things, going back to his days as one of the chief architects at IBM," Feigin observed. "He owned a Rolls-Royce, but only as a result of making a deal with Rolls-Royce—'You buy my computers and I'll buy your car.' Larry was different. I can't imagine he was ever in it because he thought the world needed a better database system."

Many of Silicon Valley's early entrepreneurs struck gold because they wanted to build something their current employers weren't interested in. Or because they wanted a better toy to play games on. "With Larry," Feigin says, "there was never that twinkle in his eye." Who could blame him? Was the tedium of information management ever going to match the thrill of making gobs of money? For a man who fancied himself a student of history, art, war, molecular biology, and a few other topics, relational databases were not the stuff of intellectual fulfillment.

Ellison represented a new kind of post-silicon attitude in the Valley, born of its changing values and the realization that high-tech beckoned opportunities that not even California could imagine. But the technology was beside the point. The marketing tail would wag the product dog. Hackerdom gave way to hype. Instead of being a Homebrewed sneakered subversive—his double-breasted European-tailored suits and black silk Japanese sport shirts are a source of great amusement to Jobs the slob— Ellison was a swaggering capitalist bent on becoming a billionaire. "He got what he wanted," Feigin rues, "leaving a lot of us millionaires, and friends, behind."

Ellison's new company, founded in Santa Clara in the early summer of 1977, was called Software Development Labs (passing up Oates's more whimsical Nero Systems—"We Fiddle While You Burn"—and Miner's UrAnus Systems—with a logo featuring products emerging from a sphincter). Ellison put up $1,200 for a 60 percent stake in the company; Miner and Oates put in $400 apiece. Ellison hoped to have a company someday with a few dozen employees. "No one thought of selling a million quantity of anything in the computer marketplace," Oates says. "If we could get to ten million dollars in annual revenue, we'd be fat, dumb, and happy. Larry would be the salesman, the guy out on the street. Bob and I had more of an affinity for the truth than Larry did. And good marketing beats good technology most days of the week."

From this seed, Ellison would become a multibillionaire and his co-founders would also do well.

The idea for a relational database had been around for years. In 1970, IBM—the unchallenged Big Foot of computerdom—published a white paper proposing a new way of managing data. Users wouldn't need to know how their raw data was stored or how to find it. They could simply ask a few questions and the database would retrieve the correct data—presented in tables of rows and columns, capable of limitless rearrangement. The very organization of this ultimate spread-sheet—the *relationship* between different piles of information—reflected the ingenuity of the "relational" model.

But it was a model only. The IBM paper described a functioning relational system, but hadn't set out to create one. Several years later, IBM researchers in San Jose looked at database management again and invented a command language rooted in simple English. SQL, the Struc-tured English Query Language, allowed anyone who could type to use it. In 1976, IBM published specifications for SQL in various technical journals and gave many programmers their first glimpse at how a rela-tional database might work.

Why was Big Blue so open? For starters, it didn't understand the commercial potential. In addition, its R&D culture was to share, at least in broad strokes. Part of that sprang from self-interest: Because IBM was viewed as the industry standard in so many things high-tech, it had every reason to maintain that perception, even if that meant some of the ideas it hatched were developed elsewhere; its subsequent decision, for ex-ample, to allow Microsoft to license MS-DOS to other companies was based on IBM's view that standardizing MS-DOS would benefit IBM as much as Microsoft.

But part of the IBM ethos to share research was based on a belief that the entire industry gained when breakthroughs happened—even if some other entrepreneur led the way. Even hard-driving Intel had helped competitors when it thought that enlarging the high-tech pie would yield bigger slices for all. But IBM erred in not introducing its own relational software until 1982, five years after Ellison's group borrowed the idea. IBM made a business of it all right, but it had only part of the business.

Unfortunately for its shareholders, IBM didn't learn. Four years after it published SQL, Big Blue had its prophetic meetings with Gary Kildall and Bill Gates. IBM made Microsoft's operating system the preferred one for its first PC. More crucially, it allowed Microsoft to license that operating system to any hardware manufacturer. The IBM PC became the Microsoft/Intel computer, a commodity that anybody could assemble. Companies like Compaq and Dell can trace their lineage to that masterstroke of stupidity. Ellison delighted in calling IBM's decision not to demand an exclusive license for MS-DOS "the single worst mistake in the history of enterprise on Earth." He might also have been referring to the IBM blunder that propelled him to the head of the teller line.

IBM wasn't by any means alone in its negligence during this period. Apple Computer's success with the Macintosh in the mid-1980s was rooted not in anything Steve Wozniak had done almost a decade earlier, but in Steve Jobs's fateful visit to PARC, the Silicon Valley research lab of Xerox. PARC invented the technology, but Jobs found it and built an empire around it. The Apple Macintosh, and Microsoft's subsequent versions of Windows, are just latter-day GUIs. "Good artists copy," Picasso observed, at least according to Jobs. "Great artists steal" (the irony being that Picasso lifted the line from T. S. Eliot, Lionel Trilling, or Igor Stravinsky, depending on who you believe used it first). Apple, Oracle, Ellison, Jobs—the genius of an idea is in execution, not just conception.

One of the programmers who lapped up IBM's database disclosures was Ed Oates. He believed that the technical hurdles could be overcome. But it was Ellison, once he learned of Oates's discovery, who shrewdly recognized that the very fact that IBM was contemplating relational software would legitimize any non-IBM product that beat Big Blue to market. And that's what he set out to do. He and Miner and Oates (and a fourth employee named Bruce Scott) had done well writing software on a contract-by-contract basis—it had financed the new company for a year. But freelancing guaranteed no income stream and offered no possibility of a dream product that might define a niche.

Several months later, to their own collective astonishment, Miner, Oates, and Scott finished a rudimentary minicomputer version of relational software. They called it Oracle, reprising the name of the failed CIA product back at Ampex; the name also fit the cocksure attitude of the company leader. Within two years, an operational "Oracle" hit the marketplace; slow and buggy, it was an achievement nonetheless. Ellison wasn't the programmer, but it was his will that got it written and his relentless marketing abilities that would get it sold. It was also at his insistence—after customers gave him the idea—that the software be adaptable to all business computer systems, not just dominant ones like IBM's or Digital Equipment's. This "portability"—or, at least, perceived portability—across different hardware and operating systems made it invaluable in computer networks managing large databases. Eventually, the software became available for a range of minicomputers, as well as mainframes and PCs. "Promiscuous software," Ellison knowingly called it, because it did the trick "with anybody." The problem was that some partners were better than others, but what the company lacked in technical excellence it made up for in image.

To try to boost its profile, Software Development Labs changed its name to the equally turgid Relational Software Inc., which didn't become Oracle Corporation until 1982. Ellison was chairman and CEO. The federal government in the late 1970s provided Ellison with his first customers—the Wright-Patterson Air Force Base in Ohio, Navy Intelligence in San Diego, and the company of companies, the CIA (not that it was ever identified in Oracle's prospectus years later). Oracle made a bundle from its software, initially charging each customer $48,000. These early successes brought out the hypercompetitive ways that became Ellison's signature: This was the man who buzzed his opponents after a regatta and made a weekend bike ride with Feigin into the Tour de France. Expounding his zero-sum business philosophy to *The New York Times*, he said, "It is not enough to succeed. Everyone else must fail." No wonder he became best pals with Steve Jobs.

In a memorable print ad from 1987, the company featured an Oracle F-15 shooting down a red triplane emblazoned with the name of a hated competitor, Ashton-Tate. The ad, Ed Oates said, "was classic Oracle attitude." Oates is relaxed and retired now ("on leave" through the year 4711, the last year programmed into Oracle's human resources database), but he saved an enlarged version of the ad for his study. In another ad, a fighter pilot gives the thumbs-up sign from the cockpit of his Oracle jet. Like notches in a gunslinger's belt, the plane's wing listed the crossed-out names of Oracle's competitors. In tiny print, a mock disclaimer read: "Oracle is a registered trademark of Oracle Corporation. Our use of [the competitors'] trademarks in this advertisement is the least of their problems."

That macho, maniacal attitude also made Oracle, like Apple, a difficult place to work. More than even the typical start-up, Oracle was an extension of its creator—alternately maddening and exhilarating, inscrutable and blatant. Ellison's makeup was part unadorned: Why wasn't everyone as driven as he was to succeed? But part of it was contrived. He wanted Oracle to look and sound the part both to employees and customers. Ellison believed in self-fulfilling prophecies: If his company boasted of the best relational database—crash-proof, portable, fast— then by God those folks paying thousands of dollars for the product just had to believe it. Whether it was true was a different matter. Sometimes, the software destroyed information. And sometimes the software didn't exist at all—despite Ellison's claims, it was still being worked out in a programmer's imagination. It was that pesky problem of "tenses" that Ed Oates and others spoke of bemusingly. "We have great software" might really mean "We hope to have great software sometime before you wonder what you paid for months ago." If the customers ever picked up that Oracle's software was shoddy, they might not be customers anymore.

Or maybe they would. According to *The Difference Between God and Larry Ellison,* a senior executive, referring to those early days, asked Ellison if any irate customers ever wanted their money back. "No, I don't think so," Ellison replied. "But I do remember that people used to call up and say, 'Could we please have our data back?'"

The Oracle way of the 1980s was to sell the product aggressively, no matter what the bugs, no matter how customers were misled, no matter how exorbitant the pricing. The point was to keep pumping up sales numbers. The motto around Oracle was "G.T.M., G.T.F.M.—Get the Money, Get the Fucking Money." "I was more interested in making software that I could be proud of than in making money," says Stuart Feigin. "For Larry, software only needed to be 'good enough.' The funny thing is that Larry thought I was the one who lacked ambition."

Oracle was Silicon Valley's version of Drexel Burnham Lambert before Ellison's friend Michael Milken took a fall. Not only did all that enthusiasm result in huge commissions to its salespeople—for a time, payment was made in gold coins—but it made Oracle the hot place to work. Oracle's recruiting style reflected the company's brand of hauteur. Ellison seemed to bring in "arrogant men and pretty women," recalls Roger Bamford, one of the early engineers. "The favorite interview question on university campuses was, 'Are you the smartest person here?' If the person said no, the recruiter would ask, 'Well, who is?' Then they'd go after that person." Who knows what they asked the pretty women? Probably not anything about senior management, where Oracle lags behind other Valley companies, which have not been particularly hospitable to women in any case.

Whatever their ethical shortcomings, the sales practices paid off and Oracle's database program became the standard. Ellison's early notion of a manageable number of employees disappeared. There were under fifty in 1984, as he once had planned; by 1990, there were more than five thousand, half of them working in foreign countries. Through most of the 1980s, sales doubled almost every year—an incredible run. In 1982, revenue was $2.4 million; four years later, $55 million; and in 1990, $970 million. In less than a decade, Oracle went from a million-dollar baby to a billion-dollar giant. In the middle of that dash, Ellison cashed in, as the founders of Intel and Apple had already done, and as so many others would.

On March 12, 1986, Oracle Corporation—the purveyor of software few consumers had ever heard of—went public. It opened at 15, and at the end of the day closed up 6⅞—an increase of 45 percent, which was pretty good in those days. Ellison found himself worth $93.5 million. It wasn't that he had been a pauper the night before—he knew his large stake in a company doing $55 million in annual business made him a multimillionaire. But not until the public markets put a value on private stock can an entrepreneur truly measure his own worth. Keeping score means letting someone else in on the game. By the end of IPO day, the insecure, forty-one-year-old Larry Ellison knew he had hit a home run.

His fellow founders—Miner and Oates—also were loaded, though Oates far less so because of a divorce several years earlier. Before the IPO, Feigin couldn't get a mortgage. The Bank of America told him he didn't make enough, even though Feigin pointed out that he had earned more than the bank the prior year. After the IPO, Feigin snickers, "they were suddenly my best friend."

The frustrating part for Ellison, as it turned out, was that another tekkie took his company public the next day. Up in Seattle, the skinny kid who also got going by hoodwinking IBM watched his personal worth soar to $300 million, leaving Ellison in the dust (and becoming high-tech's first billionaire two years later). The kid, of course, was Bill Gates, and the stock market's love affair with his Microsoft never waned; around Easter, 1988, Microsoft passed Lotus as the top software company in the world. Ellison never said that his personal battle with Gates flowed out of their dueling IPOs, but if it didn't, the coincidence was remarkable.

Ellison, like Gates, did as well personally as he had because he retained such a huge chunk of the company—more than a third, which was unusual for a founder. Typically, venture capitalists, who give a start-up its early cash, control a big slice. They didn't with Oracle; Ellison didn't much like them, but more to the point, when he wanted access to their wads, they had turned him down. Even more unusual, Ellison held on to each and every share for years after the public offering. Most entrepreneurs will sell small amounts of their stock, both to diversify their holdings and to get the spending money necessary to buy houses, cars, planes, and groceries. Ellison was too vain about his company to part with his stock; it was the longest-running relationship of his life. Down the road, that devotion would magnify his personal financial problems.

Oracle's offices showed just how far Ellison's company had come. Having moved around the Valley throughout the 1980s, Ellison wanted permanent space, both to accommodate growth and to present the right corporate image. In 1989, Oracle bought a large parcel just off Highway 101, a few miles north of Redwood City—a few minutes' helicopter ride from Woodside. There were already two cylindrical glass towers, and in time, Ellison would erect four more, complete with their own utility substation, surrounding a fake lake. These buildings were not like the Valley's other bland shoe boxes; they were shimmering elliptical monuments—sort of like Larry himself. ORACLE read the space-age lettering atop the main building, visible to anyone on the highway or landing at nearby SFO.

Inside, the culture was vastly different from the unpretentious, egalitarian feel of other high-tech companies. It wasn't just the 36,000-square-foot gymnasium or the tailored salesfolk or even the regal suite Ellison created for himself, full of expensive art and samurai toys. The parking lots were full of German imports and hot rods. "The place never felt like the typical Silicon Valley company," says Roger Bamford. "It was more yuppified—expensive furniture, designer offices for favored programmers. Part of this was Bob Miner showing respect for us—we weren't just people who put stuff in the machine. But part of it was also about appearances." Oracle's opulent emerald towers might better fit into the Southern California scape. Or a land called Oz.

The wizard himself was an enigma to most employees. Any CEO can be late to meetings, but very few fail to show up at all. Nor do they disappear for days on some unannounced excursion. The worst part was that employees felt they couldn't connect to Ellison. Steve Jobs, at least, would be in your face and tell you what a dope you were; Ellison avoided situations—and people. Stuart Feigin remembers the time he was coordinating the move of Oracle's data center. It was a massive project scheduled to take three weeks. Feigin and a crew of 150 did it over a weekend. "Larry came by with a girlfriend," Feigin says. "He apparently had nothing to do that day. The data center guys had never met him before. 'Larry, this is great. We're done!' I told him. 'Why don't you stick around and have some pizza with us?' He declined. We had just saved him a ton of money, and I tried to explain that he should recognize what these people had done. He wasn't interested." Andy Grove wouldn't have been this dumb—though he might not have picked up the tab either.

The leaders of most companies make a point of walking the halls, just "to be seen." David Packard practically invented the style. Ellison does just the opposite. "Larry hides," Feigin says. He frequently works out of his home, amid the bonsai trees and waterfalls. Or so he says. Most employees have never even seen Ellison. When they do—often in an elevator ride—they call the encounters "Elvis sightings."

Ellison's personal life has been no less odd. Three marriages hadn't worked out by the time he was forty-two. The last, to Barb Ellison—Stanford graduate, early Oracle employee, and now the empress of Woodside—produced a wedding scene out of *As the World Turns* and then, fittingly, an ugly divorce. Hours before their 1983 marriage—they already had a baby—Larry presented her with an eleven-page prenuptial agreement that attempted to insulate most of his holdings from her. What a romantic! She would be entitled to no more than $1 million, no matter how long they were married. As the guests waited, the couple's lawyers negotiated—Barb's father and Larry's brother-in-law. The marriage lasted barely three years, until shortly after Oracle went public. The divorce never went to trial, but enough of the details came out to embarrass Larry, including a resplendent account of the prewedding legal festivities. He wound up paying Barb a lot more than he had hoped—enough for her to keep their Tara estate in Woodside, take care of fifteen horses, and host the best parties in town.

With the IPO, Ellison had made good on his promise of wealth for himself and his employees. In a rare moment of reflection, he warned some of them of the legacy of forty-year-old Dennis Barnhart of Eagle Computer—the "poor bastard," as he put it, who in 1983 cracked up his red Ferrari on IPO Day and was killed. Then Ellison went out and got his own anyway. At one point after the IPO, according to Mike Wilson, he was keeping track of just how many Oraclers were millionaires. There were Miner and Oates, Stuart Feigin, and a few others. "The first thing that hit me," says Jenny Overstreet, the executive assistant who ran Ellison's life—from scheduling appointments to selecting the right peanut butter for his sandwiches to finding the best cars—"was that I could buy pretty much whatever olive oil I wanted." She noticed, too, that her garbageman took to talking to her about *his* fifty shares of Oracle stock.

By 1998, there were hundreds of Oracle millionaires (though none compared with Ellison's multiple billions). How well had Oracle stock done? If you bought $17,500 worth of Oracle at the closing bell on March 12, 1986, you had $1 million a decade later. Most loan sharks don't do so well.

But they have better accounting. Remember that the paradox of Oracle's immediate prosperity was that it doomed the company's longer-term prospects. The robust sales figures every quarter made it all the more imperative to *top* them in three months. At a certain point, that wasn't possible: There were only so many businesses around the globe that needed state-of-the-art database management and had the means to pay for it. For a big customer in the 1990s, between the software and consultation services for it, as well as the hardware, Oracle could receive $100 million. To keep the revenue engine humming and make sales quotas, Oracle needed not just salespeople who stretched the facts, but accounting practices subordinated to the bottom line.

A creative bean-counter could book revenue that had been received even if Oracle hadn't delivered its goods, which became problematic if the product never was delivered. At a normal company, that might not amount to a big issue. But at Oracle, the cavalier sales staff was selling products that hadn't been created and, in some cases, never would be; "vaporware," the wags called it. Salespeople logged in deals for which contracts were never signed. If next year's take was better than this year's, the balance sheet would still look fine. But sometime this pyramid scheme would collapse, no matter who was at the helm. Given how consciously ignorant Ellison was about the financial side of the business—and how he frittered away his time on such ventures as starting *Buzz* magazine in Los Angeles—Oracle was a corporate steamship in search of an iceberg.

It found one in 1990, when the accountants saw no way to navigate around the stack of uncollected bills. Oracle was still taking in lots of revenue—an annual rate of a billion dollars—but not much more than the prior year. The reason was a $15-million shortfall caused by the bad debts. When that news was divulged, the stock market that had long adored Oracle showed no mercy. The stock fell 31 percent. Ellison lost more than $300 million on paper. He was no longer a billionaire.

Six months later, the reckoning got worse. In one quarter alone, Oracle's losses amounted to $36 million—its first reported loss—and it had to admit it had overstated profits for preceding quarters. By the beginning of November, the stock price—which hit 17½ in the earlier crisis—sank to 4⅞, lower than it had ever been, even in the first year. Oracle had lost 82 percent of its market capitalization; the total value of its stock, as calculated by Wall Street, was only $700 million. Ellison was down to the lower ranks of the centi-millionaires. Good-bye, red

O Z
145

Testarossa. While Ellison never faced personal insolvency, his company was in peril. Ten percent of Oracle's labor force was laid off; there was speculation Ellison would be forced to quit; and the perceived invincibility of his database kingdom was gone forever. "Oracle is run by adolescents," he said at the time. "And that includes me."

Shareholders filed nineteen class-action lawsuits for securities fraud and the SEC launched an investigation into a range of "pervasively inadequate" accounting and billing practices, which was a nice way of calling Oracle a bunch of crooks. Steve Jobs suggested that Oracle's main phone number answer with a recording: "To sue Oracle, please press 3. To sue Larry personally, press 4." Settling the litigation years later, Oracle paid out a total of $24.1 million. The CEO, if not humbled, was embarrassed at his reversal of fortune. "There is no more costly schooling than to give somebody a billion-dollar company as a laboratory and teach him how to be a CEO," he explained to *Business Week*.

But Ellison, a fox and a phoenix, had the last laugh. His company recovered—hiring new financial managers, reining in abuses, and finally paying attention to customer support—and surpassed its earlier dazzle. As Ed Oates said, "Ellison wouldn't allow himself to fail," even if it meant he had to admit past errors. A billion dollars in revenue became $2 billion in 1994, $4 billion in 1996, and then roared by $7 billion in 1998, with nearly a billion of that as profit. Employees around the world totaled more than 29,000. Relational databases, with Oracle in the lead, were the core of information management in the government, in banks, in commercial transactions ranging from credit-card purchases to complex manufacturing orders. The old Oracle arrogance returned, illustrated by the bickering billboards along Highway 101 at the Oracle exit. One competitor, Informix, posted an ad picturing Oracle's headquarters beside a yellow sign, WARNING: DINOSAUR CROSSING. Another showed Ellison with devilish horns growing from his head. Oracle countered with an ad ridiculing Informix's lawsuit against thirteen of its engineers who defected to Oracle. It read, "Informix: Hiring Lawyers Experienced in Suing Programmers. Oracle: Hiring Experienced Programmers."

In the 1990s, Oracle's stock price soared higher than ever and Ellison's wealth at one point made him the fourth-richest American on the Forbes 400 list—behind only Bill Gates, investment star Warren Buffett, and Paul Allen—with almost $10 billion. More impressive than the act of creating Oracle, Ellison had resurrected himself—much like Steve Jobs would do with Apple Computer. Humbled warrior, unappreciated hero, Ellison was a paragon of what Silicon Valley offered.

Then, in the middle of December 1997, Oracle stock hit its biggest bump since the balance-sheet debacle six years before. The cause this time wasn't internal mismanagement, but economic turmoil in Asian markets. In the course of eight hours, in the greatest individual volume ever recorded on the Nasdaq, Oracle stock fell 29 percent—from $32\frac{3}{8}$ to $22\frac{15}{16}$. (The stock had split many times over since the IPO, the only reason these numbers were still in double digits.) While not as severe a plunge in percentage terms, Ellison took a bigger hit in the wallet. His net worth dropped more than $2.1 billion—reportedly the biggest single-day personal loss of anybody in history. It didn't seem to bother him extraordinarily. "If you've got five billion left," he says, "it's not so bad. After a certain threshold, cash doesn't mean anything other than in points—a way to keep score." As Dorothy said about her home outside Wichita, Ellison says, "There's no place like Silicon Valley, where your talents can be magnified and the projection of that magnification is cash. It's unbelievable.

"Would another billion dollars make me happy? I mean, it beats losing a billion dollars, yeah. But at this stage it doesn't make that much difference to me." He can't let it, because if you find yourself ultimately sated by the money, you might as well just sit by the koi pond. There's not much left to prove. How could Larry Ellison have nothing left to prove?

Ten days after Oracle's stock skydive, Ellison skipped the company's twentieth anniversary party. Shaken employees—those who hadn't been at the company forever and whose stock options suddenly became worthless—were looking to the CEO for a boost of morale and perhaps a few kind words. Ellison was off to the Caribbean for a cruise on his wonder yacht. Three weeks later, the stock price fell another 23 percent and his fortune went down another $1.2 billion. As with other big stock dips, a bunch of employees quit. The kind of corporate loyalty that Hewlett-Packard or Intel engendered was unheard of at Oracle. However bad the occasional taste, Ellison was a meal ticket when times were good. When the ticker started heading south, it was a lot harder to put aside your revulsion or laughter.

But Oracle came all the way back, as it had before. Easy come, easy go, easy come. Oracle's two towers became six, each with its own themed gourmet restaurant (today, seared ahi tuna; tomorrow, a savory Cobb salad). There's on-site dry cleaning and no employee need walk more than thirty yards to reach a $5,000 espresso machine. In his inner sanctum, atop 500 Oracle Parkway, Larry Ellison dreams of grander times than ever, when Oracle will no longer be a mere database giant but a leader of the Information Age to rival, well, Microsoft. And no longer will he be the CEO of a company that few consumers know, but the kind of demigod who can get on as many *Fortune* magazine covers as Bill Gates himself. Oracle may be the second-largest software company, says Roger Bamford, "but for Larry that's the kiss of death. Being No. 2 drives Larry crazy." A few years ago, *Fortune* ran the big headline, LARRY ELLISON IS CAPTAIN AHAB AND BILL GATES IS MOBY DICK. It only reinforced Ellison's inferior status.

Perfectly pressed, immaculately creased, in an office befitting a wizard, Ellison describes a future where PCs no longer rule, replaced by simple "Internet appliances" running on Oracle software. It's a tune he's been singing since Labor Day, 1995, when he proclaimed at an annual industry forum in Paris that the PC was "a ridiculous device." Instead, he proposed an "NC," a network computer that would merely be a terminal tied to a centralized database managed by professionals. Ellison was clever. Noting that keepers of the PC flame "made fun" of old-fashioned mainframe computers—whom could he be alluding to?—how could he be criticized for taking on the PC? There had been only one ENIAC back in the 1940s. It gave way to tens of thousands of mainframes in the 1950s, then hundreds of thousands of minicomputers in the 1960s and '70s, and finally hundreds of millions of PCs in the 1980s and '90s. Now, Ellison said, dawned the post-PC age, when billions of cheap information appliances would harness the world of data networks that is the Internet.

The next speaker just happened to be Bill Gates. He scoffed at the idea that computer users would ever accept a device that couldn't store programs and data, all the more so if the replacement was a Big Brotherish machine (assuming, of course, that Ellison ever could be imagined as Orwellian rather than Dickensian). "Larry's hype has expanded to fill his ego," Gates later laughed. But for one of the few times in his life, Gates had been forced to acknowledge something that came out of Ellison's mouth.

Ellison was talking about an entirely new species of machine—a cheap, stripped-down box that would hook up to the Internet for all of its needs other than printing. No disk drive, no gargantuan memory, no bulky "bloatware" applications, no upgrades necessary—the NC would electronically retrieve software applications and personal files, as needed, from a common source—the equivalent of a digital filing cabinet. Who would manage the network and all that data? Oracle, of course—sort of like the Ma Bell of information, a private utility serving the public good. If you believed Ellison, the PC Era would give way to the Network Era.

At a cost of several hundred dollars, the NC promised savings to businesses, computer access for the masses, and the end of the "Wintel" monopoly—Microsoft's Windows and Intel's chips. On that more level playing field, the notion went, Oracle and the other companies aligned against Microsoft would have a fighting chance. (This cabal later became known in Microsoft's camp as NOISE—Netscape, Oracle, IBM, Sun Microsystems, and Everybody Else.) Using a computer would no longer mean paying Microsoft for the privilege; there would be no more need for its operating system. Nor would Microsoft have a lock on applications dependent on its operating system (things like its Excel spreadsheet or Word word processor). The deck also wouldn't be stacked anymore against Netscape's Internet browser, for example. Sun's open programming language, called Java, could flourish because it allowed software engineers to develop programs for the Internet that worked on any computer's operating system—or none at all. Microsoft exemplified a proprietary—"closed," as they said in gobbledygeek—model of software. With dependence on the Windows system thus smashed, Apple or any other company outside the Wintel cartel would have a better opportunity to compete. For a time, before Steve Jobs returned as CEO, Ellison considered a takeover of Apple.

The NC was a compelling alternative to the PC, not that Ellison or Oracle had done any work on creating or marketing such an appliance.

In that regard, it was classic Ellison: Talk first, execute later. Over time, he perfected the pitch: PCs were too complicated, whereas network computing would be almost invisible. "The TV, the phone, water, electricity—we don't see the complex network behind *them*," Ellison told industry gatherings, interviewees, and even *Oprah* on daytime television. "You never have to 'back up' your phone or TV. . . . After aqueducts, almost no one built a well. . . . Using a PC is like me flying my jet or helicopter to the store." By the year 2000, Ellison predicted, NCs would outsell PCs by nine-to-one. And, after that, there would be other network-based appliances, like a digital-screen phone, pager, car navigation system, or even a microwave oven that looks up cooking times on the Internet based on a food package's bar code. Mrs. Swanson goes digital.

Before one dinner crowd in Silicon Valley in late 1997, he performed for an entertaining forty-five minutes, barely pausing for a swig of Evian. "The computer is the only device in the home today that's smaller than the instruction manual required to operate it." The PC is something "only an engineer could love—faster and faster, feature-rich, but full of gratuitous complexity." If it was a car, "you would only be able to drive it around your backyard and you'd still have to buy Steering Wheel 7.0." Ellison calls the PC "transitional technology," akin to the steam engine that kicked the industrial age into gear. "When the internal-combustion engine came along," the steam engine became largely obsolete. Anybody who uses the Internet, he noted, already relies on a network. After all, if you had to install your own software to buy a book online at Amazon.com, it would take a gazillion floppy disks; Amazon.com software lives exclusively on the network. In the advertising extravaganza for the Super Bowl, Oracle in 1998 took out its first big spot, using the slogan "Enabling the Information Age," and trying to reinvent itself as a consumer brand.

Ellison's riff on PCs and NCs was prophetic and anticipated the arrival of sub-$1,000 desktops, some costing only several hundred dollars. The problem was, it wasn't his NC racking up the sales—traditional Windows-based PCs were. At the end of 1998, almost a third of new PCs sold in the United States cost less than $1,000. Gates was ultimately right about the "hype." Aided by better, quicker development and an established position in the marketplace, the old machines—with lower price tags—co-opted the NC. So, too, did the Microsoft-owned WebTV, which connected the Internet to a television set. It also didn't help that Microsoft negotiated a licensing deal to ship Java, then altered part of the language so that it would run only on Windows. (Sun sued, but whatever the outcome, the damage had been done.) Ellison's testosterone-fueled dream of unseating Bill Gates as industry titan and billionest billionaire is as far away as it's ever been. The total market value of Microsoft is roughly ten times Oracle's, even though Microsoft's sales aren't three times those of Oracle; a few years back, the spread in market value wasn't quite half that. CAN LARRY BEAT BILL? asked a 1995 headline in *Business Week* and a dozen others since. Not likely. And Bill's little company is even making incursions into Larry's core database business, developing its own database software and trying to raid Oracle talent like Roger Bamford. No wonder Larry wants the MiG.

If you believe him—and who can tell what to believe from a man who sold software that didn't exist—what Ellison, now ancient at fifty-four, really wants is to be less misunderstood.

Pay no attention to that man behind the curtain. No matter that on the May afternoon in 1998 that he and I talk there's a Kansas-like funnel cloud within view of his emerald windows. Forget that his office is a corporate bunker—a forbidding zone in the Oracle complex that few employees dare go near. The decor is formal Japanese, the doors are locked, there are guards on duty, and Ellison's main assistant (one of three) is so unfriendly that Stuart Feigin avoids going by when she's there. But put all that aside, Ellison tells me—and realize he's just a simple Valley CEO trying to build a business and help society at the same time.

"Of course I'm doing this for myself," he says. "I have a hard time with all the hypocrisy here—everybody's doing everything for the other guy, there's no self-interest. It strains credulity." So why's he so keen on reforming public education and making sure "every kid in America" someday has a computer? "That would mean I changed the world. It would say something about *me*. I do everything for my own happiness, but the best way to pursue happiness is to do something for others. It's this nasty little trick."

How does he wish people would look at him? "I wish the world would see me as the man who cured cancer," he says. "Unfortunately, I haven't, so it's a problem."

Good point, but what about a real answer? "Oracle is the second-largest software company in the world and by far the most successful in Silicon Valley," he says. "Not bad, right? But did you want egocentricism or false modesty? I can deliver either one."

The cure-for-cancer line has a basis, since he started his own medical research company and has invested millions of his own money in it. But the world, he says, doesn't note that kind of thing, preferring instead to fixate on "every time I show up at a party with a good-looking young blonde." (Imagine what we'd write if he showed up with anything less.) This would be the "falsely modest" Larry Ellison.

Why does Ellison shoot his mouth off so fast that sometimes his syntax can't keep up? If he really wants respect, why can't he seem to help himself? His decision in 1996 to market the NC on *Oprah* was brilliantly conceived. It confirmed his celebrity status way beyond the arcana of relational databases and it gave him a vast audience. But in the end it proved humiliating. In a videotaped introductory clip, Ellison showed off his house, planes, and art collection, then lamented the single life. "I'm looking for a woman who is smart and funny and compassionate," he said. "And great-looking, of course." Ellison emerged from *Oprah* as America's Most Eligible Bachelor. The next day, Oracle's switchboard lit up with thousands of calls and faxed photos from prospective brides; a year later, he was still getting propositions. Bamford joked about installing one-way glass outside Ellison's office and a runway going by for nubile young women. Ellison would have a red button he could push to let the desired ones in, while the others—such was the culture at Oracle—would parade down the hallway in front of other executives.

It is because of frolics like *Oprah*, not his ravishing party dates, that Ellison invites curiosity into his private life. He did the same in his ignominious legal battle a few years ago with a former Oracle employee named Adelyn Lee. It was a classic love story for Silicon Valley in the 1990s: Boy meets girl in work elevator, they kiss, they date, she asks for an Acura NSX, she gets fired (by an Oracle vice-president), she sues. Oracle settled on Ellison's behalf for $100,000. Unluckily for Lee, she was subsequently convicted of perjury and forging incriminating E-mail allegedly sent by Ellison, and sentenced to a year in jail. (Then, in an unrelated incident while appealing her conviction, she got herself convicted of brandishing a gun at a neighbor.) While Ellison was exonerated of any legal wrongdoing—claiming he'd had nothing to do with her dismissal—the entire tabloid affair highlighted his profligate lifestyle, his dating of Oracle subordinates, and Oracle's overall treatment of women. It hardly helped Ellison's image.

With a steady girlfriend the last few years, Ellison says he'll remarry and attempt to stay married next time. Friends hope so, lest he grow old alone—except for his cats, Maggie and Big Daddy—and have nobody for whom to buy sports cars. Barb Ellison, his most recent ex is less enthralled. "I'm a good ex-wife," she's told friends. "Why does Larry need more of them?" Of course, she's the one who put pins in a voodoo doll representing her former husband. Heh-heh.

Ellison's eyes apparently still wander. In late 1998, an attractive East Coast high-tech marketing consultant (she prefers her name not be used) sent him a provocative note and photograph of herself in a hat and veil. Ellison loved it. He sent back this E-mail:

Where do I begin? There seems no obvious place. But then you are not the least bit obvious. You have undiscovered depth. Confronting yet concealed. The veil: what does it yield, what does it withhold. Liquid eyes riveting. Luminescent lips beckoning. Raven hair curving wantonly. Strawberry milk completion [sic] daring to be touched. But there is no touch or scent. Passion and desire remain a mystery. Ambition to all but unknowable. How do I lift the veil? What dragons must fall. What riddles solved. Tell me where to begin and I will. Larry.

P.S. Dinner in New York next Tuesday works for me.

"Mr. Sweet E.," replied the East Coast woman, proposing an "affair of the heart and mind." They discussed the possibility of meeting in Hawaii. "Volcanoes, verses, vino and one little size 6 vixen," she wrote. "Very private. I still prefer the City of Lights and Isle of Capri, but I'll take Hawaii this time. When?? Where?? I'll need time to see my travel agent. The Chinese dress and the eelskin high heels await." Oh, please stop! No, please continue. "Take me to your kingdom by the sea where

I can bask in your garden of delights. My buds of beauty are ripe for plucking. Nourish me your red gooseberries . . . There'll be no parting of sweet sorrows."

They didn't go to New York or Hawaii, but they did ultimately get together in San Francisco just before Thanksgiving. Ellison arranged for an Oracle stretch limousine to pick up the woman at the airport and take her to his San Francisco pied-à-terre for a romantic dinner. But before even the first course, she says Ellison got a phone call and suddenly excused himself, citing a family emergency. She never saw him again and was shuttled off to a hotel. Could it be that notwithstanding the florid messages, Ellison found out she was fifty years old and not the package he hoped for? She continued to call and write him for two weeks after the encounter. What was going on? Was he not getting her messages? Could they get together another time?

Finally, Larry answered. "I'll pass," he said, and that was that.

In the winter of 1998, Ellison spoke at the annual TED conference in Monterey, California. The letters stand for "Technology, Entertainment and Design," but the agenda is basically whatever's on the mind of the host, Richard Saul Wurman, an intellectual prima donna who looks a little like the actor Wilford Brimley. Wurman charges thousands of dollars for conferees to hear from a strange mixture of performers that in 1998 included filmmaker Oliver Stone, animal keeper Jim Fowler, TV newsman Forrest Sawyer, the singer Hazel Miller, and Ben Cohen of Ben & Jerry's Ice Cream. Wurman's oddest pairing was the final speakers, Larry Ellison and the Reverend Billy Graham. "The Oracle and The Oracle." Get it?

Ellison preached the usual—the network computer, the evils of Microsoft, the importance of education, and he got in some lines about his plane and helicopter. When Ellison finished his thirty minutes, Graham took the stage. Visibly infirm and verbally halting, he offered kind words about Ellison and then reflected on prior technological eras dating back to the time of King David. "But in the end," the seventy-nine-year-old Graham said, almost sadly, no technology had ever solved the "three great problems" facing mankind—human evil, human suffering, and death. "Technology perpetuates the myth of control over our own mentality." But it shouldn't, he said, because "science and religion aren't antagonists, but are sisters." Graham closed by noting that while he would not live to see the epoch his high-tech listeners were so breathlessly predicting, he hoped they would think about spirituality and greater truths.

It was a moving presentation worthy of contemplation for anyone, but especially someone who professed to be growing up at the age of fifty-four. Unfortunately, Ellison hadn't heard a word of it. He'd taken off the second he finished talking about himself.

Chapter VI
Money

You drive a mile east just out of Woodside and two more south on Interstate 280. You make a left and head toward Stanford University. This is Sand Hill Road. Running between the town of Menlo Park and the Stanford campus, it looks like just another four-lane strip through a brown landscape in the shadow of the Santa Cruz Range. On the right side is SLAC—the imposing Stanford Linear Accelerator, where the Homebrew Computer amateurs used to meet. On the left, in a series of little office parks, are about a dozen square, two-story buildings, most unnamed, all unimposing. The enclave is the capital of capital in Silicon Valley—$22 billion of it—the densest concentration of investment resources in the world (and by some estimates, a third of all U.S. venture assets).

Places like Stanford offer the Valley its brainpower, and companies like Intel and Hewlett-Packard represent what passes for institutional longevity. But Sand Hill provides the buckeroos. It is the Fort Knox of the Valley. When a venture capitalist asks "What's new?" he's not making small talk; he's looking for the next great technology, even if that means pilfering somebody else's idea. They call it "venture capital," for its connotation of "adventure," the Wild West. In that sense, while Sand Hill VCs date back only three decades, they can lay claim to the same spirit of the Gold Rush days—where the presence of capital, along with technological knowledge, helped make California. Even the term "VC" is relatively recent, only being used since the 1970s when people stopped thinking it meant "Vietcong."

Forget Rodeo Drive in Beverly Hills, the Miracle Mile in Chicago, even Wall Street or the strip in Las Vegas. They don't compare to Sand Hill Road. From here, the venture capitalists—"vulture capitalists" or "velociraptors" in the vernacular of the Valley, except when it's *your* hand sticking out for the money—rule the high-tech roost. Indeed, they own sizable chunks of most of the successful companies that make up the Valley's mother lode. Venture capitalists name their dogs "Midas" and put $$$ signs on their license plates. One firm, Institutional Venture Partners, displays in its lobby a four-foot blowup of a $1,000 bill, its edges charred and blackened, as a way of telling would-be entrepreneurs that money burns up fast.

The dirty little secret of succeeding here is that the romance of the true garage start-up is largely gone. It's not just the need for money, though that's a large part of it. The power of a few VCs to choose products and create industries—and to decide who will run them—is virtually unknown to the public. These are not stuffy old family firms or well-known publicly traded investment banks or the subsidiaries of big corporations. Unlike the entrepreneurs they finance, they typically go out of their way to avoid publicity. Sequoia Capital, Sierra Ventures, the Mayfield Fund, New Enterprise Associates, Technology Venture Investors—who's heard of any of them? Unlike most investment vehicles, the private funds that VCs run—about five hundred of them nationwide—are exempt from almost all SEC reporting requirements. In the end, they amount to a mysterious force in the financial universe—the invisible hand of capitalism ruling the Valley—accountable only to the laws of economics. And for the most part, over the last twenty-five years, they've done stupendously well—for their investors and for themselves.

If there's one firm that outmuscles the rest, both in how it manages its deals and its image, it is Kleiner Perkins Caufield & Byers. Though other venture capital operations were around the Bay Area before it—Arthur Rock's shop, for example—Kleiner Perkins was the first to institutionalize the business and it remains the model. It is so successful, other VCs say, that rooting for it is like rooting for the Yankees. Founded in 1972, Kleiner Perkins has raised $1.4 billion in capital and, with it, funded more than three hundred companies. At any given moment, the firm has about ten active partners (who've always been men). But these days, KP is dominated by one person—the centi-millionaire from Woodside, John Doerr. He's the one who helped out with *the largest legal creation of wealth in the history of the planet.* If he got a dollar for every time he's used the line, he'd be rich on that basis alone. At the end of 1997, if you trust their arithmetic, KP-backed companies were worth $125 billion in stock, produced revenues of $61 billion, and employed 162,000 people. (These are numbers that any KP partner commits to memory, always ready to recite, much the way a baseball slugger knows his home-run total and where it ranks in the pantheon.) KP partners alone have reaped more than several billion dollars from the game.

The firm is managed smartly by Brook Byers, also from Woodside, but it's the forty-eight-year-old Doerr whose deals are the cash cows. He gives KP the visibility that infuriates its competitors along Sand Hill Road, and causes resentment in Byers, the last of the name partners still working. Most VCs still like to go by the Taoist maxim: "Those who know, do not talk. Those who talk, do not know." Paul Saffo, an industry consultant whose offices are on Sand Hill, says that in the old days "venture capitalists' power used to be in inverse proportion to their visibility." Jim Clark has known Doerr since 1979 and used him to gear up Netscape. "There is no doubt he's the leading venture capitalist in the world," Clark says, "but I've seen him change with the increased attention. He's spread much thinner and has gone through a personality change." But wouldn't you, if Martha Stewart took you to dinner at some fancy Palo Alto bistro?

Doerr professes not to like the publicity—he's the closest thing that venture capital has to a rock star, a combination of banker, fixer, mentor, spy. But it sure didn't stop him from being quoted in newspapers and magazines 626 times in 1997 and 1998 alone (according to the Nexis online database). The shortest measurement of time in our world is the interval that passes between a friendly journalist beeping John Doerr and his return of the call. Nor does Doerr seem to mind shuttling those friendly journalists around the country aboard his comfy private jet when it suits him. ("It's not for publicity. I'm just trying to educate people about the Valley," he says.) When *The New Yorker* ran a glowing profile of him in 1997, Doerr heard that it might refer to him as the "Michael Ovitz of Silicon Valley." He was petrified and lobbied for days to keep it out. The piece didn't include the reference.

Doerr probably protested too much. The Valley has lots in common with Hollywood and he's its super-scout, the guy who hobnobs with Clinton-and-Gore more than anyone in the Valley; if Doerr's house in Woodside was bigger, the president might stay there instead of at Steve Jobs's spare place. A venture capital firm is like a film studio. Each has its overwrought stars, its new flavors every season, its parochial world-view. In Hollywood, directors squabble with actors; in Silicon Valley, directors squabble with CEOs, while the VCs fight with the twenty-two-year-old entrepreneurs. The businesses aren't that different. Each has a portfolio of projects—movies or start-up companies—that it's trying to assemble a production crew for. Some will be big hits—*Titanic* and Yahoo—and many will be multimillion-dollar disasters—*Waterworld* and Shockley Semiconductor. In Hollywood, the remains are on the backlots; in the Valley, they're represented by the red ink on a VC's ledger.

As Kleiner Perkins sees it, the Florence of the Renaissance had the Medicis, the American steel industry had the House of Morgan, and Silicon Valley in the late twentieth century has Kleiner Perkins. The partners will eagerly tell you that it's not about money—it's about "new paradigms" and "changing the world." "We're not *financiers*," Byers insists, mocking the word he and all his partners disdain. "We're *company builders*. We don't throw out some seeds and see which sprout. We don't buy a stock today and sell it tomorrow if we don't like it. That's not our mentality. We're not even investors—we're part of the entrepreneurial family of the company. I'm just in the service business."

But in their written sales pitch to KP's wealthy, select clientele—who, after all, are *investors*—the partners come clean just a little. "We are driven by entrepreneurial passion, the intellectual challenge of venturing, a genuine love of science," and, oh yes, "the financial rewards of this business." Who would've figured? When he hears VCs claim, "We're in it for the long haul," Jim Clark says it reminds him of the salesman who promises, "I'll take care of you." But, Clark says, "if he says it more than once, I really get worried."

While KP's formidable financial accomplishments are undeniable, its image is paramount. Without that image, it would just be another VC firm on Sand Hill. Thus, Doerr and his partners actively cultivate "the Kleiner Mystique," part of which is that no one partner is pope. Several years ago, the high-tech trade magazine *Red Herring* asked Doerr to cooperate for a story on a-day-in-his-life. "Does this violate the Kleiner mystique?" Doerr asked his partners in an E-mail. No, but accidentally also sending the E-mail to *Red Herring* sure did.

Byers keeps methodical track of KP's newspaper clippings and must notice that Doerr's name appears more often than his; he longs for the days of the 1980s when his biotech start-ups were flying high. Maybe Byers needs to hone his negotiating prowess. A couple of years ago, he wanted to buy a parcel of land adjoining his property, but he was out-maneuvered by another Woodsider. The following week, Byers called the purchaser with this stratagem: "You know, I'd be willing to pay the same amount you just paid." Byers didn't get the land. Lucky for Byers no one at KP heard about this embarrassment. Lucky for him, too, that he and Doerr both get chauffeurs and both receive equal cuts of the partnership (cuts above all the other partners).

Maybe Byers could improve his standing . . . if he learned how to drive a soapbox derby racer.

Since 1997, on a Sunday morning in mid-September, an eccentric, strutting fifty-year-old restaurateur named Jamis MacNiven has run the Sand Hill Challenge—a soapbox derby for the silicon set, held in a carnival atmosphere of overpriced T-shirts and a four-story-tall inflatable dinosaur. It's a race for any venture capitalist who'll pay the $2,500 entry fee and doesn't have the time to race Larry Ellison on the high seas. The road is closed temporarily and several dozen gravity-propelled hot rods get a chance to zoom down it. The point is to raise $100,000 or so for charity, but also to give the hard-core VCs and the gearheads they've funded a chance to build something *really useful* and have some fun, too—all without the need to make even more profits. (MacNiven's first choice was to stage a life-size game of Monop-oly—real houses, real cars—but nobody, of course, could agree whose house would constitute Boardwalk.)

The race is also a way to keep MacNiven out of trouble. For his day job, he runs Buck's in Woodside, a cross between a country diner and Morton's of Hollywood. If Matt Drudge could cook, he'd be Jamis MacNiven. A former contractor who built the Hard Rock Café in San Francisco, as well as renovated twenty-five-year-old Steve Jobs's first house ("I was a terrible contractor and Jobs was a worse client"), MacNiven is a buffoon whose utter lack of pretense makes him the ideal antidote to the blowhards and overlords of technology who fill Buck's on most weekday mornings. In his sombrero, fur shorts, and noisy shirts, the six-foot-four Jamis runs the place like an impresario, in and out the booths, greeting diners at the door, harassing the help, and making random announcements with sentences longer than this one. At one point, he spread the rumor that he served eight years at Folsom for murdering his eighteen-year-old first wife because she wouldn't bring him a beer. (It's not true.) It's all rather entertaining, even at sunrise. Coffee helps.

But while the Filoli garden omelettes and USDA-disapproved $4.25 milkshakes at Buck's are fine, the reason one eats there is to enjoy the show. Virtually every VC in the Valley, including the ones who don't live in Woodside, has a business meeting at Buck's now and then; Doerr is a regular, which is why MacNiven tells people that Netscape was founded over a Belgian waffle. It wasn't, but it's a good story. (Like Larry Ellison, MacNiven believes no story is too good not to be made up.) MacNiven will tell you that Thursday mornings at the restaurant are the West Coast version of a power breakfast, with more deals getting done before nine than in all the conference rooms in the Valley combined. The entrepreneurs are the ones talking with their hands; the VCs have dollar-filled comic-strip balloons over their heads. Maybe MacNiven's picture is true and maybe it's not—all the $135,000 white Mercedes S600s out front and cell phones inside suggest it is—but it makes for good shtick regardless. The occasional $100 tips aren't bad either.

Above the front door at Buck's are a series of notches in the molding for each television camera crew that's visited to do a slice-of-the-Valley segment. In 1998 alone, the total neared thirty; if he kept score of all the mentions in the local papers ("venture capitalists met yesterday morning at Buck's" is redundant), the molding would crack. It's MacNiven watching the observers watching the scene: Look! Over there in the corner booth! It's *Doonesbury*'s Garry Trudeau and director Robert

Altman planning their Fox TV series on intrigue in the Valley. And there's Chelsea Clinton, an undergraduate at Stanford, who might be unnoticed except that Jamis points her out and offers appetizers to the Secret Service guys. Bill Gates has even eaten here—and legend has it that he wasn't much of a tipper.

The restaurant reflects its owner's bizarre sense of humor. On the walls are a life-size black-velvet painting of Elvis, photos of turn-of-the-century inmates at San Quentin, an Ernest Hemingway fishing rod, detritus from a flea circus, the ashes of a softball friend who dropped dead on the field one afternoon, and the hipbones from Willy the Famous Circus Lion, who used to be buried on Mike Markkula's property before the bulldozers found him. (Willy was the proud possession of Frank Buck, the famous animal trainer, who tooled around Woodside with Willy in the back of a Duesenberg convertible.) MacNiven even instigates some of the kitsch. By the men's room is a copy of his 1993 letter to the Republic of Russia, asking to purchase Lenin's body and offering "a payment in the low six figures." The Russians wrote back: "Tell us what is exactly meant by 'low six figures.'" Who knows if this stuff is authentic? Does it matter? Over the bar hangs a fake robotic buffalo head named Buck; thanks to a hidden microphone, Buck talks, though some folks think it's MacNiven throwing his voice. The annual Christmas display includes Santa and his reindeer getting pulled over by customs agents.

With his history, a soapbox derby orchestrated by Jamis MacNiven isn't going to be a sedate affair. The pre-race parade includes army tanks, bagpipe bands, cheerleaders, and as many Nobel laureates from the area who are willing to be seen in the same publicity shot with Jamis. Then come the official entries: There is a car made out of real French bread, one that's a queen-sized bed, and another that's a giant skateboard. There's a Big Fish Car from *Red Herring* magazine and a Shark Car from a law firm. But there are also high-end machines made of carbon fiber, Mylar, and titanium. In 1997, one firm hired the design team that built an America's Cup boat; another found a national biking champion living in Woodside. Three Arch Partners consulted with roller-coaster designers and experts on "mechanical violence," whatever that means. The fancy cars are built by the likes of the Stanford Linear Accelerator and *Scientific American* (1,445 pounds shaped like a miniature Titan missile).

"I'm celebrating the Valley's spirit of entrepreneurship and invention," MacNiven says, "and I'm hoping they'll remember what it was like to be a kid." Rich kids.

The inaugural race, held in front of ten thousand spectators and a few dogs, came down to a final half-mile-long heat between Mohr Davidow Ventures and Kleiner Perkins. Both drivers don their crash helmets. Officials check the hay bales and netting that line the course that goes right past the driveways of all the VC firms. Mohr Davidow's entry—a low, narrow 150-pound orange-red vehicle resembling an Oscar Mayer with wheels—is driven by Larry Mohr, on crutches from an ankle injury but undeterred. It cost $4,000 in parts and reflected 1,600 hours of engineering time at IDEO, the Palo Alto industrial-design firm. KP's car is a little longer, looking like a blue suppository. It cost $10,000, was built by a local wheelsmith, and has a young partner, Doug Mackenzie, at the wheel. Both cars go *zoom! zoom!* down Sand Hill Road. But Mohr Davidow's is three seconds faster, topping out at 47 miles per hour (2 mph over the road's speed limit), which is pretty good considering they didn't cheat. As everybody in the Valley knows, gravity can take one a long way. Mohr Davidow's edge: They recruited two former running backs from the National Football League to give their car its push off the starting line. Hey, there was nothing in the rules against using professional athletes. It may all be for charity, but in an atmosphere of testosteroned egos, it's still about winning.

Bill Walsh, the Woodside resident and former coach of the San Francisco 49ers, presides at the awards ceremony. Mohr Davidow gets the two-foot-high, brass-plated Coveted Perpetual Trophy of Toast; why it's toast is known only to the demented mind of MacNiven. He then gives the runner-up prize to KP. It's two inches small and reads simply LOSER. Everybody has a good laugh. Everybody except the partners of Kleiner Perkins Caufield & Byers, who never seem to take a day off. John Doerr and Brook Byers, red-faced, are furious and leave, Brook babbling to himself. Their friends are embarrassed for them. "I was surprised by the reaction," MacNiven reflected later. "Here I was trying to bring out the kid in everyone."

The partners at Mohr Davidow can't resist rubbing it in. The next week, they send over a teeny-weeny trophy display case to their vanquished competition. Not everybody at KP is amused. The VCs vow to get revenge the next year. But they lose again, slipping to third place. Mohr Davidow ditched the football players and shipped in U.S. bobsled champions for the push-off from the starting line. At least KP didn't suffer the tiny trophy this time.

The history of KP dates to the early days of the Valley. Gene Kleiner was the fourth of the Traitorous Eight to leave Fairchild. But before that, in 1956, he had left Western Electric in New Jersey upon the entreaties of William Shockley to come to the Bay Area. As he drove west on Route 66 in his Buick convertible—"please make sure to say it had red vinyl upholstery!"—the thirty-four-year-old Kleiner thought he was a little crazy. He grew up in a wealthy Vienna family, then fled the Nazis in 1940 and came to Brooklyn, where he went to Polytechnic University. His father owned a shoe factory in lower Manhattan and Kleiner got his first job as an apprentice toolmaker. It was not a normal career move to abandon an established company on the East Coast, with a wife and baby in tow, to work for a brand-new one three thousand miles away.

Now he was leaving Fairchild after six years and after having run every department except marketing. "I had essentially been working for Gordon Moore," he says, "but decided I would not work for Andy Grove," who was coming into the company. "He was simply too intense for me, even then." Moreover, Kleiner recognized that he had come out way ahead from the other two job exoduses, and he had seen at Fairchild what entrepreneurship could unleash. What could he now accomplish on his own? His idea was to manufacture "teaching machines," an early form of interactive learning in which students had to push buttons on a gadget linked to questions from a teacher. Kleiner called his company Edex, for "educational excellence." And while it didn't represent an electronics breakthrough like the transistor or chip, Edex racked up millions in profits. In 1965, he sold out for $5 million to Raytheon, the old East Coast defense contractor. Kleiner stayed on for a while, technically as an employee in weapons research. "I was supposed to be selling to the education market," he remembers, "and here my business card said 'Missiles Division.' That ended most sales conversations."

Worse for Kleiner, Raytheon wanted to transfer him to Michigan City. Kleiner, whose thick Austrian accent and European manners mask a sense of humor, thought it was ridiculous. "You just don't move from Palo Alto, California, to Michigan City, Michigan. The place is best known for having a penitentiary." So he quit and, financially secure, took his family abroad for two months. With time to reflect, and the recognition that he'd been in the business of starting companies for half his adult life, Kleiner decided to do it for a living. Not for just one company anymore, but for a bunch. This was how he got into the business of midwifing businesses, perhaps entire industries. He would institutionalize what the Traitorous Eight did on a whim. To this day, Kleiner keeps a framed photograph of the Eight, signed by all, in the study of his home in the hills of Los Altos.

"Venture capital" didn't really come to Silicon Valley until the 1960s. The term was probably coined by Arthur Rock, who backed Fairchild and then Intel. "It's not like there wasn't any money around," says Rock, now in his seventies and still doing a few deals by himself from a modest, anonymous office atop a San Francisco tower. "We just kind of organized it out here around high-tech." In those early days, according to Jack Wilson's *The New Venturers*, the VCs were called the San Francisco Mafia. Rock's $300,000 investment in Intel alone was worth about $700 million at the beginning of 1999, even though he'd sold about half his shares and bought such things as a piece of the San Francisco Giants.

Before the emergence of John Doerr, Rock was the object of the national media's VC adoration. *Time* once put him on the cover, dressed in greenbacks, beside the headline CASHING IN BIG: THE MEN WHO MAKE THE KILLINGS. Rock hated it. "I don't like people to count my money," he told the *Time* correspondent, who used the quote in the story. It was the virtual entirety of the interview. The correspondent was a young Brit named Mike Moritz, who would go on to be the venture capitalist who backed Yahoo years later. The cover got it about right. Rock is private about his money, but is the first to acknowledge he's not in the altruism business. "I wouldn't have done what I've done for nothing," he says.

Part of the reason venture capital hadn't taken root earlier was because large institutional investors risked liability if they took on too much risk. It was one thing for a Rockefeller or Fairchild to bankroll a new company—for them, it was as much a sport as an investment. Old Money had little hunger. But pension funds, college endowments, insurance companies—the institutions that controlled much of America's financial assets—couldn't take the chance. Most state courts had long followed the "prudent man" rule; fiduciarily, the big investors could only put money where a prudent man would and venture capital wasn't in the ballpark.

The irony was that many so-called safe investments weren't so safe and many venture capital deals paid off handsomely. The neat thing about making a bet on, say, Intel or Apple is that the downside is so much smaller than the upside. You can lose all your money only once, but you can make ten or a hundred or a thousand times your investment. When a company goes public or sells to a larger player, the rewards can be astronomical and the dead can be buried with barely a notice. It's the absolute reverse of what traditional (read: safe) banks do. If your loans do well, you make 5, 10, 15 percent a year; but just one bad loan can wipe out ten good ones. That's why First U.S. Begrizzled Bank & Trust is so risk-averse—and would never invest in start-ups, lacking as they do any collateral or earnings. Yet it is the lemming-like banks, which wouldn't go near an entrepreneur, that so often seem to be on every stupid lending bandwagon (Manhattan real estate, California real estate, oil and gas, Latin America, Mexico, Southeast Asia). That realization—along with the good returns that a lot of early venture capitalists were showing—was largely responsible for easing the restrictions on institutional investors. And that's what allowed venture capital to become part of the Silicon Valley system. More big money came in, which led to more company formation, which led to more successes, which led to more money coming in—a cascade of riches.

After Fairchild's success in the late 1950s and early '60s, Arthur Rock left Hayden Stone in New York and started his own shop in San Francisco. "The money was on the East Coast, but the opportunities were here. Given the way these things worked, that meant the opportunities here didn't get funded." The very fact that the Traitorous Eight had to look east to fund Fairchild convinced him there was a financing vacuum in the emerging El Dorado. But Rock, ever cool and contrarian, admits he was "a little bit tetched" in coming to San Francisco.

In 1961, Rock hooked up with Tommy Davis, a graduate of Harvard Law School who had served during World War II as an OSS operative behind enemy lines in the Burma jungle. Recovering afterward in a San Francisco hospital, Davis fell in love with the area and decided to stay. Davis & Rock, as the new firm called itself, raised a pool of $5 million from wealthy individual investors ("limited partners"), which it used to fund start-ups. As the general partners, Davis and Rock got 20 percent of the profits between them for the seven-year run of the partnership—establishing the standard cut for VCs, until Kleiner Perkins Caufield & Byers came along.

Davis was the ebullient, exuberant one; Rock more often than not said nothing at meetings. Their operating philosophy, as Davis put it in a speech, was to "back the right people." The product and the market were not to be ignored, but personal drive was the critical factor. "When I go to the racetrack," Davis would say, "I try to pick the horse that wants to run." One of their first investments was in Scientific Data Systems, which became a significant maker of minicomputers in the 1960s. When SDS was acquired by Xerox in 1968, the price was nearly $1 billion. Sixty million of that went to the Davis & Rock fund, which until then had anted up only $250,000. Over its seven years, Davis & Rock returned to its investors a profit close to forty times what they put in.

Unfortunately, on a personal level, the two VCs broke up in acrimony. Rock never spoke to Davis again. After backing Intel with just his own money, Rock teamed up with a young Harvard MBA named Dick Kramlich, after Kramlich had sent him a handwritten note inquiring about a job; Rock was impressed with the personal touch, though according to Kramlich, Rock sent the note to a handwriting analyst just in case. That association lasted eleven years. On the day Kramlich left, he got a call from Tommy Davis. "It's time you and I had lunch," Davis said. They went to the Velvet Turtle, where Davis unloaded his bitterness about Rock. "I'm only going to say one thing about this and I'm not ever going to talk about it again," Davis told Kramlich. "I just don't think one human being should treat other human beings the way he did." Kramlich went on to form New Enterprise Associates, turning down a KP job offer along the way.

Gene Kleiner knew about the venture capital business in large part through Davis & Rock. He was a limited partner in its fund, putting in $100,000, and was one of the investors Rock called when he was rounding up money for Intel (another $100,000). "I liked the work," Kleiner says. "It was fun being involved in starting companies—the variety, not being tied down to a job that would become routine." The only problem was that he was doing his investing alone. He didn't like being lonely.

In early 1972, Tommy Davis was approached by a friend, an East Coast scion interested in getting a piece of the California action. Henry Lea Hillman was heir to one of the last robber barons, J. H. "Hart" Hillman, the Pittsburgh coal-and-gas industrialist from the early twentieth century. Henry Hillman, fifty-three, ran the family's business, worth $3 billion even back then and one of the ten largest family fortunes in the United States. The Hillman Company had boatloads of cash, and Hillman shrewdly wanted to redeploy it out of the old smokestack industries and into the new world of electronics; the semiconductors and microchips of Silicon Valley were just the steel and coke of his father's era. Hillman also wanted to invest quietly—he hated publicity and lived by the motto: "The whale only gets harpooned when he spouts."

Davis wasn't looking for other investors when Hillman called. But Kleiner was, and Davis, along with Rock, recommended him. "I went to see Hillman," Kleiner remembers. "He offered me a job in Pittsburgh, which I didn't think was any better than Michigan City, where I turned down the Raytheon job. I told him no."

"What if you can stay in California?" Hillman replied. "I'll give you $4 million to invest, if you can raise another $4 million."

That was a big chunk of change in those days—not for Hillman, but for the still-nascent venture community and its investors, who were made up, in effect, of financial dilettantes. To raise that kind of money, Kleiner sought out his friend Sandy Robertson, one of the area's early investment bankers. Robertson had come to San Francisco in 1965 to man an outpost for Smith Barney, leaving four years later to start his own corporate-finance firm on Montgomery Street, which would be an institution for nearly thirty years before selling out in 1997 to BankAmerica. Robertson agreed to help Kleiner and suggested he meet another entrepreneur-turned-financier looking to raise capital—a former Hewlett-Packard executive named Tom Perkins. "I may not be able to raise the money for you separately," Robertson told Kleiner. "But I might be able to raise it for you together."

Tom Perkins, then forty, was an MIT-trained engineer and a Harvard MBA who studied under Georges Doriot, one of the earliest venture capitalists. Doriot was a former brigadier general in the U.S. Army, who liked to invite entrepreneurs to his study and play French military songs to inspire them. Perkins was raised in the northern suburbs of New York City and hadn't traveled west of the Hudson his entire life. "California and the West were basically a mystery," he says. In the summer of 1956, between his two years at Harvard Business School, Perkins worked for General Radio, near the university. General Radio made various testing instruments, the kinds of devices the public later associated with Hewlett-Packard; the primary customer was the military. Perkins's job was to run quality-control checks on HP's products, as a way of scouting out the competition. At that point, HP was still a relatively small company with $25 million in revenues—it wouldn't go public until the following year—but General Radio was taking notice. "That's when I first heard of HP," Perkins says.

Though he assumed all along that he would go to General Radio upon graduation, Perkins was intrigued by the notion of these two upstarts going up against his employer. He was curious enough that when he heard that Hewlett and Packard would be at an electronics trade show at the New York Armory, he went. "There they were, setting up their own booth!" Perkins recalls. He talked to Packard for most of the afternoon and was offered a job. "I had little to lose and figured I'd try California for a few years before returning to the East Coast." Perkins was not just impressed with the intellect and ambition of HP's founders, but their "down-to-earthiness" and skepticism about his fancy MBA. It was only after he arrived at the company in 1957—after a leisurely drive across the northern part of the country with his girlfriend—that he realized just how skeptical they were. Despite its one thousand employees—the most in Palo Alto, but still dwarfed by a company such as Lockheed in San Jose—an MBA from Harvard was like a creature from another planet. "Typically at HP, you worked as an engineer first and then proceeded into management. The idea that an MBA would go directly into management was totally strange.

"So they came to me and said, 'If you don't mind, we're going to put you to work in the machine shop on a lathe.'" It was not a promising beginning.

Single and charming, Perkins spent much of his free time in San Francisco surveying the social scene. During his first summer, he dated Ellen Davies, part of one of the city's bluest-blooded families. At a dinner with her mother one evening, he was told, "So, Ellen tells me you're a machinist. You have to understand we're of a certain position here." Perkins got the message. He didn't see the daughter again. If only Mrs. Davies had known the riches that awaited him.

After several months in the machine shop, HP put Perkins in charge of all the independent company salesmen. In time, though, he got restless, deciding that even HP wasn't growing fast enough for his tastes and he wasn't learning enough. For much of the next six years, he seemed to bounce around jobs, none a complete bust but none exactly a success. In retrospect, his wanderings were not so much a lack of direction as a certain conviction that he knew best. He quit to work for a national consulting firm in San Francisco, then left for Optics Technology, a new fiber-optics venture that Packard and Hewlett had funded. When Perkins and the CEO crossed swords, Perkins issued an ultimatum to the board of directors. It didn't choose him. Packard asked Perkins to return to HP to help organize a central R&D department, which included the new computer division.

Perkins agreed and in 1966 began a second stint at HP—with one caveat. The deal he struck with Packard allowed him to work his own commercial venture on the side. His idea was to build a smaller, cheaper laser. "It may have been foolish, but I couldn't get the idea out of my head," Perkins says. "Packard had no problem with me doing it while I was at HP." The foolish part was talking his wife into putting their life savings—$15,000—in the company, which he called University Laboratories, for its location along the main road in Berkeley. It would be the first of many times that Perkins's risk-taking psychology would show itself: He was always willing to push all his chips to the center of the table. Within six months, the laser invention worked. The device could be used, for example, in the construction of sewer lines. Excavators no longer had to survey every inch of a path in order to ensure a straight line, but could use a laser. Rather than explore a public offering, Perkins took the quicker, surer deal and sold the company to Spectra Physics, the dominant player in lasers. He came away with just under $2 million—"my first home run," in the baseball parlance of venture capital—and got himself a far finer house, overlooking the Bay, than his $15,000 in life's savings could ever have purchased.

Meantime, at his real job, Perkins was given the unpleasant task of jump-starting HP's recent entrance into minicomputers. It was a political morass, affecting the sales force, the people in finance, and the HP Way itself. Dave Packard never had been interested in discounting prices as a way to build market share, and Perkins's protests notwithstanding, he wasn't about to start doing so simply because of a new kind of product. Packard believed that if a company had a good product, market share would take care of itself. "The problem," Perkins says, "was that too many at the company thought of themselves as engineers rather than merchants." In a famous meeting, when Packard and Perkins faced off on their disagreement, Packard got so mad that he walked out, though his views prevailed for the time being. "What now?" Hewlett asked Perkins. "I need more ammo to make my case," was the reply.

More intense than other HPers, Perkins instilled a competitive ethos. "Tom taught me the rule about not letting the phone ring three times," says Joe Schoendorf, now at Accel Partners, one of the Valley's leading VC firms. But it was the manner of the lesson that Schoendorf remembers. "If the phone ever rang a third time in the sales department, you were in trouble. The way he let you know is whenever he heard two rings, he would start counting, then leap from desk to desk—while he continued counting—until he got to the phone that was ringing. If he got to the phone before you, you didn't want to have to explain why. You don't have to do that much before the three-ring rule becomes ingrained." Even now, Schoendorf shows a special zeal when he bounds from the dinner table to get the phone.

Because of Spectra Physics, Perkins had enough wealth that he didn't have to work. So when he tired of office politics, he quit Hewlett-Packard again, this time for good. Much like Gene Kleiner and his success with Edex, Perkins had gotten a charge out of founding a company several years earlier. Like Kleiner, he wondered if he could create companies for a living. It was one thing to found a single company, or even two; Perkins aimed to make a habit of it, getting the adrenaline rush over and over again.

One of the people Perkins turned to for advice was Sandy Robertson, who knew that more venture capital deals would produce more companies that might go public that he could then underwrite. "It wasn't until then that I realized that Gene Kleiner was going through exactly what I was," Perkins says.

On a summer morning in 1972, Kleiner and Perkins finally met for breakfast at Rickey's in Palo Alto. Although they were different in temperament as well as technical expertise, their styles seemed to mesh, as in other great collaborations of Silicon Valley: Hewlett and Packard, Jobs and Wozniak. Kleiner was ten years older than Perkins and had a formality antithetical to the Valley; Perkins had flair and supreme confidence in his own ability, leading one of his VC colleagues to remark that Perkins probably was self-assured in his birth crib. Kleiner knew manufacturing, Perkins understood hardware and management. Kleiner had modest tastes; Perkins liked classic yachts and vintage cars, at one point owning the world's best collection of antique roadsters, including a Duesenberg once owned by a maharajah. (In the early 1990s, he sold the collection for tens of millions of dollars.) In the venture business, Perkins was the Valley's big engine, finding many of the ideas in the first instance; but it was Kleiner who provided reflection and skepticism.

The two of them hit it off and spent most of the next two days together. Thus was born Kleiner & Perkins. The name was arranged so for two reasons—alphabetical and, as Kleiner noted, "I had the commitment for $4 million from Henry Hillman." With the additional $4 million they had to raise, the $8 million fund they planned was unheard of in those early days of venture capital. It was one thing for companies like HP to have large engineering R&D budgets. But Intel aside, restaurant chains were what was hot. The recent bear market on Wall Street and unfavorable capital-gains laws, too, made a venture fund look unattractive. And so it took Robertson, and the two engineers he arranged in business matrimony, four months to come up with the other $4 million.

Their roadshows to potential investors were an adventure. One company, on whose board Hillman sat, agreed to have its investment committee meet the trio. Recalls Kleiner: "The guy in charge asked us a series of questions like, 'What kind of companies will you invest in?' We said, 'Successful ones.' He kept pressing for specifics. We finally mentioned a word-processing typewriter.

"He threw us out. Didn't even give us lunch. He thought we'd be crazy to compete with IBM. That was the last time we mentioned a specific idea."

Kleiner and Perkins got their money eventually—from fewer than a dozen investors, including $1 million from Rockefeller University, nearly that much from two insurance companies, and the balance from wealthy individuals and trusts. "People who could easily part with their money," as Perkins put it. And always "legitimate sources," which meant money without any taint of organized crime or gambling (not that any VCs took money from those sources). "We were too risky for the mob," Perkins laughs. He and Kleiner themselves each put up $100,000. Kleiner & Perkins, as they called both their partnership and their inaugural fund, set up shop at 3000 Sand Hill Road, the first VC tenant of real-estate developer Tom Ford.

In the rich tradition of the shopkeepers to the 49ers, Ford figured out early on how to make a buck from the new gold-diggers. Venture capitalists could dream of all the money they wanted, but they still needed office space. Ford, raised in Youngstown, Ohio, had been stationed in Northern California during World War II. Escaping private law practice, he came west again to work for Stanford, running its land development office, including the Stanford Industrial Park. A decade later, he went into business for himself. A 1969 project to move *Sunset* magazine to open space on Sand Hill fell apart, but it gave Ford the chance to think about the parcel, which not only was surrounded by a golf course but was adjacent to the planned eight-lane freeway, Interstate 280, that would run through the Valley, from downtown San Francisco to San Jose.

It was also just a canter's ride from Woodside, which many of the venture capitalists were beginning to call home.

Ford bought the land, putting up $10,000 of his own and getting more than $1 million from an Eastern life insurance company. He then built four woody low-slung office buildings, heavily landscaped and centered around a subsidized restaurant overlooking the range. In the main bricked thoroughfare sits a brass disk displaying an image of the sun. Ford didn't foresee 3000 Sand Hill becoming the nucleus for venture capital, nor did he think he would become the Donald Trump of Silicon Valley. But one firm moved in after another, the better to see who's talking up whom.

Following the VCs came the scavengers of business—lawyers, head-hunters, accountants, corporate turnaround specialists, mobile fitness centers, and the one shop that all VCs like to envy. That's Kohlberg Kravis & Roberts, the leveraged-buyout firm immortalized in *Barbarians at the Gate* and that Henry Hillman himself bankrolled in the 1970s. KKR partners, who prefer to dismantle companies rather than form them, take in even larger sacks of cash than VCs. If money is a way to keep score, finishing second counts for little except a lifetime supply of ostrich salami. VCs talk about KKR the way Red Sox fans do about the Yankees; in comparison, the Sand Hill branches of investment banks like Goldman Sachs and Morgan Stanley get grudging respect.

Larry Ellison, too, moved in for a while with the only start-up on Sand Hill. Ford remembered meeting this "tall man in sneakers and jeans" in 1978 and being appalled by his rusted-out Mercedes with Fred Flintstone holes in the floor. In addition to rent, Ford ended up with a piece of the Oracle fortune. A couple of months before the Oracle public offering, Ford bought 10 percent of Stuart Feigin's shares for $50,000. Feigin was the company's fifth employee and eager to cash in some of his equity. He thought Ford was crazy—who could predict when an IPO might happen and whether it would be a hit?—and demanded Ford sign a letter indemnifying Feigin. After Oracle went public, of course, the value of those shares skyrocketed.

The Sand Hill complex has the feel of a country club and the dues to match. Ford, who bought out the life insurance company, continued to buy nearby property. Today, managing about 400,000 square feet of office space, Ford Land Company charges more in rent—up to seventy

dollars a square foot—than anyplace in America, including Wall Street.
And why not? Venture capitalists make more than the Wall Streeters.
The Sand Hill vacancy rate is zero. Location, location, location—this is
the gold standard. Tom Ford, who died in late 1998 at seventy-seven,
was the only person in the Valley who could hold the VCs up for money.
They paid him back the ultimate compliment: They let him invest in
their funds. To his credit, Ford became one of the Valley's leading phi-
lanthropists, quietly giving away millions, including setting up an endowment that provided $1,000 to every student in the poor community
of East Palo Alto who went to college. (Ford also helped bring Bill Walsh,
the former San Francisco 49ers football coach, back to coach at Stanford.
Ford himself paid part of Walsh's salary.) To honor his VC tenants, Ford
in 1997 endowed an engineering professorship at Stanford. Instead of
naming it after himself, he called it the Kleiner Perkins, Mayfield and
Sequoia Chair in Engineering. Where did that order come from? Ever
the image cop, Brook Byers insisted the names be alphabetical.

Kleiner & Perkins had big dreams and an
$8 million pot, but initially could find little to invest in. The days of
young engineers rushing in the front door with business plans was still
far off. For the first two years, K&P found almost nothing in high-tech,
instead putting money in the kinds of deals that wind up being ridiculed
in the center of the front page of *The Wall Street Journal:* a combination
snowmobile-motorcycle that came out during the first oil embargo and a
business that would retread sneakers. Those two turkeys alone ate up
$650,000, though Perkins had a grand time with the recreational vehicle
in the Sierras. A waste treatment company blew a million dollars in three
years, only to be left, in VC lingo, "at the bottom of the ocean." Another
troubled company, Dynastor, taught the partnership the perils of luck
and timing. Dynastor aimed to revolutionize the market for floppy disk
drives. K&P put nearly $500,000 in the company, which made technical
advances but was never quite fast enough to keep up with competitors.
The moral to Kleiner & Perkins was to get out of bad deals sooner; even
though Dynastor returned a profit of a few hundred thousand dollars, it

had wasted a lot of time. K&P maintained an "ICU List," made up of companies in need of emergency treatment. But the intensive care unit wouldn't stick with a patient forever; when the plug needed to be pulled, the partners prided themselves on not showing any sentiment.

The point might have been obvious, but the VC landscape was filled with examples of good money chasing bad. Kleiner & Perkins aimed to formalize certain principles of VC-ing—raising the money, evaluating proposals, finding management talent. Whereas entrepreneurs were supposed to be fanatics ("drinking the Kool-Aid" is a favorite metaphor), a venture capitalist was supposed to be agnostic. He didn't care so much if the widget changed the world—though that was great—as long as the investors had a chance for the moon. Better to double your money on mediocrity than lose it all on a dream. An entrepreneur would be dancing away on the *Titanic* even after she hit the berg; the VCs would be first in the lifeboats. "Venture capitalists are bankers—they're after only one thing and that's a return on their money," says Jerry Yang of Yahoo, who in the 1990s made the VCs very happy. "You wonder if they once had religion, then lost it, and why something is now missing."

What Gene Kleiner and Tom Perkins tried to do is create a dispassionate system for winning more often than losing, even if instinct ultimately governed. In the process, they created a new vocabulary of money, even if the system sometimes looked like a crapshoot to outsiders. Most important, they would be active in managing the companies in which they invested, a hands-on preference based on their backgrounds as technologists and managers—experience that even a brilliant financial analyst like Art Rock didn't have.

It was Rock whom the young firm wanted to distinguish itself from— and Doerr later saw him as his arch-rival. "The other VCs would turn over the money to the entrepreneur and then just watch in the grandstand, only asking questions after a company went south," Kleiner says. "We weren't going to be checkbook investors." That meant serving as architect, hustler, strategist, and, when necessary, executioner. ("Redeploying the founders," as the euphemism goes.) That kind of control is why K&P, like other venture-capital firms, demanded a substantial percentage of the company, as well as a seat on the board of directors, in return for its deep pockets. It could turn out to be a Faustian bargain.

VCs will say their allegiance is to the founders and the CEO, but they're there primarily to protect their investment. Most times, all three coincide. When they don't, there's bloodshed.

Even after a company went public, K&P didn't go away, holding on to its board influence for a decade or more. It was one thing to be a trusted adviser, but the times often came when the VC had to be the tough guy who would fire the entrepreneur-turned-CEO or sell out to the highest bidder—thus, the tag "vulture capitalist," coined by the normally mild-mannered Gordon Moore. Moore was never bothered by Art Rock or anybody else, but hated when new start-ups and their VC backers would try to raid Intel of talent. (How to learn the names of mid-level engineers nobody had heard of? It's called "Dumpster diving"—rummaging through a competitor's trash for personnel records and phone lists.)

The "system" of venture capital also meant knowing how to play rough with entrepreneurs—playing on their inexperience and emotions, playing founders against each other, negotiating tough prices and generous ownership stakes that in hindsight infuriated the entrepreneurs. If entrepreneurs attempted to play competing venture firms against one another, then the VCs might band together in a deal; that meant a lower-dollar-amount investment, but also kept the overall price low. Not a particularly free market, but pretty smart when they got away with it.

K&P's guidelines even took on a kind of mythology within the Valley. There was Kleiner's First Law: "When the money's available, take it," which in other tellings became, "When they pass the hors d'oeuvres around, take two." Kleiner doesn't remember saying either, but then Yogi Berra says he didn't say most of the items attributed to him. And Kleiner's Second Law: "There is a time when panic is the correct response," which was another way of Perkins saying to get out of bad deals quickly. And Kleiner's Third Law: "Never sell unless there are two buyers." Or: "If a decision is incredibly difficult, it doesn't matter what you decide, because it isn't going to turn out that way anyway." Perkins's best-known law is: "Market risk is inversely proportional to technical risk," which is quite insightful, because it counterintuitively places value on products that look impossible to develop. If the product is easy, presumably others will come up with it and crowd the marketplace (as KP-funded Netscape learned from Microsoft). As a rule, Kleiner & Perkins never wanted to face both significant market risk and technical risk in a start-up. It almost never paid to be a follower; you don't ride the wave— you try to make the waves. An old Valley joke: "What do you get when you cross a lemming with a sheep?" Answer: a venture capitalist.

Kleiner and Perkins also tried to establish rules of business conduct for their firm, as a way of avoiding ethical allegations that had hovered around other VCs. First, they promised to avoid conflicts of interest: If they invested any of their own money in companies, they would do so only as limited partners themselves. If K&P turned down a chance to invest in a deal, Kleiner and Perkins personally would not put their own money in the start-up. This would prevent Kleiner or Perkins from becoming distracted by competing deals or "cherry-picking"—saving the best deals for themselves. K&P also would not take its cut of the profits until the limited partners got their initial investment back. Nor would K&P reinvest any of the profits; this ensured that investors weren't committed forever and that every fund had a finite life, about a decade. Finally, K&P would issue audited quarterly and annual reports to its investors (though never to the public).

Due largely to its inability to find high-tech ventures to fund, K&P decided to incubate its very own younglings. Rather than be reactive— waiting until an entrepreneur arrived with an idea—Kleiner and Perkins aimed to anticipate where the world was going and how to invest there. It was this self-assurance—some called it arrogance—that became a hallmark. In later years, Doerr wasn't kidding when he said he ought to have a list of five companies—*industries*, better yet—in his back pocket that needed to be started; for some of them, he or another partner might even serve as CEO for the first few years.

In 1973, the firm had hired Jimmy Treybig, a Hewlett-Packard electrical engineer by way of the Texas Panhandle and Stanford Business School. Treybig, who once worked for Perkins at HP, would be the first of the young "associates" brought in by K&P to bird-dog deals and pursue their own entrepreneurial ideas. Treybig, who accentuated his Texas heritage and loved to note that in Houston "you start playing poker at the age of six," wanted to sell a foolproof minicomputer to businesses that would preserve data even when it broke down. The trick would be dual microprocessors so that even if one crashed the others would back it up. Banks and airlines and stock exchanges, for example, were desperate for a reliable "nonstop" fail-safe computer. Treybig got K&P to kick in $1.5 million, for 40 percent of the company and Perkins as chairman of the board. In three years, when Tandem Computers of Cupertino went public, that investment had gone up more than eight times; by 1981, it was worth more than $220 million. (Tandem ultimately sold out to Compaq in 1997 for $3 billion, though in 1996 Treybig got sacked as CEO by the board. "I still own a bunch," says Kleiner, chortling that his cost basis is "still twenty-two cents." Perkins made $70 million in the deal.)

K&P then hired another young entrepreneurial gun named Bob Swanson. Like Perkins, he had degrees in both engineering and business management; he had come west to run an investment fund for Citibank. But twentysomething Swanson didn't want to manage other people's money, and he certainly didn't want to move to Hong Kong as Citibank wanted. He wanted a company of his own and approached Kleiner & Perkins, whose incubation of Tandem had achieved some notoriety around the Valley. Perkins suggested he join the firm for a year or two to work through his ideas. Swanson settled on genetic engineering. Biotechnology was another form of information technology—the information just happened to be DNA and it was stored not on chips but in genes. In the mid-1970s, gene-splicing was largely theoretical and the locus of research was academe, not industry. Kleiner and Perkins knew nothing of the new science and were skeptical.

Sniffing around, Swanson heard that other companies were working on a way to commercialize gene-splicing—moving a piece of DNA from one bacterium to another. With the correct coding sequence, the second microorganism reproduced and passed on the new traits to its offspring. Presto—a genetically altered creature! The ultimate object would be replicating a human gene. But mass-producing recombinant DNA simply was not part of the academic mind-set and provided Swanson his opening. He asked K&P to back him in a genomics start-up, to the tune of $1 million—for a lab, technicians, equipment. In keeping with its policy of trying to overcome technical hurdles early and as inexpensively as possible, the firm balked at that much money. "Wondering if God would let you make a new life-form struck me as rather high-risk," Perkins recalls. "Especially in an $8 million fund where we had already had some very significant losses." Instead, Perkins proposed subcontracting out various parts of the experiment to existing companies. Swanson effectively quit the firm, but kept a desk there and for a time worked as a lone venture wolf.

He called around to the various universities to talk to researchers working with recombinant DNA. One of them, who agreed to meet briefly—"ten minutes on a Friday afternoon"—was Herbert Boyer of the University of California at San Francisco Medical School, one of the early patent holders and the "genie of genes." They wound up talking three or four hours and about that many beers—academic and businessman, the ideal complement for an industry they would create. With an initial $100,000 from K&P—in return for 25 percent ownership of the new company—Swanson and Boyer founded Genentech in early 1976, with Perkins as chairman. Perkins prevailed in his suggestion to have Swanson and Boyer farm out discrete parts of the first experiment. The University of California, Caltech, and the City of Hope Medical Foundation each got a separate research contract; none of them was told about the overall project. Within a year, Genentech had manufactured somatostatin—a human brain hormone—and then human insulin for the treatment of diabetes.

K&P had put in another $100,000 to go with major university funding, but Swanson knew by now he had a viable company and knew that K&P no longer had the best negotiating position. K&P's decision not to back Swanson at the very beginning had cost it sweeter terms. But success in a good venture deal is all relative. After Genentech went public in October 1980—the hottest Valley IPO since Intel, making the front page of the *San Francisco Examiner* in the banner headline GENENTECH JOLTS WALL STREET—Kleiner & Perkins's original $200,000 stake was worth $160 million—an eight-hundred-fold explosion. Perkins had highschool friends he hadn't talked to in decades calling him to get in on the deal. (Unfortunately, none of them were K&P limited partners.) In its first ninety minutes of trading, Genentech went from 35 to almost 89. Like other K&P-brewed companies to come, it nicely captured the public's fascination with a new technology, even with its absence of profits. Critics mocked Genentech as "a science-fiction stock," but even that put-down imbued the company with a futuristic glow. Swanson was Mr. Green Genes.

Because of the aggressive Tandem and Genentech deals, the first venture fund put together by Gene Kleiner and Tom Perkins was a gigantic hit. "It took us a while to realize the magnitude of the money," Kleiner recalls. Perkins could buy all the Bugattis and Alfa-Romeos he wanted; he even convinced Kleiner to splurge on a Jaguar. In six years, the two general partners only managed to invest $7 million of the $8 million they had raised. Yet ten years after they launched K&P, the fund was worth a prodigious $400 million—an average annual return of 47 percent, even taking into account the profits paid to KP's partners—with most of it returned to the investors. That was better than the rule of thumb that venture investors tended to follow: Get back ten to twenty times your money within ten years.

Of the seventeen deals in K&P's fund, seven had lost money; other possible deals—like a personal computer start-up by Steve Jobs and Steve Wozniak, which went public two months after Genentech—never happened because the partners weren't interested. "Not a great batting average," according to Perkins, "but the two home runs more than made up for it." A few big winners always erase a pile of losers—everyone loves a long shot—and there was nobody in the house setting the odds. For the same amount of time and effort, why not go for the moon? It was a pretty good investment scheme and made K&P the first brand name in venture capital. The home runs obscured the reality of failure that faced most entrepreneurs; K&P provided a seal of approval for a start-up, but was hardly a guarantee, a fact that disillusioned a lot of dreamers who struck out. Investors had the luxury of seeing their risk spread out among deals, but entrepreneurs had all their chips in one basket.

K&P's success attracted enough business that it was clear the firm needed to grow. Kleiner and Perkins wanted partners who were as ambitious as they were—as ambitious as the entrepreneurs they would back.

Brook Byers grew up in Georgia in the 1940s, a little nerd who loved ham radio so much he aimed to make contact with every country in the world so he could fill in the ham hobbyist's equivalent of a bingo card. By Byers's count, that was 320 countries. (He

finished in 1996, at the age of fifty-one.) His interest in science led him
to Georgia Tech, where he worked his way through to get a degree in
engineering; at school he was a concert promoter and he also did antenna
design for the Federal Communications Commission. Byers decided he
didn't want to do engineering and went off to get an MBA from Stanford.
"I liked the science-and-engineering case studies they seemed to be
doing there," he recalls, "and you could just tell that something was
happening at the Stanford Industrial Park right next to campus, with
Hewlett-Packard and Varian."

In 1970, during his second year at the business school, Byers says he
had "an epiphanous moment." He attended a brown-bag lunch hosted
by three Silicon Valley venture capitalists. One of them was Franklin
"Pitch" Johnson, who had been with the venture crowd since 1962. Byers
was fascinated with the VCs and wrote a paper on venture-capital returns, the first ever at Stanford Business School. Two years later, he heard
that Johnson was looking for an apprentice. Johnson was running his own
firm, mostly investing his own money. "How bad do you want to work
for me?" he asked Byers.

"Real bad," Byers answered in his best Southern drawl, more boyish
than his years.

Johnson recognized his eagerness and offered him a pay cut, down
from $16,500. There was another condition. Byers had to invest in any
deals he came up with—putting his money where his mouth was. Johnson was teaching Byers about risk. The only difficulty was that Byers
hadn't yet paid back his student loans and had what he called "a negative
net worth."

But Johnson arranged and guaranteed a loan at Wells Fargo, and
Byers went to work for him for the next five years, the same period when
Gene Kleiner and Tom Perkins were getting established. There were only
about thirty VCs around the Valley in 1972 and Byers got to know them
all, a networker before the term was invented; every month, as the Western Association of Venture Capitalists, they gathered at the University
Club in San Francisco for lunch and entrepreneurial gossip. Johnson
wasn't interested in raising a large fund like K&P's. Byers was, the better
to attract good deals and to fund them without having to resort to other
competing sources of capital. Perkins was particularly impressed with

Byers because he had invested $10,000 of his own in Tandem—the best riverboat gamble of Byers's life, making him a millionaire—and gotten Johnson to go in as well.

In 1977, Byers left Pitch Johnson and accepted Perkins's invitation to join K&P as an associate; Byers was the consummate schmoozer—he even became bachelor roommates with Bob Swanson in San Francisco. Byers represented the exception to the firm's rule—he was hired despite never having run a company. So, a year after joining K&P, Byers took a leave of absence to incubate Hybritech, a biotech company out of San Diego that aimed to clone antibodies for the immunological treatment of disease. Hybritech naturally thought Perkins, and not his associate, was going to be K&P's point man. But Byers did okay. From an investment of $1.7 million, Kleiner & Perkins reaped $28 million when Hybritech went public in 1981, before Eli Lilly & Co. eventually acquired it. Byers headed the company only briefly, a decision that flowed out of the most fortuitous event of his life.

On September 25, 1978, Byers was supposed to take his regular Monday-morning flight to San Diego. He spent weekends up at Lake Tahoe and usually caught the first plane from Sacramento, but on this day he missed it. Pacific Southwest's Flight 182 never made it, colliding with a small plane over San Diego. All 135 people aboard were killed (along with thirteen on the ground and two in the small plane)—then the worst accident in American aviation history. "Fate worked out well for me," Byers says. "I'm kind of superstitious, so I don't like recalling it." (Apparently so, since when I first asked him about the flight, he answered simply, "What a strange question," without acknowledging the incident.)

Byers's replacement as Hybritech CEO happened to come from a competing biotech start-up. Not only did K&P get themselves a new CEO, they wrecked a rival in the process. The Hybritech experience crystallized Byers's interests in biotech, and he would lead K&P's efforts in "bugs and drugs." Perkins and Kleiner stuck with "bits and bytes." Even today, those two areas—biotechnology and electronics—are the firm's specialties.

As Byers worked with Hybritech, K&P took on another partner, thirty-eight-year-old Frank Caufield. The son of a career Army general, Caufield combined both intellectual discipline and a bon vivant's love of the pleasures money can buy. As an eight-year-old growing up in Spain—

his father was with the American embassy—Caufield became a fan of bullfighting and still knows the best matadors in Seville. If you want to know the best club in San Francisco to hear the blues, ask Frank Caufield—he owns it with Boz Skaggs. Like Perkins, Caufield had a sharp wit and little patience for fribbling chatter. They even both married Norwegians. Unlike Kleiner, Perkins and Byers, Caufield had no experience in electronics. What he brought to the table was traditional financial analysis. He was a numbers man but had a layman's common sense.

After spending much of his childhood overseas, Caufield attended West Point and then wound up in the Army as a captain. He went on to Harvard Business School (just as Perkins had) and then to Manhattan as a management consultant. It was there that he learned about the perils of poverty, at least relatively speaking. "I hadn't really made the connection between having money and living well," Caufield says, recalling the lesson most of us learned when the neighbors got a Cadillac while we were still driving a Chevy. "Because my father was an Army officer, we never had much money. But we lived very well. In Europe, we had a maid for the upstairs, a maid for the downstairs, a cook, a chauffeur, and a gardener. In the States, we had the biggest house on the post and the officers' club. We lived great. When I got to New York City, I was twenty-eight, married, had two children, and I thought I was going to be rich beyond the dreams of avarice with my salary of . . . exactly $16,500. Within three months, I was cutting my own hair with one of those weird combs with the razor in it. I said to myself, 'Jeez!' " Frank Caufield didn't like the idea of cutting his own hair.

Naturally, he had to seek out a career in venture capital. Caufield had an inkling that high-tech was the financial place to be and might also allow him to be independent; when he was in high school, he bought thirty-three shares of Texas Instruments with Christmas money that eventually paid part of his way through Harvard. Caufield convinced his Manhattan consulting firm to move him to its San Francisco branch. "It was just like, 'Go West, young man,' " he says. Once there, he landed a VC job. Following seven successful years managing a small venture fund that invested in many of the same deals as Kleiner & Perkins, he moved again in 1978. "You make your best friends in your worst deals," Caufield says, and that's how he and Tom Perkins came together. The company they both had tried to rescue was a word-processing venture called

Office Communications. Savin Business Machines bailed the company out, allowing the investors of both K&P and Caufield's firm to escape with some self-respect. The time that Caufield spent in the bunker with both Kleiner and Perkins convinced K&P to make him a partner.

With two new members, in 1978 K&P renamed itself Kleiner Perkins Caufield & Byers, the name it still goes by. The firm moved its offices to a tower in San Francisco, in the Embarcadero financial center built near the old, notorious Barbary Coast of Gold Rush days, upon the wrecks that had sunk in the Bay. The move was to accommodate the commute of Tom Perkins, who lived on Belvedere Island, north of San Francisco; several years later, the firm returned to Silicon Valley, where the action was. With its initial fund nearing an end, the four partners went about raising a new one, totaling $15 million, which they called KPCB I (for some reason, VC firms, like Super Bowls, like Roman numerals). Two years later came KPCB II at $65 million. These would be the first of eight main funds the firm would raise over the next two decades. There were two reasons for establishing a series of new funds— rather than just replenishing capital in the old ones. It allowed KP to make changes among the investors, either because they wanted out or because KP thought some newer faces deserved entry into the club. The new funds also allowed KP to reallocate the profits among its general partners. Typically, at least one of them would retire to make room for fresh blood.

Somebody like John Doerr.

Chapter VII

Profits

John Doerr—"JD" to Valley insiders—joined Kleiner Perkins in 1980. With the retirement of Gene Kleiner and the expanding base of investment capital, KP needed more partners to manage the money, find deals, and sit on corporate boards. There was a limit to how many companies—a dozen or so—that any one partner could keep track of. The twenty-nine-year-old Doerr was another Midwestern transplant, having grown up in St. Louis. His father was an entrepreneuring engineer himself, owning the largest supplier of sulfuric acid pumps in the world. Doerr attended engineering school at Rice in Houston and, in KP tradition, got his MBA from Harvard in 1975. Like Byers before him, Doerr gravitated to venture capital immediately.

"He cold-called me a few years before we took him on," Byers says, "explaining how badly he wanted to go into the venture business. I told him to go to work for a brand-name company and get some experience." Doerr approached Dick Kramlich as well, who also suggested getting some time at a company, perhaps at Intel.

Doerr went to work at Intel soon after it introduced the advanced 8080 microprocessor. He became a marketing ace and the company's top-ranked salesperson. He wasn't selling doughnuts to Dunkin', but he was known to do anything to make a sale; he closed one deal in the heartland by throwing in a decidedly low-tech lawn mower. His only liability to Intel came when he got behind the wheel. He drove like a rabid dog, wrecking company car after company car, forgetting a light here, missing a turn there, swerving in and out of lanes as if the white lines were mere suggestions. He was proof that a short attention span is a handicap on the highway. Netscape's Jim Barksdale once said that Doerr "wouldn't qualify to drive for FedEx"; Barksdale would know because he used to be No. 2 at that company.

These days, Doerr's wife tries to keep him off the road. Doerr is chauffeured around Silicon Valley in a white Chrysler minivan that's stocked with an extra set of rumpled khakis, button-down blue oxfords, and the other tie he owns. Doerr sits in the rear, a jumble of cell phones, bookbags, notepads, two laptop computers, and a two-way pager that never leaves his belt. He's got at least nine phone numbers to his name. Within seconds, he can be hooked up to any satellite in Earth orbit; within seconds, anybody with Doerr's very own 800 number can reach him. "I try not to stand too close to him and all his electronic devices because I may still want to have more children," Scott McNealy likes to joke. Even for the three-mile dash from his Woodside house to KP's Sand Hill offices, Doerr uses the driver, keeping the California Highway Patrol happy and allowing him to work the gadgets in the backseat. How else could he read 150 E-mails a day?

After five years at Intel, Doerr called Byers again. "You remember me?" he asked.

"No," Byers told him, "but I run every day at Stanford at 5:30 in the afternoon. Do you?" Byers was planning the rivalry that would mark their relationship even years later. "I wanted to see how motivated he was. We had an opening only once every several years and we got an awful lot of calls."

Doerr was smart enough to be at the Stanford track. They ran and talked and met over the next few weeks. Byers thought Doerr was a Kleiner Perkins prototype: He combined technical capability and business savvy. He understood sales, systems, and customers. "In checking references on him," Byers recalls, "I'll never forget what someone at Monsanto, an early job of his, said. I was pushing to find some fault of John's. They told me, 'Well, when he gets going on something, he doesn't sleep. Literally. He'll just work himself to exhaustion.' That sounded pretty good to us." (On good nights, Doerr sleeps three to four hours; his best E-mails are sent at two in the morning.) Andy Grove urged him not to leave

Intel—suggesting VCs were little better than real-estate agents, even if one named Rock had given Intel its start—but Doerr could not be swayed.

The partners brought him in as an associate, and then made him the firm's fifth partner in 1982. The letterhead did not change, something Doerr says never bothered him, though his friends say he mentions that fact rather frequently.

Doerr's manic style, no less than his intellect, quickly became apparent. He was (and still is) the embodiment of pulsating energy, frenetic to the point of distraction. Colleagues and entrepreneurs take turns trying to figure out what creature to compare him to. Why? It certainly wasn't the sandy hair, toothy smile, or baritone voice, which makes him quite charming. It is his weird physicality, the worst case of the fidgets since Mark Fidrych pitched for the Detroit Tigers. Simply put, John Doerr rarely sits still. In meetings, he dangles on the arm of a chair, his legs twitching like a thoroughbred's at the gate. Then he'll jump up and pace around "like a whippet in a cage," as Onsale's Jerry Kaplan described it. Another friend called him "the Energizer Bunny on steroids." The fact that Doerr's so thin—a beanstalk with glasses—makes him seem even more wound; caffeine might actually calm him down. Caufield says he has the "metabolism of a hummingbird." McNealy, the head of Sun Microsystems, refuses to jog with him. "He doesn't run," McNealy told *Forbes ASAP* a few years ago. "He *bounces*."

In his book *Startup*, Kaplan described his first telephone conversation with Doerr—at 4:30 in the morning. Kaplan was in Boston. Doerr was in San Francisco, on his way to New York. Doerr wanted a meeting, someplace. "What airline are you flying on?" he asked. "TWA," Kaplan told him. Before Kaplan could catch his breath, Doerr had zipped into action. "Okay, they have a hub in St. Louis. I've got an *OAG* [*Official Airline Guide*] here. Change your flight to the 9 A.M. tomorrow from Logan. That will arrive in St. Louis around 10:55. There's a flight to San Francisco at 12. That will give us an hour. I'll arrange a connection through there as well, and meet you at the gate." And so he did.

That kind of energy had a flip side and it made his partners wonder about him. Perkins remembers asking him one Friday afternoon if he wanted to go sailing in San Francisco Bay on Perkins's yacht that weekend. "I'm not sure," Doerr replied. "I may have to be in Tokyo."

"But it's Friday, John, really Saturday in Japan," Perkins said incredulously. "At this point, don't you know where you'll be?"

Gene Kleiner, of all people, today says he's not sure he'd have hired Doerr, if he had to do it over again. "If it's about money, I would," Kleiner says. "If it's about temperament, I would not. Part of it is intentional on John's part and perhaps every group of ten should have someone like him. But he sometimes is too driven, too intense, and gets himself into deal making too quickly."

Early on, Doerr showed he was not a manager. One time, Perkins found out Doerr had scheduled two business dinners simultaneously. "At least have them in the same restaurant," teased Byers. "John risks veering into self-parody," Caufield says. "But the decision by Tom and Gene and Brook and me to hire him has made more money for us and the investors of Kleiner Perkins than any other single decision."

Self-parody? Maybe he's talking about the customized ski helmet Doerr had made for his Aspen weekends. It's got a cell phone built right into it—earplug and mouthpiece. Dipsey Doodle and Little Nell are fun runs over Christmas, but you still have to do deals, right?

In standard KP fashion, Doerr incubated his own company, Silicon Compilers, which used computers to design complex microprocessor circuitry. But Doerr, who knew he was no manager, soon returned to Kleiner Perkins, pocketing millions when Silicon Compilers was sold. In contrast to the other partners' early tendencies to explore ventures beyond their expertise, Doerr was wise enough to concentrate on what he knew best. Having worked at Intel, he was in the best position to anticipate the beginnings of the personal-computer era and capitalize on it. Doerr also appeared to be fearless. One of his first deals, Seeq Technologies, involved raiding talent from Intel, which got KP sued. (The case went nowhere.)

Spending time around Stanford—he lived in a nearby apartment—Doerr could see what new technologies the talented university engineers were incubating: sophisticated computer networks, high-performance computer workstations, and 3-D computing. This is where Cisco, Sun

Microsystems, and Silicon Graphics were born, where Jim Clark was still a professor and Scott NcNealy an MBA student. Doerr's gift was in anticipating and conceptualizing the marketplace before the marketplace did. As if he were running an entrepreneurial assembly line, Doerr got Kleiner Perkins in the 1980s to fund Sun (the maker of souped-up computer "workstations"), along with such other companies as Compaq Computers, Symantec, Quantum, Cypress Semiconductor, and Lotus. Compaq cloned the IBM PC and then passed it by to become the largest personal-computer manufacturer in the world. Lotus developed the electronic spreadsheet package that overtook VisiCalc and, in the words of their classic ad, "gave IBM PC owners a reason to use their machines." For a time, Lotus was KP's most profitable deal ever and the largest software company in the nation before Microsoft got rolling. These new companies reflected a sea change for venture capital. In 1972, when you funded a start-up, potential investors were skeptical—an unknown product from an unheard-of company wasn't very appealing. Now, that attitude had reversed—new meant better.

Most of the fledglings were no longer selling technology to other high-tech companies but themselves becoming part of consumer markets. The personal-computer revolution created a new wave of businesses—hardware and software—with growth rates few could imagine. An industry that didn't exist in 1975 was worth $100 billion fifteen years later. The best VCs, like Doerr, rode the wave and allowed Kleiner Perkins to deliver double-digit returns to investors most years, even during lean times when other venture funds got pasted. Only in the late 1980s and 1990 did KP post some lousy numbers—0.3 percent for one fund in 1988, 3.7 percent for another in 1990—and these still were better than most competitors. It is these unimpressive numbers, however, that cause the firm to be so taciturn about its fiscal performance in any given year—even when it's been better than 100 percent, like in the early 1980s or some years in the mid-'90s. It's much more dramatic to give a rate of return over ten or twenty years, which hides the barren times.

The chief beneficiary of the good numbers was John Doerr. His string of investment hits—from a VC in his mere thirties—began to establish the Doerr mythology. Today, he gets introduced at speaking engagements as the "Bill Gates of the Valley." It's meant as a compliment.

Doerr's philosophy of venture capital amounts to his own version of Kleiner's Laws. The days of random genius are long gone: Innovation and rapid change are part of the VC infrastructure. "Forget the idea of new industries being born in a Hewlett-Packard garage," he says. "Companies are too attuned. Tandem didn't have competition for six years. Today, within six months, a competitor will have a cruise missile headed right up your tailpipe." The solution? Find a good venture capitalist.

The Doerr formula of venture capital has been presented to audiences the world over in his famous Slide Show. From his little laptop, projected onto the big screen, it's not much on fancy graphics—"I like to work in Helvetica Bold," says Doerr, ever the control freak—but entrepreneurial wannabes at Stanford, high-tech executives at $4,000-a-head conferences, and Al Gore himself have all raved about the performance. If you want to learn about *the largest legal creation of wealth in the history of the planet!* this is the show to attend. Same for "The best way to predict the future is to invest in it," "The Web changes everything—there's no wait and it's always on," and "It's just possible the Internet is *under*-hyped." How so? Because every graph on growth of the World Wide Web is "exponential"—home pages, users, Amazon.com purchasers. "It's not about Moore's Law anymore, but Metcalfe's Law," which holds that each new person connected to a network—like the Internet—increases the power of the network exponentially. (Bob Metcalfe, founder of 3Com, in the 1970s had invented the Ethernet, a new way of linking computers together electronically.)

On a Saturday afternoon in May 1998, Doerr is performing before the Monte Jade Science and Technology Conference in San Jose. It's a gathering of eight hundred Taiwanese-American engineers who want to be *rich* Taiwanese-American engineers, and Doerr has invited me along for the fun. In the minivan ride down to the local convention center, Doerr adds some updates to the Slide Show, plugging a hot KP company or two. Otherwise, it's identical to the version he gave to KCBS radio and to the Russian president a few weeks earlier. ("Largest legal creation of wealth in the history of the planet" translates into any number of languages.)

The centerpiece of the Slide Show is a bare-bones list, a comparison of the old world order and the utopia Doerr hopes to create. It reads:

OLD ECONOMY	vs.	NEW ECONOMY
a skill		lifelong learning
labor vs mngmt		teams
bz vs. environment		growth
security		risk-taking
monopolies		competition
plants		intelligence
standardization		customized choice
sues		invests
status quo		speed, change
standing still		moving ahead
top-down		distributed
wages		ownership, options

Doerr might have included the best part about the New Economy: He's made lots of money from it.

Silicon Valley, according to Doerr, is at the heart of the revolution. "You hear the conventional wisdom around the country that the American Dream is over," he tells the crowd. "I have a problem with that. Look at the four pillars of the New Economy—microchips, PCs, the Net, genomics. Forty percent of GDP growth is from technology in the Valley. We have unemployment under 3 percent, high wages, every segment is moving up!" (Especially those with KP investments.) Who's the biggest employer way off in South Dakota? Doerr asks. Not some agricultural combine, but Gateway Computers. "Silicon Valley is a state of mind," to be exported across the nation and around the globe, he says.

The New Economy means not just better living, but a new political order, presumably with folks like Doerr pulling the strings. And his TechNet, a bipartisan political action committee, has successfully lobbied Washington on such issues as securities litigation and software encryption. "Move over, Soccer Moms and Angry White Males," he declares, announcing tomorrow's "Digital Citizens," who are libertarian, who vote, and who "have faith in the future." First came the PC wave, then software, now the Internet and its World Wide Web. The digital universe is just beginning—"like ten milliseconds after the Big Bang." Our children will look back on today's early convergence of media—computers, telephones, cable—"the way we remember manual typewriters." (For all his political success at the national level, it was particularly galling to Doerr that he couldn't get his own town of Woodside to approve a mundane bond issue in 1998. Doerr spent much of the spring on personal calls and E-mail to his neighbors, appealing for passage of a school bond vote that would raise $10 million for the Woodside Elementary School—the one that his kids and Larry Ellison's attend, the one that raised $439,000 in the charity auction the following month. Doerr's cause was just—the school needs major capital improvements, like heat—but it seems that a lot of the locals didn't like being pressured by Doerr.)

Who are the rebels driving the New Economy? They just happen to be a lot of Kleiner Perkins companies—and they figure in practically all of Doerr's "forecasts." The @Home Network, from which KP investors made more than a *billion* dollars (from an investment of only $7.4 million), will one day wire the country for high-bandwidth Internet access via TV cable. Reforming health care certainly is a national priority—that niche belongs to Healtheon, the most recent start-up of Jim Clark, late of Netscape. The model of online commerce? That would be book-seller Amazon.com (which earned KP investors at least another half a billion). It was the type of business that could exist only in cyberspace—millions of items available that no one store could accommodate. "Buy Amazon!" advises Doerr merrily, ever the salesman; since that date, it more than quintupled in value during 1998.

Audiences soak this up. Many conventioneers at Monte Jade try to write everything down, though Doerr's light-speed delivery makes it difficult. A few appear to be reciting his words to themselves, as if at a revival. After the show, he's deluged with business cards and business plans; they ask him questions about Netscape the way a ten-year-old pleads with Reggie Jackson to describe one more time how he hit that home run in the bottom of the ninth. For an agnostic VC, Doerr knows how to work the congregation. To true believers, he's the high priest in a way that perhaps only Steve Jobs could also pull off. And Jobs isn't out on the entrepreneurial hustings looking for talent, which is what really gets an audience going.

Once he's convinced he's found talent, Doerr explains, the key to any KP investment is to "identify the risk up front and get rid of it." This, he says, is Kleiner's First Law (or, at least, *his* Kleiner's First Law). "Every VC will tell you he's looking for the same thing—technical excellence, outstanding people, and a large, rapidly growing market. But the question we ask is whether we can get rid of the risk up front—long before there's a management team, before there's a product, before others have poured in money, when it's just somebody with an idea. There's technical risk: Can we split the atom or map out the human genome? And there's market risk: Will the dogs eat the dog food, or will the fish jump out of the tank? The way to succeed isn't to cut a tough deal with the entrepreneur, but to ruthlessly evaluate the risk." Removing that market risk can be expensive, though KP prefers to dribble out its war chest until the risk is eliminated and then pour in larger sums when a start-up invades the marketplace.

Doerr says he worries most about market risk, because it's out of his control. You can invent the world's best electronic fork, but it doesn't mean anybody'll buy it. Technical risk, however, is KP's forte. The partners are scientists and engineers by background, who like being close to the technical part. Other venture firms will invest in later stages of a young company's life cycle, once there's product and management; for that, they'll have less risk, but they'll also pay more to own part of the company. "We prefer high risk and high rewards," Doerr says. "Many people just don't want that constitutionally. It leads to gray hair and anxiety attacks." Thus the mantra of venture capitalists is that they "sleep like babies"—they sleep for two hours, wake up and cry, sleep for two hours, wake up and cry. Doerr's hair is still blond, and he doesn't sleep much anyway.

The paradox in venture investing is that, for all the analysis that goes into it, decisions are ultimately made by hunch. KP employs no economists or bright-eyed analysts to crunch the numbers of a potential deal. "I'm still looking for a bunch of burrito-eating Stanford kids who walk in with no idea what a business plan is, we write them a check and work intensively with them for several months, and—*boom!*—you've got a company with a market cap of $2 billion," Doerr says, racing in cliché overdrive.

Most Monday mornings, the partners of Kleiner Per-
kins assemble in the conference room of their sun-washed headquarters
at 2750 Sand Hill Road. KP has the best look-at-me offices along venture
capital's gilded avenue. Almost all the interior walls are glass, giving the
office the feel of a giant aquarium, except that none of the specimens
are guppies. Renovated a few years ago for $2 million, the single-story
space has a patio and exposed beams for each partner, and is paneled
with a light-colored, highly grained, frightfully expensive exotic wood
from western Africa called avodire. The soaring ceilings give the space
an air of a house of worship—quite fitting in this the cathedral of capitalism. Around a conference table large enough for a Cuban boat lift
are the real seats of power in the Valley. Any entrepreneur who's ever
been in that room feels compelled to comment on the table, which is the
point of the table.

Mondays at nine, as Jerry Kaplan described it, are "as sacred as
Sunday is to the Vatican." This is when the partners of KP, as well as
most other VC firms, assess existing investments and consider new deals.
This is how they get their juices going for the week. If you can't be there
in person, that's why one of the propeller heads you financed invented
video conferencing. There's a joke that any partner on his deathbed
should still be able to muster up a speakerphone. In forty-five-minute
sessions, entrepreneurs are invited in to pitch their idea and are then
grilled by the partners whom they're asking to write a big check. There's
little tolerance of chitchat, and the investment decisions are made the
same day. The setting is a cross between a Ph.D. dissertation defense
and Star Chamber, with some great catered fruit slices on the side. If a
partner is bored, he tends to show it; T. J. Rodgers remembers giving
one presentation on semiconductors and noticing Tom Perkins doodling
a sketch of his latest ketch.

"I am sure there are entrepreneurs that have never recovered from
making a pitch to us," Perkins says, though T. J. Rodgers is not among
them. "We all have operating experience, and we get into their plans
right away, rapid-fire. Some are taken aback by how much we push
them." Perkins remembers one fellow who was making a presentation in

the old San Francisco offices when a 5.0 earthquake struck. "You know, we all were grabbing the table and holding on, and this guy was so intent on getting through his forty-five minutes that he didn't even notice. He asked, 'Why are all of you so quiet?' " Impatient with the free-flowing, open-ended, Californian comfy meetings he had seen at Hewlett-Packard, Perkins ran a tight ship, even as he was drawing one. There was an agenda, there were time limits, and there was a denouement to each issue—a KP tradition that continues.

Every year, KP gets about two thousand business plans, most unsolicited, most preposterous, like enclosing Los Angeles in a dome to avoid the smog. "We don't invest in real estate," Doerr deadpans charitably. The plans arrive by box, letter, E-mail, and fax. All at least get skimmed, though KP cares more about the résumé of the proposer than the proposal. West Coast entrepreneurs are favored, so KP partners don't have to travel far to monitor their money. So, too, are known quantities— almost every KP deal involves an entrepreneur the firm has dealt with before. Several hundred of the plans annually make it to the next level of review—a meeting or a call—out of which forty or so, "The Hot List," make it before the Monday sessions. The entire process takes just a matter of weeks, compared with the more leisurely pace twenty years ago. On occasion, an idea makes it to the partnership based on a single conversation and no business plan at all. That's what happened in 1994 with a start-up that became Netscape.

Each of KP's partners usually takes on only a couple of new companies per year, so most of the forty will be rejected. Typically, one or two partners will sponsor a particular proposal, and it takes a majority to approve an investment. Nonetheless, a partner who objects strenuously enough can veto a deal. With unanimity required, it prevents finger-pointing when a company flames out. And most of them will, though it's all relative. Of every ten KP-backed start-ups, roughly five will be out of business in a few years; four will yield a decent return; one will fly to the moon. No wonder Doerr likes to call himself a "co-conspirator" in "the largest legal creation of wealth" . . . you know the rest. KP now invests about $60 million a year in new companies, several million per start-up.

While Kleiner Perkins says it doesn't negotiate the last dime, it tends to be the toughest bidder in the Valley. "There's always a give-and-take, but you don't really make money in the venture capital business by squeezing the entrepreneur a little tighter," Caufield says. "On the other hand, we think we add a lot of value and that we shouldn't have to pay any more than we have to. The fact is, when you have such a wide range of outcomes, on those companies that are huge successes, we grossly underpaid for our share of ownership. But on those companies that are failures, we grossly overpaid. Basically, the entrepreneur who's a big success—for whom we'd have paid much more in hindsight—is effectively penalized for all those deals where if we knew then what we know now, we'd have paid nothing. The successful entrepreneur who complains we got too much is right—but he forgets about the deals gone bad." Another Kleiner's Law: "If the negotiation is so tough and the relationship starts out that way, it isn't worth it." Perkins called it "slicing the salami too thin." But then again, that's a pretty good negotiating ploy, to the extent the entrepreneur is dying to do the deal with you. Byers says only one in six turns down KP's money. Doerr says the firm hopes to see about half the good projects in all of high-tech.

In return for its investment, KP gets a 20 to 30 percent ownership stake—more than anyone else, including the founder—as well as a seat on the board of directors, which gives it unbridled right to meddle. What the entrepreneur gets, in addition to funding, is KP experience—and the brand name; Doerr's imprimatur alone warrants a mention in the press. There is a credibility associated with KP money, as if the partners' vote of approval automatically means you'll be living in Woodside in no time. The number of start-ups turned crack-ups belies that, but even failure depends on who's left holding the bag. The company that looks like a disaster to the public may have done fine for KP. *Fortune* ran a graphic in 1998 showing that of the seventy-nine companies KP had taken public since 1990, fifty-five had tanked, with their stock then trading below the closing price back on the first day. Five more had registered gains below the market average, leaving just nineteen as winners.

But that 76 percent "failure" rate pertains only to people who might have bought shares in these KP start-ups on IPO day. KP still made money from most of them, because the price it paid at the moment of creation was virtually nothing—a small fraction of the initial public-offering price. Once a company reaches the public market—and many do not—the private VCs who were there at the beginning make out like bandits. Combine these "singles" and "doubles" with the "home runs," and VCs get rich. It's a pretty good system. Even companies that remain private don't have to be total losses, if a merger partner or corporate white knight comes along. But there have been some spectacular busts—meteors that flamed across the Northern California sky—and it pains KP when their names are resurrected. It's well and good to celebrate the Valley's tolerance of failure—without it, nobody would take the risks necessary to reap the rewards—just as long as it's somebody else's crater.

The two most infamous Kleiner Perkins implosions are GO, which didn't, and Dynabook Technologies, for which ex-director T. J. Rodgers keeps his stock certificate of twenty-two hundred shares framed on the bathroom wall right above the toilet paper. "Same difference," he snorts.

Tom Perkins liked to say that "it took two bad deals" to educate a KP partner about the thin line between success and failure. John Doerr's initiation just happened a decade into his tenure. In a weird way, it demonstrated his amazing powers of persuasion: He was so good, so carried away by his own intuitions, that he was able to sell his KP partners on duds. Dynabook Technologies was one of the early companies seeking to make slim, fast laptop computers—ones that didn't seem to weigh more than you did. Launched in 1987 by Doerr and his younger KP partner, Vinod Khosla—they dreamed up the machine themselves, according to the classic incubator formula—Dynabook was backed by chipmakers, designers, and $37 million spread throughout the Valley (including Goldman Sachs, Prudential, and Sequoia Capital). KP alone sank in $8 million and projected first-year sales approaching $100 million, a lot in those days. But the technology—Doerr's sine qua non *technical* risk—never worked. Dynabook's attempt to simplify chip circuitry was a great idea but only that. Even the screen kept malfunctioning. With manufacturing problems and management turmoil, the company didn't have a chance.

In the summer of 1990, *The Wall Street Journal* ran a brassy front-page story, complete with mugs of Doerr and Khosla, under the headline

COMPUTER GLITCH: VENTURE-CAPITAL STAR, KLEINER PERKINS, FLOPS AS A MAKER OF LAPTOPS. It was an utter humiliation for the firm that had brought Genentech and Tandem into the world, an article that seared Brook Byers's clippings scrapbook. And it came on the heels of another nasty piece, in the first issue of *Upside*, a monthly trade magazine about the Valley; KP partners, it suggested, were becoming "more interested in their sabbaticals than their investments." (That piece was written by Nancy Rutter, who several years later became Mrs. Jim Clark, the wife of the man who would help found Netscape with KP money.)

You don't hear about Dynabook in the relentlessly upbeat Slide Show of John Doerr. The *Journal* article quoted criticism from Gene Kleiner himself and contained gripes from company executives who had come and gone; it also noted that Doerr and Khosla lied to the press and Dynabook employees about the status of the CEO they were firing on the eve of its product's debut. When it finally came out, stores were more interested in selling Compaq's new line of laptops. "Kleiner Perkins Caufield & Byers, perhaps the nation's most prestigious high-tech venture-capital group," the piece concluded, "is being haunted by the ghost of a new machine."

Weeks later, Unisys Corp. claimed the Dynabook carcass for only a few million dollars. Rodgers still cringes at the mention of the company. "I would've quit the board," he recalls, "but they went broke first."

There was another embarrassment for KP. A few months before Dynabook was founded in 1987, Jerry Kaplan was sharing a cross-country flight with Mitch Kapor, founder of Lotus Development, where Kaplan was the technology guru. They struck upon what they believed was, in Kaplan's words, "the modern scientific version of religious epiphany," a realization "startling in its raw power and purity." The key to the next wave of computing, they surmised, "was to create a handheld device that worked like a notebook instead of a typewriter." No more *tap, tap, tap* on a keyboard—just write out what you want on a little electronic pad and the silicon brain would understand, all at a price under $1,000. Thus was born the dream of "pen computing"—part of a new generation of so-called nomadic devices—which the wildly successful Palm Pilot came to fulfill in the mid-1990s. Trouble was, the Kaplan-Kapor epiphany came about a decade too early, which they learned too late. So, too, did Kleiner Perkins.

Kaplan pitched his new kind of computer to Doerr, and KP funded it for approximately $5 million. Other investors put in at least $30 million more; IBM and AT&T loved the idea. Doerr predicted a market of "more than $100 billion" by the year 2000.

The name of the company was GO. "As in GO forth, GO for it, GO for the gold," Kaplan explained to Doerr.

"As in GO public," Doerr replied.

The son of a New York garment executive, Kaplan was and is a marvelous salesman—quirky, funny, earnest, with little of the guile of another showman, Steve Jobs. Kaplan kept a plaster bust of himself at GO, but his employees were more amused by him than anything else.

As with Dynabook, GO was a good idea that never got properly executed. The engineering hurdles were never overcome, Doerr's obsession with technical risk notwithstanding. And GO didn't help itself by getting into a fight with Microsoft over pen-computer operating systems. "The real question is not why the project died," Kaplan later wrote resignedly in his autobiography, "but why it survived as long as it did." The saga of GO, as Kaplan described it, was "six years of hell." At the end, what he had to show for it was an entire carton of smart-looking leather cases embossed, Gucci-style, with the "GO" logo. "Looks terrific," Kaplan says now. "No pen computers, though."

Kaplan cataloged his hell. GO (and its spin-off company, EO) was bought out by AT&T in 1994 and eventually shut down. In the declining months of GO, Kaplan could be found going around to colleagues with a tape recorder in hand. The result was *Startup*, which bemusingly chronicled the life and death of his company and how it burned through more than $30 million. The book did well—Danny DeVito had a film option on it—and Kaplan made some money. But nothing compared with his next venture, Onsale, the online auction house. When Onsale went public in 1997, Kaplan suddenly was worth more than $100 million; GO's legacy of failure hadn't impeded Kaplan from finding the mother lode. KP invested in him, though Doerr conditioned it on Kaplan not writing another book about the Silicon Valley game.

KP made tens of millions of dollars from Onsale, thank you very much, Jerry. And while GO and EO went down the drain and hundreds of employees lost their jobs, KP managed not to lose a dime, despite Doerr's public claim that KP "lost a ton of money." It was AT&T left holding the bag in the end. AT&T had been an investor early on and it bought out GO/EO with its corporate eyes fully open, then decided to give up. If KP had shrewdly sold AT&T the Brooklyn Bridge, AT&T couldn't complain that it didn't know about the potholes.

Dynabook and GO were exceptions for Doerr. His triumphs—along with those of new partners such as Jim Lally (another Intel manager), Floyd Kvamme (an executive at Fairchild and Apple), and Khosla (a cofounder of Sun, who had vowed to become a millionaire by age thirty)—allowed Kleiner Perkins to ramp up its new funds, KPCB III, IV, and V. In 1982, 1986, and 1989, each fund raised capital of $150 million (and invested in roughly forty new companies apiece), a staggering amount in the universe of venture capital. It might have been more.

In 1980, Tom Perkins had considered starting a fund of half a *billion* dollars—the New Industries Fund, he called it—and Morgan Stanley agreed to raise the money. At that level, KP could never directly invest all the capital and would keep for itself a far smaller percentage of the profits. What Perkins had in mind was a scaled-up "fund of funds." Together with about a dozen other VC firms in the United States and overseas, KP would use its expertise to passively manage investments in other venture funds, sort of the way some mutual funds buy stakes in other mutual funds. Perkins called Henry Hillman in Pittsburgh to see how much he'd go in for. Hillman, after all, had contributed half the capital for the first two of KP's limited partnerships. Much to Perkins's surprise, Hillman disapproved of any change in the firm's strategy.

"Just keep doing what you're doing," he urged Perkins. "Maybe yours is a good idea, maybe it's a bad idea, but I wish you'd just keep doing the same thing. It seems to be working pretty well." Hillman said he'd invest something in the mega-fund—though obviously not half—but he made clear how foolish such a venture seemed to him.

"I found what he said to be persuasive," Perkins recalls. "So I called Brook and John and the other partners and told them, 'I have some good news and I have some bad news. Henry Hillman objects to a fund of funds. But he has another idea: He wants us to raise the "carry" from

20 to 30 percent! Can you believe it?' " And he volunteered to put $50 million into the next fund. "Persuasive" indeed. The partners agreed within twenty-four hours.

The "carry" was VC jargon for "carried interest," which in English translated to "our cut." In a blink, Hillman had proposed that the KP general partners get a lot richer, potentially by many, many millions. This was Hillman's way of motivating KP not to switch gears; the other investors were likely to go along, because they knew KP was producing great returns for them. Now, for every dollar the funds made for investors, the KP partners would get thirty cents, up from twenty cents, which is what every other VC firm on Sand Hill got. If a fund of, say, $150 million, returned a profit of $50 million, that meant the partners would get $15 million between them. Of course, the partners were also still reaping their take from earlier funds. So, let's say it's 1985. The first fund has essentially run its course and its dividends have been distributed. But KPCB I, II, and III are all active; between them, there's about $100 million out there in investments. If each fund is returning about 30 percent a year—and some did far better—that's a pretty swell living, when only a few partners are divvying up the pot. All the more so when their downside was simply getting nothing, since they weren't the ones putting up the capital.

And it wasn't quite that bad. In addition to their slice of the profits, the partners every year—regardless of returns—received 2 percent of the money that investors had committed. For later funds, it didn't matter whether the investors had actually sent in all their capital; Kleiner Perkins, unlike other VC firms, still got its 2 percent. If Henry Hillman was in for $10 million, for example, he didn't have to send in that amount at the outset, because it might take the fund several years to invest it. Nonetheless, KP was annually entitled to 2 percent of the full $10 million.

Unlike the "carry," which was based on performance, the 2 percent "management fee" was a tollgate. So, even though the inaugural K&P fund went through a million dollars for a waste treatment company that yielded nothing, the partners still took home $20,000 a year in spending money for that boo-boo alone. If Kleiner Perkins in a typical year in the 1980s had $200 million committed to its portfolios—and it often had more—that meant $4 million in the pockets of the partners. In the 1990s, with more than half a billion dollars committed, the partners got a guaranteed $10 million between them. At ten partners—not very different than a decade ago—that was a million dollars per. "A great venture capitalist is equivalent in talent and responsibility and productivity to the CEO of a Fortune 100 company," says Byers. "The 2 percent fee is analogous to salary, and the profit participation is then like a stock-option package and bonus."

Ironically, Henry Hillman, that beneficent billionaire who invented the 30 percent carry for KP's partners, bowed out of KP's funds in the 1980s, choosing to dabble directly in start-ups. It was a train wreck. His venture investment company, wrote *The Wall Street Journal* in a front-page story, "quietly piled up what may be the biggest losses in the history of U.S. venture capital"—more than $80 million. His ineptitude led to the joke that he was more philanthropist than venture capitalist; any deal that was going to eat up vast amounts of capital, with no hope of returns, was said to be a "Hillman deal." It was no great tragedy—Hillman didn't need the cash and had other investment successes—but he would've done a lot better if he'd stayed with Kleiner Perkins.

In the 1990s, the amount of capital in KP's successive funds continued to rise, as both the price and number of deals increased. And so with them did the take of the KP partners. Perkins and Caufield had retired, leaving Byers as the last name on the door still working, even if Doerr was the rainmaker and prima donna. KPCB VI, in 1992, raised $173 million. Two years later, KPCB VII started with $255 million. And in 1996, the current fund, KPCB VIII, raised $328 million—more than forty times the money that Tom Perkins and Gene Kleiner spent months scraping together in 1972. It's become a bit easier.

NOW, the only thing harder than getting money *from* Kleiner Perkins is trying to give it *to* them. In fact, you don't actually ask—you wait to *be* asked. A Kleiner Perkins investment fund is like a country club with really high dues and no golf course (though lots of white men). For every investor KP invited in, half a dozen waited for the call that never came.

The success of KP's funds has not only turned the notion of a risky investment on its head, but made membership in the KP club a badge of honor. "We piss everybody off because we won't take their money or won't take enough of it," says Perkins, who retired from the firm in 1986 but who still attends strategic meetings and serves on several corporate boards. (At one point, he was juggling seventeen, as chairman of all but one.) "We'd like to take more, but there's no way to scale up our kind of venture capital, unless you completely change the recipe."

KPCB VIII, with committed capital of $328 million, will have no more deals than the prior six funds—they'll just cost more, given both the competition among proliferating VC firms and the bull market of the 1990s. In 1998, there were a thousand funds in the United States, double the number from five years earlier. No need to weep, though. In the age of the Internet, when companies can go public in just a year or two, the few extra million are table scraps in the case of an IPO hit.

As with each previous fund, the KP partners in KPCB VIII fan out on a road show to talk up the investors. In the days of Kleiner and Perkins, the show was necessary to get money. These days, it's just for show—to make eye contact with the investors, to demonstrate respect, perhaps to come home with a new idea. On separate planes—you can't risk having the firm go down all at once—the partners spend several weeks around the country meeting with university endowments, pension funds, foundations, and corporations.

Armed with colorful promotional binders and a slide presentation— not as good as Doerr's—they review KP's past profits and explain the returns that the high-tech future may bring. The KPCB VIII theme: "The Marathon Continues." (KP likes corny sports metaphors. A few years back, the pitch was "Our Home Run Strategy" and the firm distributed baseballs with the KP VII logo.) Byers, who takes the most important visits, explains that his business "isn't about sprints, but the long haul." Charts boast that the firm has "106 years of venture experience" and "169 years of technology management experience"; has been the founding investor of 10 companies in the Fortune 500 and an investor in 279 more companies; has been part of 94 mergers and acquisitions; has done $4 billion in IPO financing and arranged another $4.5 billion in private financing from other institutions.

Most important, Byers tells the prospective investors that the KP funds have typically returned more than 30 percent every year since 1989, at

---

or near the top of all venture funds. (KP's most recent funds, bet heavily on Internet companies, have produced returns of more than 70 percent in some years.)* After his presentation to investors, Byers hands out a "Term Sheet," which summarizes the 70/30 split, the 2 percent management fee, and other housekeeping details. Byers and the other partners are preaching to the converted. Nobody says no; most have said yes long before KP arrived at their doorstep. A few are even afraid to ask questions, lest they offend the fountainhead of so much profit. "My view is that they can take whatever cut they want and just let me have my little piece," says one awed institutional investor whose "little piece" has amounted to millions over the years. "KP knows it doesn't even have to show up," says another endowment manager who doesn't want his

---

*The nomenclature VCs use to describe their investment returns is somewhere between unfathomable and impenetrable. Rather than a simple term like "annual return"—how much your money, on average, appreciates in a given year—the VCs use "internal rate of return." IRR is an economic term of art that seeks to account for the fact that the $1 million you commit to a fund isn't actually all there on Day 1 (instead being "drawn down" by installments). You may have sent in $100,000 at the beginning, another $200,000 in six months, and gotten back $53,000 in dividends at the end of the year. IRR takes into account these different cash flows.

I still have no idea how to compute IRR. After talking to a business professor, an accountant, two VC financial officers, three investors in KP funds, and one of my neighbors, I'm convinced IRR cannot be explained in English, which is my first language. The trouble with IRR—a mathematical equation that a leading corporate finance textbook describes with eight variables—is that it doesn't indicate the bottom line. Invested a million dollars in KPCB VIII? Given an IRR of X percent, how much did you make? The correct answer is, "It depends."

Yet in describing VC returns, virtually every business publication—from *The Wall Street Journal* to *Business Week*—takes the IRR of venture firms and describes it as just "annual return," which is misleading. Because, while an IRR of, say, 80 percent sounds impressive, it probably isn't. If it's based on only a single investment of a small chunk of your million dollars, then you haven't actually made many dollars, the high IRR notwithstanding. Some venture firms will go out of their way to front-load their wins—for example, by selling a stake in a company—so that early on in an investment fund's life they have a high IRR that, in turn, allows the firm to go out and round up new investors. IRR isn't part of some nefarious scheme to inflate VC success rates. Using it is standard practice in the industry. But its use does suggest that VCs can exaggerate their numbers by not forcefully distinguishing to investors and business journalists between IRR and the simpler measure of success—what did I make in total dollars over the life of my investment? I asked one of KP's investors whether he was frustrated by the nomenclature. He said, "I gave up years ago. All I know is I get back a lot more than I put in." Fair enough—after all, that's the point. What I've tried to do here is avoid the definitional morass and just look at overall returns in layman's terms.

name used. "Many other VCs in Silicon Valley don't bother. It's another way Kleiner markets itself so well."

The KP investors also receive a thick "Offering Memorandum" that isn't shy about KP's fiscal ambitions. "There has never been a better time to practice the hands-on, technology-based venture investing that KPCB pioneered over the past two decades," it states. Eager entrepreneurs and glossy magazines may be told by Doerr that the firm extols patience and the long haul, but investors hear that KP requires that ideas "yield products in a venture time frame to earn a venture return." Indeed, "even in ventures based on Nobel Prize–quality science," "new paradigms" must put out "revenues and earnings that build value." The best part of the prospectus is a four-page appendix detailing the returns of KP's investments. It's the equivalent of fiscal pornography: The charts could satisfy the desires of even the greediest investor.

But what happened to all the talk of changing the world? "We lose very few of those [deals] we pursue," KP further tells its investors. Better yet, "We look for new industries where tens or hundreds of billions of dollars of value can be created over a decade." Me, too! I'm just not so good at it, nor can I claim a pedigree that "blends experience and a proven track record with broadened expertise and mentored enthusiasm," "a unique blend of seasoned executive" and "young, rising superstar." (No wonder KP goes to such lengths to keep these materials out of the public eye. They'd be overrun by fans.)

The most remarkable aspect of KPCB VIII's $328 million is that 11 percent of its capital—$37.4 million—comes from the KP partners themselves. That's a far higher percentage than every other major venture fund in Silicon Valley. It signifies just how well KP has done over the last decade and it signals to outside investors that the partners will ante up when the betting with other-people's-money begins. In the first K&P fund back in 1972, Gene Kleiner and Tom Perkins kicked in $100,000 apiece, plus furniture. Split among the seven full partners in KPCB VIII—a few of the other KP partners only have "junior" status—that's an average of $5.34 million. Not much for Doerr or Byers, who could fund far more. (Some years ago, one of the then-younger partners couldn't write a check for his full buy-in and had to take out a loan from the partnership. Even though he made millions, he's complained about it ever since.) Conceivably, the KP partners could one day eliminate their outside investors altogether, since they don't really need their money anymore.

The major outside investors for KPCB VIII—ninety-one of them—include many usual suspects, friends of Doerr and Byers, and some folks and institutions with deep pockets that the world has never heard of. Like members of any exclusive club, they prefer you don't know they're in a KP fund; why's it any of your business where they invest their millions? In that spirit, as well as its own predilection for secrecy, Kleiner Perkins doesn't release the names of the investors or the amount of their investments. But the firm does distribute both lists to all the investors themselves. There's no legal need for KP to do so and every prudential reason for it not to. Why, then, does KP do it?

Seeing and knowing who else made the club only makes belonging that much cooler. So does KP's dire warning to the investors that they should never, ever photocopy the lists. But the pecking order within the tiers of investment provokes intense rivalries among the participants, all blue-chippers. And a few of them go to great lengths to hide their identities; AT&T's pension fund, for example, cloaks itself in an entity called Leeway and Company that, in turn, exists only within the State Street Bank in Boston. The sport involved in tracking down these identities is probably similar to the game that the principals play in setting up the shells in the first place.

The five largest investors in KPCB VIII, which put in $20 million apiece, are Yale University; AT&T's pension fund; the University of California; the Horsley Bridge Fund (a pool of pension money from Xerox, Kodak, Exxon, John Deere, and three universities); and Harvard ($19 million from the university endowment and $1 million from the faculty retirement account).

Next, at $15 million each, are Stanford University; the Ford Foundation; and the Common Fund (a pool of endowments from dozens of schools, at both the secondary and college levels, that includes Cornell, Dartmouth, Emory, Juilliard, Exeter, Hotchkiss, and Sidwell Friends).

At $10 million are Duke University, MIT, General Motors, the Andrew Mellon Foundation, and the Hewlett-Packard pension account.

At the $5 million tier: the University of Michigan; Vanderbilt University; the foundation of Bill Hewlett, cofounder of HP; and Michael

Dell, founder of Dell Computer. Dell is the individual holding the largest piece of KPCB VIII and had asked to put in $20 million. Apparently, he didn't want the other investors to know he was in with them: His identity in some of KP's records is masked through the use of a Texas holding company called Kiralexa. Now they know about Dell's $5 million. Jim Clark, cofounder of Netscape, wanted in for that kind of money, but was turned down—so he didn't bother to invest at all. By contrast, Steve Jobs doesn't want anything to do with venture capital—KP's or anyone else's.

Georgia Tech (home to Brook Byers) got to put $4 million into KPCB VIII. Rice University (John Doerr's alma mater) got to put in $2 million, as did Notre Dame and Rockefeller University. The pension fund at Monsanto, where Doerr once worked, added $1.7 million (its identity alternately cloaked in something called How and Company and "Chancellor F/B/O #24," which stands for a management organization). The absence of many big companies is consistent with KP's ability to be picky, and to choose those institutions it thinks can be helpful in entrepreneurial gossip, deal referrals, and, not unimportantly, the ability down the road to finance a developing start-up directly (in a way that a KP fund's limited resources often cannot).

All of those buy-ins—$2 million to $20 million—constitute roughly 55 percent of KPCB VIII. Doerr, Byers, and the KP partners put in another 11 percent. The remaining third comes from RRPs—Really Rich People. They tend to be individuals who've been part of a successful KP deal or somehow managed to become tekkie personages without KP's help. Sorting out the roster in these lower ranks of KPCB VIII is more interesting than at the institutional level. Who's in and who's out gives a snapshot of KP's view of the all-stars of Silicon Valley (and who has a million bucks in spare change)—a pecking order for a place that never stops keeping score.

Investing $2 million are Steve Case, CEO of America Online; and Bill Joy, cofounder of Sun Microsystems and Doerr's Aspen buddy. Among the people buying in for $1 million:

Marc Andreessen, wonder boy of Netscape

Andy Bechtolsheim, cofounder of Sun

Jim Barksdale, former CEO of Netscape

Howard Birndorf, cofounder of Hybritech and CEO of Nanogen, which mixes semiconductor technology and DNA diagnostic techniques

Frank Caufield, one of the name partners of KP

John Chambers, CEO of Cisco Systems

Andy Grove, Intel's chairman of the board (and who pooh-poohed venture capital when Doerr quit Intel once upon a time)

Eric Hahn, former chief technology officer of Netscape

Tom Jermoluk, CEO of @Home Network

Jeremy Jaech, founder Visio Corp., which makes graphical tools

Mitch Kapor, founder of Lotus Development and backer of GO (and now a competing venture capitalist)

Tim Mott, cofounder of Electronic Arts, the computer-games company

Gordon Moore, cofounder of Intel and Fairchild Semiconductor

Tom Perkins, cofounder of KP

Bob Pittman, president of AOL and former president of MTV Networks

Michael Schulhof, former president of Sony USA

Wesley Sterman, cofounder of Heartport, which makes cardiovascular surgical devices

Bob Swanson, cofounder of Genentech

Then come the stragglers. The $500,000 investors include: Eckhard Pfeiffer, CEO of Compaq; Scott McNealy, CEO of Sun; Jerry Kaplan, cofounder of Onsale and GO; Mike Homer, executive vice-president of Netscape; Les Vadasz, senior vice-president of Intel; Morton Meyerson, the right-hand man to Ross Perot; and John Stevens, cofounder of Heartport and a professor at the Stanford Medical School. Below that are Bill Campbell, former chairman of Intuit and former CEO of GO, at $400,000; Jerry Yang, cofounder of Yahoo, at $250,000; Ted Leonsis, a senior executive at America Online, at $250,000; Peter Currie, chief financial officer at Netscape, $250,000; Scott Cook, cofounder of Intuit at $200,000; Kirk Raab, former Genentech CEO, at $200,000; and T. J. Rodgers, founder of Cypress Semiconductor, at $200,000. The inclusion of Jerry Yang is notable, because KP didn't back Yahoo and did back Excite, a rival.

Even though KPCB VIII stands at a bountiful $328 million, the KP partners have managed to preserve their 30 percent cut of the profits, even though other leading funds still receive only 20 or 25 percent. Given that 30 percent of the profits on a $328 million fund are rather more generous than a corresponding percentage on an $8 million fund, or even a $150 million one, and given that the size of the partnership has stayed roughly constant for years, it means the KP partners are making a fortune as mere "service providers," in Byers's modest words.

For Doerr and Byers, the returns are especially good. They like to emphasize that the KP partnership is a democracy where every voice is heard and no one dictates the deals. It may be so, but some partners are more equal than others. Doerr and Byers get a larger helping of pie than the others—Doerr because he's entitled to it and Byers because he's got seniority. In a single good year, between his KP partnership and unrelated investments, Doerr has made close to $100 million, according to sources familiar with his finances. He needs something to pay for his two small jets (one shared with Netscape's Jim Barksdale, the other with Netscape's Marc Andreessen) and the chauffeur service that drives his pets to Aspen in the summer. (The planes aren't *that* big.) Byers only has to worry about that magnificent hillside château in Woodside—the kind of place Gene Kleiner derisively calls "a KP house," where wealth can alter values. Even after he got get rich, Kleiner maintained a frugality that amused his partners. When Kleiner saw that Crest was marked down at the drugstore, he bought an entire case. "Consider yourself a commodity speculator," Frank Caufield ribbed him. "You're 'long,' Crest."

Caufield, who retired from KPCB in 1989 but, along with Perkins, keeps an office in its small San Francisco branch, knows how much the VC business has changed. "VCs way back made a good living—hundreds of thousands of dollars," he says. "But not like today. There was less money around altogether. People invested less, we made less. What the VCs can make now on one deal is what a VC used to make in an entire career. The velocity of money has changed." With his $1 million in KPCB VIII, as well as earlier funds, Caufield may take home more now than he did when he was busting his hump twenty years ago. Perkins once told an eager lad, "If you really want to get rich, you should be an entrepreneur rather than a venture capitalist." The thirty-two-year-old wannabe was Mike Moritz, a former reporter for *Time* who had started a financial-publishing company and was now looking to become a VC. He ignored Perkins's advice and went on to become a partner at Sequoia Capital, where he discovered the two graduate students who founded Yahoo.

In addition to its mega-funds, KPCB in the 1980s launched a series of smaller funds designed not so much for the accumulation of working capital as to assemble a well of consulting talent. Whoever controls information is most likely to hear of the Next New Thing that will yield the Next Biggest Deal. Getting invited into these "Zaibatsu" funds is like being made an offer you can't refuse. Says Perkins: "We think and I think they think we're doing them a favor by bringing them in. It's not about their money." The Zaibatsus are just another way KP reinforces its role as the Valley's chief stable of thoroughbreds. Just as the old Hollywood moguls achieved power by locking up actors and directors, KP, too, tries to hoard stars—notwithstanding the putative independence of Valleyites. Michael Ovitz indeed. KP may be the spider at the center of the Valley's web, but it still relies on others to bring in food.

The buy-ins to the Zaibatsu funds are almost always a mere $100,000 or $200,000—pocket change in today's Valley. "Zaibatsu" is Japanese for "money clique" and refers to the old family-controlled banking and industrial conglomerates like Mitsubishi and Sumitomo; Mitsubishi, for example, had its own bank, its own heavy-industry division, and its own carmaker—the ultimate in vertical integration. The Japanese government traditionally gave the handful of zaibatsus a privileged position in the country's economic development through subsidies and tax breaks. At Kleiner Perkins, that's a bit base, too suggestive of a cartel; they prefer to call the Zaibatsus a "close Family of Friends." Zaibatsu funds confer status on their members. In return, the investors only have to give the KP partners half their usual 30 percent share in the profits, though they still pay the full 2 percent management fee—business is business, even among friends. The Zaibatsu funds always mirror the investments of the big funds, though for a lot less money.

The first fund, Zaibatsu I in 1987, had $6 million in it. In 1995, Zaibatsu Life Sciences had $5 million and Zaibatsu Information Sciences had $12 million. ("Biotech" and "high-tech" weren't sufficiently classy titles.) The Zaibatsu rosters, like the higher-rollers in KPCB VIII, reflect the several dozen Valley players who matter to KP most; in 1995, for example, that meant a lot of Netscapers, given the company's glorious debut. But given Netscape's more recent problems, it's a good bet that a lot of them will be gone in the next Zaibatsu. Many of the names overlap with KPCB VIII—people like Andy Grove, Scott McNealy, Marc Andreessen, Frank Caufield, Scott Cook, T. J. Rodgers, Jerry Kaplan, Bill Campbell, Tom Jermoluk, Bill Joy, Mike Homer, Ted Leonsis, and Peter Currie.

But some of the names are not part of KPCB VIII, suggesting either they were not offered buy-ins for the big fund or weren't willing to put up the money. They include: John Malone, chairman of TCI, the telecommunications giant; Eric Schmidt, chairman of Novell; Gordon Eubanks, CEO of Symantec and former lieutenant to Gary Kildall; Bill Harris, CEO of Intuit; David Beirne, headhunter extraordinaire and now a competing venture capitalist; Trip Hawkins, chairman of 3DO, the online games company; David Dorman, former CEO of Pointcast, the online news service; Larry Sonsini, the Valley's super-lawyer; John Sculley, former CEO of Apple and Steve Jobs's nemesis; Randy Komisar, a consultant to WebTV and other start-ups; Todd Rulon-Miller, former head of sales at Netscape; Eric Benhamou, chairman of 3Com; Richard Schell and James Sha, Netscape vice-presidents; Bruce Ravenel, a Doerr friend from Intel who went on to run TCI's Internet business; Eric C. W. Dunn, Intuit's chief technology officer; David Cole, head of Internet efforts at AOL; and Naomi Seligman, "queen" of the Research Board, a small, super-secret fraternity of technology executives from Fortune 500 companies.

There is one particularly notable omission from the recent Zaibatsus. The initial fund, raised back in 1987, includes a fellow named Bill Gates, kicking in his $100,000. But he's nowhere to be found in the newer Zaibatsus, nor is he part of KPCB VIII. Remember KPCB VIII includes the likes of Michael Dell, another force in the computer industry but one who has relatively little to do with KP. Could Gates have been excluded for symbolic reasons and the fact that so many of KP's start-ups loathe Microsoft (at least until they want to sell out to it)? KP claims it has offered Chairman Bill a spot in some of its recent funds. Maybe he's just saving up for something else.

Two curious investors in the Zaibatsus are pundit-publishers-"industry observers" of the high-tech landscape. Esther Dyson, who is in Zaibatsu Information Sciences for $200,000, is her own cottage industry—publishing a respected computer newsletter; running PC Forum, the preeminent industry conference; advising government officials; getting mentioned in *Vanity Fair*'s racing form for the "New Establishment." Dick Shaffer, in the same Zaibatsu for $50,000, is head of Technologic Partners (cofounded with Mike Moritz), which publishes several influential high-tech newsletters, including one for the VC industry that frequently discusses KP. He's also a columnist for *Fortune*. Every year *Upside* magazine includes Dyson and Shaffer on its "Elite 100" list, which begs the question: Does KP invite them to invest because they made the list or did they make the list because of KP?

Dyson and Shaffer are universally well regarded. No one questions their integrity. But their inclusion in KP's "Family of Friends" raises the question of how they can appear to be impartial when commenting on the venture business, or about any companies within KP's portfolios or competing against them. If Dyson or Shaffer own, say, Microsoft or Intel stock, that, too, might be problematic. But at least those behemoths' fortunes are unlikely to be affected by isolated observations; that's not the case with a start-up company or industry, whose very success may be affected precisely by the sort of buzz that a Dyson or Shaffer can create. In his *Fortune* pieces, Shaffer discloses any financial holdings related to his topics. But his newsletter includes only boilerplate language in the masthead stating that any staff could own stock in covered companies. Dyson says that many of her comments only involve private companies, so there's no public market for her to possibly affect. Whatever one's conclusion on the potential for conflict of interest, the Zaibatsu status of Dyson and Shaffer surely underscores KP's tactical judgment, as well as the incestuous nature of the Valley. Too bad KP couldn't entice anybody from *The Wall Street Journal*.

KP's references to its "family" make sense. It has created a latticework of corporate alliances, interlocking directorships, and personal links that looks like an Arkansas family tree. Kleiner Perkins even has a name for it. In addition to the exotic-sounding KP Zaibatsus, there's a KP "Keiretsu." Kleiner Perkins, as do other Silicon Valley organizations, seems to have a fascination with things Japanese, notwithstanding that Japan never stops trying to outdo the Valley at its own high-tech game. Originating in postwar Japan, a "keiretsu" is a regulated, strategic network of companies—typically suppliers and manufacturers—linked by mutual obligations like endorsement or licensing arrangements. Keiretsu, in short, is synergy in warp drive; far more than Wall Street's notion of collegiality, the Keiretsu implies some formal relationship, ranging from shared information to overlapping ownership. In Japan, the center of the network is a bank; for Kleiner Perkins, it's Kleiner Perkins. A cynic would call the Keiretsu a syndicate that squeezes out competition and binds its members to the leader, whose gravitational pull is irresistible.

KP's Keiretsu, by its own count, consists of more than 175 companies and thousands of executives, organized around "bits and bytes" and "bugs and drugs." KP's most recent investment pool—the $70 million Java Fund, raised in 1996—is nothing more than a consortium of KP partners. Its purpose is to promote Java, Sun's programming language for the Internet that doesn't need Microsoft's operating system to run. The theory of the Keiretsu is that many separate reeds are strongest when sewn together in one bundle; you meet people, swap gossip, license products—all in the name of shared self-interest. The metaphorical alternative is that all you get from putting three dogs together is a litter. Sharing strengths invariably means propping up weaknesses. It also can become disingenuous when KP has funded start-ups that wind up competing with each other. Doerr liked to say that if there was "no conflict," he had . . . "no interest."

KP believes in reeds and makes no secret of the basket it has spun. It says there is no such thing as a truly independent high-tech company anymore. As Doerr gushes in his Slide Show, the Keiretsu is one of the things that makes his firm work so well (and, as he'll at least admit privately, establishes Doerr as leader of the Valley's anti-Microsoft cabal). If Kleiner Perkins is the franchise, its Keiretsu is the marketing mantra. Look at Netscape, the breakthrough Internet software company that remains Doerr's best-known hit. Founded with KP money in only 1994, it has had partnerships with at least a dozen other KP companies, including Intuit, Sun, America Online, @Home, and Excite (one of Yahoo's main competitors that, in 1999, agreed to merge with @Home). Doerr is on Netscape's board of directors, as he is on the boards of Intuit, Sun, and Amazon.com; other KP partners have similarly interlocking board memberships. Netscape's first full-time CEO, Jim Barksdale, serves on the board of @Home; Netscape's cofounder, Jim Clark, is cofounder of Healtheon, another KP start-up. Frank Caufield is on the AOL board. And so on. The point is that after Netscape's launch, KP set out to nurture or create other Internet companies. From an infrastructure start-up like @Home (which uses cable-TV wires to provide fast access to the Internet) to online businesses like auctioneer Onsale and book-seller Amazon.com, Kleiner Perkins tried to establish new businesses to profit from and assist existing ones. Onsale and Amazon.com (each of which yielded delicious ten-to-one returns for KP investors) are card-carrying members of the Keiretsu; Jerry Kaplan is part of the 1995 Zai-batsu, and you can be sure Amazon's Jeff Bezos will be a big-time part of the next one.

Fortune magazine in 1998 set out to diagram the KP Keiretsu. The details of the two charts—one for the KP corporate partnerships, the other for directorships—don't matter so much as the total dizzying effect. KP is an octopus.

Sometime in 1999, unless the Internet market dries up completely, Kleiner Perkins will begin to raise KPCB IX—"IX in 1999," they've started to call it. The $400 million to $500 million it will likely raise isn't a lot of money by most standards, though it far eclipses the meager $8 million that Gene Kleiner and Tom Perkins began with. The firm's success and the competition it's created has made deal making far more expensive. (Former entrepreneurs who've gotten rich and don't want to try again themselves now just set up as VCs and compete with the folks who once backed them—"angel investors," they're called, as in Lucifer.) You won't see KPCB IX advertised in the paper and its representatives won't be calling you at dinner hour looking for a contribution. But KPCB IX will be the transcendent private in-vestment fund in America.

Just how far has KP come since its founding in 1972? Its wealth is extraordinary, but its reach shouldn't be measured in dollars alone. While Doerr is among the richer people in Silicon Valley—most tycoons limit themselves to one plane—he's peerless in terms of influence. Steve Jobs is a scourge, Larry Ellison's a clown, Andy Grove's a respected manager but still of just one company. There are other premier venture firms like Sequoia and Mayfield and Accel Partners. But these firms aren't KP and no one there is like Doerr. Doerr is everywhere; and by extension, so is KP. Quite apart from his access to unlimited institutional capital and a range of friendships in Washington that lead to "Gore & Doerr in 2004?" jokes, Doerr represents the next tsunami. When Doerr speaks at Stanford Business School or the Palo Alto library or on an industry panel, it's always standing room only, as if the high-tech future emanates from his ebullient wisdom. In the Valley, what comes tomorrow counts far more than yesterday's success.

But in the end, the KP magic may best be measured not in the forty-eight-year-old Doerr, its grinder, but in Tom Perkins, its original spirit. Approaching seventy, he is still a grand figure in the life of the Valley. He keeps a KP presence, invests in his own deals and as a limited partner in KPCB funds, owns a moated manor house in East Sussex, and is a part of the San Francisco social scene. He's restoring a steel vessel used in the rescue of Dunkirk, he's been knighted by the king of Norway for his philanthropy there, and he races a 1915 Herreshoff schooner—the 110-foot aristocrats of the ocean—in Mediterranean regattas. In the rest of his spare time, he sails the globe on his 154-foot yacht, *Andromeda La Dea,* visiting such places as Antarctica. Andromeda was the wife of Perseus, who rescued her from a sea monster; for Perkins, she's the most beautiful creature in the water, billowing 10,764 square feet of sail. I serendipitously came upon her in Newport Harbor the day I went out to see *Sakura,* Larry Ellison's floating palace that's thirty-eight feet longer but has none of the majesty of Perkins's boat. (In summertime, Newport is the Hollywood of the maritime set, its harbor full of stars.) Perkins wasn't aboard, but the crew was getting the boat ready for an upcoming cruise; next to *Sakura* and the Newport Suspension Bridge, she was the largest thing around.

Deep marine blue, flying the naval ensign of Bermuda, where she's registered, *Andromeda* would cost $30 million or more to build today. (The 38½-inch model Perkins commissioned for his study alone took 4,200 hours to build and cost $126,000.) Designed by Perkins himself, as T. J. Rodgers learned at a KP partners meeting, *Andromeda* is his second super-yacht. The first *Andromeda,* twelve feet shorter and also built in Italy by Fabio Perini, didn't live up to his plans. An even earlier sailboat is now owned by none other than Doerr, who perhaps is seeking to emulate his mentor. *Andromeda* figures in one of Perkins's many private acts of grace, which include tracking down his favorite author—Patrick O'Brian, who writes historical novels about the Napoleonic Wars—and offering him free use of the big windjammer. (When Perkins contacted him, O'Brian wrote back from France: "Perhaps I accept your kind invitation with obscene haste.")

In 1994, Perkins's wife of thirty-five years, Gerd Thune-Ellefsen, died of cancer. Last year, he got remarried—to Danielle Steel, the author who's sold more books (400 million) than anyone in history besides the Lord. The very same woman who, as the merciless tabloids keep reminding us, married investment banker Claude-Eric Lazard, son of the man who made Lazard Frères; then Danny Zugelder, a convicted bank robber (she met him while visiting another inmate and they married inside the prison); then a heroin addict and burglar; then a shipping magnate. Now, Perkins is a classy No. 5 and the dedicatee in one of her latest novels; now she can resuscitate her image as an elegant Bay Area socialite and devoted mother of nine. "At least she's not after my money," Perkins cracks. She says the same thing; after all, he's probably worth more than her.

Fifty years ago, or even twenty-five, marrying a venture capitalist was hardly considered a catch. In the Steel-Perkins union, which was covered with great relish in the celebrity press, he got just as much play as she did. John Doerr, eat your heart out.

Chapter VIII
Mozilla

Planes, trains, automobiles, and boats—what does a billionaire with everything do to break the boredom? Learn to pilot a helicopter. That's what Jim Clark did.

Anybody can fly a Gulfstream or sail across the South Pacific. Everybody has an NSX—Larry Ellison bought four. But a helicopter takes singular daring, or stupidity, depending on your attitude. Yes, it can fly at tree-line—"nap the earth"—and land in the middle of a mountain stream. And it certainly has a steeper, well, learning curve. Planes have wings—they like to fly. A helicopter, by contrast, is little more than rock attached to a whirling rotor—"a thousand parts, all conspiring to kill you," the saying goes. It has no interest in staying up and none of the forgiveness of other machines. It is a deadly device that only a cowboy would want to fly. In short, it is the ideal toy for Clark, the serial entrepreneur of Silicon Valley, whose motto might as well be "If at first you succeed—try, try again." Clark is John Doerr without the airs. Silicon Graphics Inc. (SGI), Netscape, Healtheon, and a few other companies you haven't yet heard of—these are Jim Clark. If there's high-tech money to be made in the Valley, chances are the fifty-four-year-old Clark is thinking about it. He hasn't crashed-and-burned yet.

On a hot Saturday morning in September, near the San Jose Airport, Clark is ready for his latest lesson on flying his six-seater Boeing MD600, the $1-million minivan of helicopters. Tall, boyish-blond, and blue-eyed, Clark's a mismatch of *GQ* chic and blue-collar kitsch: Mephisto tennis shoes, leather gloves, green socks, black T-shirt, beige Ralph Lauren shorts, Il Moro di Venezia belt, aviator sunglasses, and a Rolex watch. "My wife says I look like a dork," Clark says, "while Larry Ellison is on *Playboy*'s best-dressed list." Funny Clark should mention the comparison. Before Clark's remarriage, the similarly aged Ellison apparently was on his mind. One night, hanging out at Gordon Biersch Brewing in Palo Alto, Clark was told by an acquaintance that Ellison's then-girlfriend, Kathleen O'Rourke, was standing at the other end of the bar—with no Larry. Until that moment, Clark hadn't paid any notice. Minutes later, he was chatting her up. "It was typical of how competitive these guys are," says the amused friend.

Even as Clark goes through a preflight inspection for his latest toy, he bounces from one topic to another, as if no single one is enough to hold his attention. Why did two of his wealthy friends buy suits wholesale? Can you believe that Eddie DeBartolo, owner of the San Francisco 49ers, associated with gamblers in New Orleans? Of course Bill Clinton fondled "that other woman," Kathleen Willey. Clark's teacher, John Quayle, a forty-seven-year-old San Jose police officer, tries to get him back in gear. "You haven't screwed up yet," Quayle tells Clark, "but you're getting a little cocky."

Takeoff is an adventure. As Clark banks right, he hits a tornadic blast of air from a Cessna that sends the helicopter into a dive. Quayle grabs his set of controls and rights the ship. *Whee!* Clark brings the helicopter into the clear blue skies over the Valley and both men resume their banter. The surprising thing is that they can talk to each other at all. Clark is an on-again, off-again billionaire, who owns a $300,000 stunt plane, a chartered Gulfstream, the most advanced cyber-yacht ever built, and big houses in Atherton and Palm Beach. (He claims residence in Florida, where there's no state income tax, though he cringes that the primary worry of many of his neighbors is taking their pills on time.) "I grew up the prototypical poor boy," Clark told me once, making no apologies for his famous tastes. "When you make it, you start to think there isn't anything you want that you can't buy."

Clark can't get enough of the rush. "There's no magic to all this," he says later, munching a burrito at Beeb's Sports Bar and Grill, near Livermore, California. "It's a skill like riding a bike, only the consequences can be more severe when you fall off." He's no good at tennis, his weak ankles don't allow skiing, and he's hated golf ever since a hardscrabble childhood in the Texas Panhandle when caddying was humiliating. So what's left to do except . . . fly a helicopter that'll do 155 knots? "Just think about hovering. Or moving around in three dimensions with precise geometric shapes. Or 360-degree turns." Or, better, "falling three thousand feet a minute with the power off." At least Clark doesn't talk about flying the helicopter upside-down.

Over the crevices of the Hayward Fault, past the house of rap star M. C. Hammer in the East Bay, safely above the electrical wires painted bright orange after another copter flew into them—Clark is exultant as he bobs his helicopter through the air like a dolphin through water. Impulsive, playful, childlike, Clark seems a little crazy. No wonder this is the wild man who brought Silicon Valley—and the world—into the Age of the Internet. What the telephone and television did for the twentieth century, the Net promises to do for the twenty-first. How un-PC.

In the time line of modern technology, the Internet was nothing new, dating to the late 1960s. But Jim Clark and others were the ones who figured out how to commercially exploit this global computer network—what once was a digital backwater of academe and the military. In early 1966, no organization relied on computers more than the federal government and no U.S. agency was more interested in them than the Pentagon. Thirty-four-year-old Bob Taylor was in charge of R&D for the Advanced Research Projects Agency. ARPA was a Defense Department response to the Russian launch of *Sputnik* in October 1957. The agency's mission was to meld science and technology with the American military. Taylor was no hacker, no early-day Wozniak motivated by whimsy. He was a Pentagon bureaucrat, trained as a psychologist. His little electronic empire on the Pentagon's third floor was an engineer's dream—computer terminals linked to major universities and technical centers. There was one hookup to MIT, another to Berkeley, and one to an IBM machine in Southern California, identified as "AN/FSQ 32XD1A," that kept track of information for the Strategic Air Command. Conventional wisdom later had it that Taylor's ARPA network was conceived by the military as a post-apocalyptic communications grid that could survive nuclear attack. Even if the Russians took out some intelligence sites, enough machines would remain. But that purported motivation for the network was more of an after-the-fact realization by the Pentagon. As Katie Hafner and Matthew Lyon showed in their history of the Internet, *Where Wizards Stay Up Late*, the point of the network was more mundane. It was meant to allow leading researchers, in both the private and public sectors, to share data and ideas.

But Taylor's terminals at the Pentagon were frustrating to use. Each had its own operating system and programming language, and each had its own way to log on. Hafner and Lyon compared Taylor's dilemma to having "a den cluttered with several television sets, each dedicated to a different channel." Taylor persuaded his agency to develop something to get the different computers talking to each other. In a twenty-minute pitch to his boss, he got a million dollars added to his budget. By luck or design, it would be one of the best projects the feds ever funded, a benchmark of what a wise investment in science could yield. The government hadn't had such a bargain since it bought Alaska from the Russians.

By 1969—with researchers around the world experimenting with network hardware and switches that used leased phone lines—Taylor's idea of an ARPA network was up and running, connecting computers at UCLA, the University of California at Santa Barbara, the University of Utah, and the Stanford Research Institute. This was the ARPANET, the foundation for the Internet and, a quarter-century later, the emergence of mainstream digital culture. The first message carried online in October 1969 was nothing as momentous as "Mr. Watson, come here, I want you." Rather, a UCLA undergraduate named Charley Kline—a friend of Stuart Feigin, later of Oracle fame—simply transmitted: "L-O-G." Geeks are not poets. The computer was supposed to be clever enough to recognize the rest—"-I-N"—but there was a bug and the system suffered its first crash. By afternoon, all was well, as UCLA and the Stanford Research Institute connected.

The secret of what became a global interactive network was a counterintuitive concept called "packet switching." Instead of sending data in one digital lump—from the sender directly to the recipient, as the telephone does—the network broke information down into uniform chunks that traveled individually along different routes of the network and converged at the designated destination. It was a new, more efficient delivery system. Imagine putting in one room all the people who want to communicate with each other. This works for a small group of people, but the room eventually gets too noisy. If you make the room too big, some people will be too far apart to hear each other. And, of course, some people won't like having everything they say heard by everybody else. The "routers" used for "packet switching" and to connect digital networks solved these three problems of capacity, distance, and security.

By the early 1970s, the ARPANET was wired to twenty-three sites, all confined at some level to the esoterica of government-funded computer research. MIT, Harvard, NASA, and the Rand Corporation were added, as was Bob Taylor's old computer room at the Pentagon. Then the first international sites were hooked up, at University College, London, and Norway's Royal Radar Establishment. By 1976, there were 100 ARPANET sites; by 1984, more than 1,000; by 1989, 100,000. The network was not a collection of spokes connected to some central hub, but a series of interwoven connections.

Taylor moved on to the University of Utah, where Nolan Bushnell had recently finished his engineering studies and was on his way to founding Atari. When Larry Roberts succeeded Taylor, he and a group of other scientists around the country—including Vint Cerf, Bob Kahn, Len Kleinrock, Ray Tomlinson, Doug Engelbart, Jon Postel, and Bob Metcalfe—guided the network's expansion.

The network faced two problems—one sociological, one technical. Some scientists resented the leveling of the playing field. Access to computers represented power, and allowing déclassé institutions to log on to the network was like letting Little Leaguers into Wrigley Field. The more significant issue for the ARPA network was how to connect to the matrix of other networks as they formed around the country and the world. Not everyone was part of ARPA. Devising universal technical standards, or "protocols," for transmitting data *between* networks was the next major development for what became known in the late 1980s as the Internet. After all, "Internet" meant an *inter*-network of communication—a network of networks that would create the ultimate in human consciousness. The linked computers came in different sizes and flavors, but it made no difference. The beauty of the anarchic network was that it could accommodate any machine—just as long as that machine conformed to the protocols, which had no political or normative content. It was kind of like the English language, as science-fiction writer Bruce Sterling suggested in an online Internet history. "Nobody owns English," he wrote. "Everybody just sort of pitches in, and somehow the thing evolves on its own, and somehow turns out workable. . . . 'English' as an institution is public property."

As the ARPANET spread to more and more sites, it finally found a compelling use for itself. But it was not the lofty sharing of libraries and databases contemplated a few years earlier, or even the computer hooked up to an Associated Press news wire out at Stanford. The "killer app" for the network was E-mail, not that anybody had made any money from it yet. An assistant U.S. postmaster general in 1976 observed, "We are being bypassed technologically." The same year, in England, Queen Elizabeth II sent out her first E-mail. Even today, more than online stock trading or the opportunity to see photos of naked women, the Internet's most popular function is message-sending. More than two billion E-mails are sent each day in the United States—for some, replacing the telephone as the most common instrument of communication—and that total is expected to quadruple by 2002.

Electronic mail combined the spontaneity of speech with the reflection of writing. Initially called "network mail," E-mail spawned the @ sign and "emoticons" like the happy-face :-), as well as discussion groups and "virtual communities," that are now an essential part of cyberspace. (By one estimate, 90,095 different mailing *lists* exist on the Internet, some of them including tens of thousands of names.) Through the miracle of networking, a popular Bay Area online group, The Well, in 1986 started a discussion group about the Grateful Dead.

Like so much else online, the conventions of E-mail arose on the network not by edict from some centralized governing authority, but by scientists and hackers experimenting to see what others would adopt. The only people in control were those using it. The computer network, Hafner and Lyon wrote, was developing into "a place to share work and build friendships and a more open method of communication. America's romance with the highway system, by analogy, was created not so much by the first person who figured how to grade a road or make blacktop . . .

but by the first person who discovered you could drive a convertible down Route 66 like James Dean and play your radio loud."

When the start-up Sun Microsystems began selling workstations in the 1980s, it included networking software for free, which further expanded the global network (and helped make Sun machines an instant hit). There were now enough people online that, for E-mail in particular, traffic congestion started to be a problem. According to *Wizards*, every site's "host" computer had its own designation, but a lot of folks wanted theirs to have the same name, like Frodo the Hobbit from *Lord of the Rings*. "Sorting out the Frodos of the Internet," Hafner and Lyon said, "wasn't unlike sorting out the Joneses of Cleveland or the Smiths of Smithville." The solution came in 1983 in the Domain Name System, which introduced an address system for the Internet. There would be seven domains—com (for companies), edu (schools), gov (government agencies), mil (military), org (nonprofit organizations), net (network service providers), and int (international treaty groups). Countries, too, would one day have their own domains, including such hot spots as Guernsey (gg), East Timor (tp), Christmas Island (cx), and the Vatican (va). Every address at each domain had a long sequence identifying numbers associated with it; the domain system provided a shorthand for these impossible-to-remember strings.

The ARPANET eventually split into one network dedicated to military information and another for civilian computer research. ARPA's coordinating role was handed off to the National Science Foundation. (In 1971, AT&T turned down the chance to take over the ARPANET, a decision that ranks in corporate wisdom somewhere alongside IBM's decision to let Microsoft freely license MS-DOS.) By the end of 1989, the ARPANET—the original interlocking network—had been unplugged and shut down. ARPANET's component sites became part of other networks, which together would now be known just as the Internet.

The year 1989 was significant not just because of the demise of the ARPANET. In Geneva, Switzerland, at the European Laboratory for Particle Physics, Tim Berners-Lee invented the specifications for a new "web" *within* the networks. He christened it the World Wide Web. Technically, what Berners-Lee did was devise a new set of computer commands and syntax for computers to talk to each other and share data. The new language was called HTML, for HyperText Markup Language, and it would be managed by new HTTP specifications, called HyperText Transfer Protocol. Berners-Lee also developed URLs (Uniform Resource Locators), which served as a universal online address system. When you're on the Internet and type in the prefix—"http://www."—you're using the brainchild of Berners-Lee, connecting to some address on the World Wide Web. After that prefix, comes the company or university you're going to—for example, Stanford—followed by the domain name. Thus: http://www.stanford.edu. Never before had *w*'s traveled in threes.

"Hypertext" was the key to the Web. Just as a domain name was a substitute for a long set of numbers that nobody wanted to type every time, hypertext, through the ingenuity of HTML, was also a shortcut. The notion of electronic hypertext was rooted in old, garden-variety printed footnotes, but it was first imagined in 1945 by Vannevar Bush, FDR's chief science adviser during World War II and the former dean of engineering at MIT. In an *Atlantic Monthly* essay, "As We May Think," which achieved mythic status, Bush described a theoretical machine that could help people organize information. Just a few decades early, he was talking about hypertext and the Internet. "The human mind," he wrote, "operates by association. With one item in its grasp, it snaps instantly to the next that is suggested by the association of thoughts, in accordance with some intricate web of trails carried by the cells of the brain." In other words, if you multiply one brain by the millions of people linked by networks, you have a worldwide web—fusing the processing power of a machine with the intuition of human beings.

Typically underlined or colored, Berners-Lee's hypertext was a highlighted word or phrase in a document that served as a link to another document or page on another computer in a completely different location authored by someone else. To go there, all you had to do was "click" on the word or phrase. You could hop, skip, and jump in endless worldwide romps through related and cross-referenced nuggets of information—say, from a Library of Congress collection on atomic history to an Oxford University wartime archive to a posting at the Los Alamos lab in New Mexico. Berners-Lee was a thirty-four-year-old British computer scientist who simply was looking for a better way to conduct his research on the chaotic Net. He hardly planned to remake it. But the Web became a new, burgeoning lane on the Internet, a trove of material that any reasonably sophisticated computer user could access. Berners-Lee openly distributed his HTTP specifications and never considered turning his idea into a business.

The World Wide Web became a new part of the Internet in the early 1990s. This was the "Cyberspace" envisioned by William Gibson in *Neuromancer,* his groundbreaking science-fiction novel of 1984. Cyberspace was a "consensual hallucination experienced daily by billions of legitimate operators . . . a graphical representation of data abstracted from the banks of every computer in the human system." The Web was as important an addition to the network as any previous online creation, including E-mail.

During the same period, the National Science Foundation removed the few barriers that remained against commercial use of the Internet. But the various versions of primitive "browsing" software that had been devised to navigate around the Web weren't intended for a mass audience. For one thing, they didn't run on personal computers, and for another, they still required knowledge of extra commands and protocols, as well as the installation of additional software. If you wanted to download a photo or other image, you had to use another program. Most browsers were text-based and couldn't support multimedia—no video, no graphics, no sound, none of the other rich stimuli that Vannevar Bush's dream contemplated. Tim Berners-Lee had no interest in these media (and was also concerned, mistakenly, that adding them to the network would overload it). The browsers were fussy, too—difficult to load on a computer, prone to crashes, and not especially smart. If, say, you called for a page from a site that wasn't processing requests at that moment, some browsers just froze there; to a machine, an eternity was no different than a millisecond. For all its gee-whizzity, the World Wide Web remained a hostile place for the uninitiated.

Even so, the universe of networked computers had grown phenomenally from the day when Bob Taylor decided to outwit his Pentagon computers. Because it was decentralized and thereby capable of limitless multiplication, and because it cost so little, the network held the potential to democratize communication and launch an information revolution—a latter-day version of Gutenberg's printing press. Compared with the accession of the network, the rise of the personal computer in the 1980s might one day be a mere historical footnote. But for the moment, the Internet was still the domain of the gods.

The Big Bang of the Internet started not with a cosmic plan, but with a pastry.

Far away from Silicon Valley, in the late fall of 1992, Marc Andreessen was attending the University of Illinois and working part-time at the school's National Center for Supercomputing Applications. For $6.85 an hour, he wrote code for the NCSA's high-end, government-funded computers that ran on Unix, an operating system developed in 1969 at Bell Labs. Andreessen could see that the Internet was accessible only to the digital cognoscenti, and he envisioned a more populist space, where any cyber-schlub could "surf" the Net and, in particular, the Web.

Then twenty-one—he wasn't even born when Bob Taylor at the Pentagon dreamed of an ARPA network—Andreessen seemed to have programming in his genes, having learned code in his teens. But his sensibilities came from a larger mold. He read several daily newspapers, scanned scores of magazines, figured out how to watch CNN on his computer screen, and wrote dozens of E-mail every night. He'd even been spotted using the phone once in a while. Andreessen, in short, was the perfect mix of PC and MTV. He could assimilate vast amounts of information from both the technical and pop-culture bandwidths, making him the ideal candidate to foresee the marriage of the Internet and mass society, the union of computing and communications.

At the same time, other smart people saw it coming. On Christmas Eve of 1992, by the fireside in Aspen, venture capitalist John Doerr was riffing with Bill Joy, cofounder of Sun Microsystems and its mad-scientist-not-in-residence. "Someday," prophesied Joy, "you'll be backing an eighteen-year-old who's writing software that will change the world." Joy was off by three years, though who could blame him? Bill Gates had dropped out of Harvard at nineteen to start Microsoft.

One evening that same December, at the Espresso Royal Caffe in Champaign-Urbana, Illinois, Andreessen decided to change the world. He knew the potential of the Web—the graphics and pizzazz beyond E-mail and text-ridden communications between scientists. The Web was still embryonic, with only a few dozen "Web sites," but the idea should have been obvious to others. Most of the academics inhabiting cyber-

space, though, didn't think beyond their specialized needs—they liked
the exclusivity of their electronic fraternity, much as the early users of
MS-DOS bemoaned the point-and-click Macintosh and Windows operating systems that came along. They were content to share information,
no matter how inelegant the access; for them, it was sufficient that the
Web made things a bit easier. At least now they could steer through the
vast, disparate network without a clumsy set of protocols.

Andreessen's notion of simplifying Web travel was one of those epiph-
anies little regarded as such by the discoverer at the time. In that way,
he was like Bob Noyce or Steve Wozniak or even Larry Ellison once
upon a time. Yet ultimately Andreessen's idea would be seen as the birth
moment of an era—the beginning of the Internet boom and a new generation of Silicon Boys. Andreessen's browser, building on the Web that
refined the Internet that grew from ARPANET, was the missing link that
completed the chain that unleashed the network.

The Valley had started as a hardware mecca, inventing and capitalizing on the microprocessor and business computers. Then came the PCs
and the rise of software. Now, in a computer lab 1,800 miles away, began
the third wave of riches—and the richest vein yet. Andreessen would
arrive in the Valley soon enough, the first of its Netrepreneurs. NCSA
management and Andreessen detractors later argued that he was just
another hack-programmer who misappropriated credit for an idea no
single individual had thought up. In their account, forty people came up
with it. A 1997 article in *GQ* magazine, smugly titled "Imposter Boy,"
served up one paragraph after another of accusation, largely citing NCSA
sources. While tales-of-origin invariably have different versions, the
NCSA's history-has-no-heroes rendition smacks of sour grapes: Many of
the other programmers, when they had every reason to say otherwise,
ascribed creation to Andreessen.

At the beginning, Andreessen's notion needed to be executed for Unix,
the operating system for high-end workstations. He turned to his NCSA
soul mate, Eric Bina, who was far more of a programmer. The thirtyish
Bina would be the Woz; Andreessen would be Jobs. While Andreessen
mused about a new order and how to market it, Bina was content to
harness the Internet for basic matters like research. But Bina thought
Andreessen's project sounded like fun—and it certainly beat the university work they were doing. For three months, they were virtually in-

separable. Working eighteen hours a day, they argued not only about code, but presidential politics, music, and junk food. Andreessen lived on jugs of milk and Pepperidge Farm Nantuckets. Bina preferred the less traditional Mountain Dew and Skittles.

It turned out that executing Andreessen's idea wasn't that difficult. Audio clips and pictures, for example, had been around before. What Andreessen did was standardize the format for content so that users could get at it more easily. As with other watershed ideas, its genius was in conception. As a program, Andreessen's idea amounted to just 9,000 lines of code—compared, say, with the approximately 8,000,000 lines for Windows 98. Andreessen and Bina called the program Mosaic—a simple, intuitive graphical "overlay" for the World Wide Web. Unlike earlier browsers, Mosaic was an all-in-one piece of software. Not only did it have point-and-click simplicity—gallivanting through hypertext without having to type strings of numbers—but the ability to display fancy graphics and other multimedia, play sounds, and resist crashes. The look of the interface was slick, with color and other touches such as "back" and "forward" virtual buttons displayed across the top of the screen. Best of all, a person didn't need to know what HTTP or HTML or "Transmission Control Protocol" meant in order to install Mosaic. It started to do for the Internet what Vatican II did for Roman Catholics: It put Unix into the vernacular.

In time, Mosaic would do for the entire network what the Macintosh had done for the PC—render it friendly. "Mosaic" was meant to suggest the full range of machines on which the software would eventually work. "Browsing" entered the lexicon as something other than looking through the classifieds in *Electronic Engineering Times*. All you had to do now was point-and-click on a hypertext link and you were transported, virtually. Mosaic processed the HTML commands invisibly and within seconds. The possibilities were endless, a data maze without a finish.

From the moment Mosaic was available to be downloaded from the Net for free—"freeware," in Net parlance—it became a sensation. It wasn't particularly fast, stable, or secure—but it was *alive*, giving dimension and vitality to a flat digital desert. By the spring of 1993, other NCSA engineers were recruited to write Mosaic for use on Microsoft Windows and Apple Macintosh machines. Jon Mittelhauser, Chris Wilson, and Aleksandar "Mac Daddy" Totic—all in their twenties—had those versions up and running online by Thanksgiving. Word of mouse traveled fast. "The first day we made these versions available," says Mittelhauser, "the server died" (referring to the central Mosaic computer from which Internet users downloaded the browser). The Web was now available to anyone with a PC.

Soon, a million people had Mosaic, allowing them to retrieve not only data from arcane libraries, but, soon, from each other, as they learned how to create information on the Web. One estimate put the annual rate of Web traffic growth at 342,000 percent. The Internet was becoming a huge organism, animated by an ever-growing number of people and machines. A list of places to buy poodles or porn, collections of phone books and street maps, a shrine to Karen Allen or other celebrities, literary musings and family albums, and lists of these kinds of lists. "Web sites" and "home pages" began to multiply. Anybody with rudimentary computer skills could connect, without economic barriers, government censors, or artistic roadblocks. It was a universal, boundless library of information, an egalitarian publishing medium, a world without editing!

Just as E-mail had been, the spectacular growth of the Web was proof of Metcalfe's Law—that each person added to a network boosts the value of the network exponentially. Skeptics of Mosaic questioned why the masses would want access to the Web because its content was so uninteresting. But the browser's user-friendliness *created* content. Just as highway planners couldn't determine the size of a bridge by counting swimmers across a river, you couldn't measure the reach of a network before it came into existence.

At the beginning of 1993, there were about fifty commercial Web sites; by year's end, the total surpassed ten thousand. People might be separated by geography, but the technology of the network united them. And since it could do so at marginally less cost for each person added to the network, the critical mass necessary for the network to take off was small. The phone system had demonstrated this decades earlier. When you bought a phone, you got the benefit of a billion other phones around the world.

As the industry gadfly George Gilder pointed out in *Forbes*, Metcalfe's maxim addressed the "law of the telecosm"—the convergence of telecommunications and the computer—while the equally prescient Moore's Law, governing the power of chips, addressed merely "the microcosm."

The *power* of the Internet—the "magic of interconnections," as Gilder put it—began to feed on itself. Moore's Law facilitated efficiencies of the old—better ways of calculating, organizing, managing. Metcalfe's Law, realized through the Internet that Mosaic popularized, promised things that were fundamentally new—new ways of communicating, educating, and governing, not to mention selling stuff and spawning fantastic new wealth.

And it did these things with a speed previously unknown. While other networks, like the phone system, constantly required updated infrastructure—wires, switches, poles, and the wage earners who install it all—the World Wide Web was largely in place when Mosaic was invented. If you had a computer, all you needed was a modem and some existing access provider like America Online or Prodigy to connect to the network. These providers predated Mosaic, but were initially closed, proprietary services operating outside of the Web; the whole point of paying to be on AOL was to talk to other AOL users. In time, though, they also accepted the open standards of the Net and provided full access to their customers. One day, say the prophets, all things in the physical world—milk cartons, airplanes, lampposts, our very beings—will be linked to the network, transmitting and correlating information. The actual machines—computers or telephones or wristwatches—won't be important anymore. What matters will be the information connected to them.

In December 1993, Andreessen graduated from the University of Illinois. NCSA asked him to stay on—as long as he quit the Mosaic team, which had expanded to several dozen members. The NCSA pinheads felt that "centralized management" was needed to take Mosaic to the next level, especially as Mosaic's notoriety spread. But this corporatespeak was a surefire way to turn off the technoids. "There used to be five of us with Domino's and Cokes at two in the morning," says Mittelhauser. "Now we had big meetings. So we practiced passive resistance—we ignored them."

NCSA seemed intent on marginalizing the engineers behind Mosaic. A long article in *The New York Times* noted the new software and featured a photograph of the NCSA's acclaimed director, Larry Smarr, who talked about Mosaic as "the first window into cyberspace." This only demoralized Andreessen further. NCSA's management stupidity proved to be another legendary blunder of American high technology. Letting Marc Andreessen get away was sort of like Boston shipping Babe Ruth off to the Yankees (all references to heft aside).

At twenty-two, and given his treatment by NCSA, Andreessen felt burned out. Without even bothering to pick up his Illinois diploma, he headed to Silicon Valley, accepting a programming job at a small company in Palo Alto called Enterprise Integration Technologies that did security for E-commerce (and also proved the rule that any company name with so many syllables wasn't going anywhere). While hardly offering Andreessen the challenge he'd had with Mosaic, it gave him a paycheck in Northern California, and put him within a few miles of the Peninsula Creamery and Grill, home of civilization's best malt. "I had some idea that I wanted to be part of a new company," Andreessen says, "but I didn't even know what a VC was."

When he arrived, the Valley was at an economic and emotional low point, very different from the Valley of prior times. The PC Era was now maturing and wasn't the fiscal turbine it once was. Few companies apart from Intel were showing the profit margins that Wall Street wanted. There hadn't been a Big Idea in years, unless you counted Larry Ellison deciding to grow a beard. Business leaders even formed a chamber-of-commerce group to figure out what to do. "Everyone seemed rather morose," Andreessen says, "kind of looking at each other and asking why nothing exciting seemed to be happening in the Valley anymore."

Andreessen was born and raised in a small Wisconsin town—his mother is a customer service representative at Lands' End—but he was no corn-fed hick. He may have looked the part, with his pudgy six-foot-four build and his sweet smile, but Andreessen didn't just happen upon Silicon Valley. He was shrewd and ambitious. He really may not have known what a VC was, but he probably had a good idea. Though his colleagues at EIT remember Andreessen as likable enough, they sensed he was on the prowl; he reminded the CEO of a "Steve Jobs in training."

Just down Highway 101, another burnout was plotting his next move. Jim Clark, already ahead of the game as founder of Silicon Graphics—the trailblazer in 3-D graphics that rendered the digital dinosaurs in *Jurassic Park*—was frustrated and wanted out. He wasn't running the company, there was conflict over his compensation, and he was sick of the Sturm und Drang. Unlike other stay-put entrepreneurs like Larry Ellison or Gordon Moore or even Steve Jobs, he wanted a new rocket to ride. The fun for Clark was in the founding—breaking free of past entanglements and starting something new. Having demonstrated that he could create a company from scratch, all that remained was to do it again—and prove it wasn't chance. The chase was better than the catch (though dividing it up wasn't bad). And besides, the venture vultures who helped him get Silicon Graphics going owned so much of the company that he had only $30 million for his effort. Clark wanted real money.

In late January 1994, Clark asked one of his SGI marketeers, twenty-eight-year-old Bill Foss, who he thought was hot in high-tech. Clark wanted raw talent, someone who, as George Gilder put it, could alter "the axes of technology and economy." Foss knew about Mosaic, and he mentioned only one name.

"How do I reach this Andreessen?" Clark asked. Foss downloaded Mosaic on Clark's workstation and located Andreessen's Web home page, which consisted only of Andreessen's ursine mug and his E-mail address. "You'll find him there."

Clark wasn't known for being contemplative. He'd left a professorship at Stanford not just to find a pot of gold but because academe seemed so irrelevant. "My original desire as a youngster was to have a really big impact on the world, to really make a contribution to knowledge or progress in humanity," he says. "I envisioned the way to do that was as an academic. But when I got into it, I saw it was so mired in politics and backstabbing and grubbing for money. People grabbed your ideas, took credit for things—it was the most political environment I'd ever been in. I found out I enjoyed business more because the metric is so much clearer: You make money or you don't. The VCs like to say that failure is okay. No, it isn't." Clark had never heard of Mosaic—he wasn't an online addict—but he sat down at the keyboard right away. "You may not know me," he typed to Andreessen, "but I'm the founder of Silicon Graphics. I've resigned and intend to form a new company. Would you be interested in getting together to talk?"

"Sure," Andreessen replied, putting together one of his longer online sentences.

They hit it off at a breakfast and that turned into a series of meetings—
"group gropes," Clark called them—over the next eight weeks. Foss and
other engineers participated, but the principals were Clark and Andreessen. "Jim looked at Marc the way a machinist looks at a tool—as a means
to an end," Foss says. They met at Clark's house in Atherton, driving
his wife crazy. They'd begin with coffee in the morning and wind up
raiding the wine cellar by nightfall (though Andreessen was still partial
to chocolate milk). "Clark just knew he was going to do something to
make a billion-dollar company, which was sort of putting the cart before
the horse," says Foss, adding, "God bless him for it." Foss would be
Employee No. 3 of the new company and made out quite nicely.

Clark initially thought of games and interactive TV, sort of an online
Nintendo. Interactive TV, movies-on-demand, five hundred channels enabled by fiber optics—these were the buzzwords of the "information
superhighway" that people were talking about in 1993 and 1994, with
techno-hip Bill Clinton in the lead. This would be the next great technology, and Clark wanted his entrance ramp onto the highway.
Andreessen wasn't keen on a Nintendo network, and in any event, Clark
recognized that the corporate alliances they would need presented conflicts of interest with his former employer, Silicon Graphics. Instead, they
got to talking about Mosaic.

"You know, Jim," Andreessen said one afternoon over a glass of
Clark's best Burgundy, "some people think the information superhighway has already arrived. It's called the Internet."

Foss had thought similarly a few months earlier, having borrowed
$50,000 from Clark to start a company that would do online real-estate
brokerage. Foss had been spending every weekend driving around the
Valley with his wife and a broker in search of a house. He thought that
the Web, using Mosaic, could truncate the process, especially by organizing photographs of properties for sale.

Maybe the future wasn't five hundred TV channels, but 50 million
Web home pages—those electronic collections of data and images assembled from a vast range of Webbies, from fourteen-year-olds in Idaho
to the chief information officer at General Motors. Already the Internet
had a base of users and was swelling because of Mosaic. The Net didn't
require anybody to build anything or any cable company to come drill

holes in your house. The proliferation of Windows and Mac versions of Mosaic produced what Andreessen called "a kind of positive feedback loop"—more users producing more content drawing more users, and so forth.

At first, Andreessen had looked to Clark as the leader, but now Andreessen seemed in that position. He mentioned his friends back in Illinois were interviewing for jobs and were about to leave NCSA. And it dawned on him that the Internet might be Jim Clark's information highway.

"You think of something to do," Clark instructed Andreessen, "and I'll fund it."

"We can always create a Mosaic-killer—do the program right this time," Andreessen answered. He'd always been frustrated that Mosaic contained so many bugs and amounted to second-rate software, regardless of its novelty. It also didn't hurt that he might get to show up the folks at NCSA. Revenge can be a fine motivator.

"God knows how we'll make any money," Clark answered back, "but okay. Let's get the other guys." Clark had an entrepreneur's instincts, and for him, part of the Internet's appeal was precisely that nobody was yet making money from it. Others would look at that economic fact and be scared off; Clark discerned opportunity, all the more so because of the Web's snowballing growth. Clark painted a picture for himself that no one had imagined—a picture filled with green. Not even John Doerr had thought of capitalizing on the Web. Beyond a company like Cisco— which made the network "router" infrastructure for the Internet— or Sun—which made servers, the powerful computers that managed networks—the venture capitalists had no idea how or where to invest.

But for all the Valley's chatter about changing the world, Clark was more genuine about his goals. "I was in it for the quick buck," Clark told me, without embarrassment. "Yes, at a certain point, the money doesn't matter. I do own an acre in Florida and an acre in California, but what if I'd like to have a hundred or a million acres someplace? I don't ever want to be down to $150 million and have to think about liquidating certain things." Clark was not the tenderfoot that Wozniak or even Jobs had been when they launched a revolution. He knew exactly what lay at the end of the rainbow.

Out of that was born Mozilla—the "Mosaic killer."

Andreessen suggested they bring his friends out to the West Coast to talk. But Clark worried that other entrepreneurs were hatching the same idea of improving Mosaic. The following week, he and Andreessen flew to Illinois themselves. No matter what Clark did, he did it with urgency.

The remaining Mosaic men had missed their leader in recent months. "We were just the plumbers," Mittelhauser says. "To build a great house, you need a great architect. Marc showed us where all the bathrooms went." Mittelhauser and the others had stayed in touch with Andreessen, but they were still surprised when he E-mailed: "Something is going down here—be prepared to leave."

Clark and Andreessen arrived during a miserable spring snowstorm. The next day, at the University Inn, Clark met individually with Bina, Mittelhauser, and Totic, along with three others who hadn't worked on the original code. Rob McCool had built a Mosaic server for NCSA and Chris Houck had worked on various projects. And then there was Lou Montulli, a student at the University of Kansas who authored Lynx, an efficient text-only browser. Andreessen wanted him in. Montulli, hundreds of miles away, had to borrow money for the plane fare and barely made it to the meeting on time. Eccentric even when compared with the other engineers, Montulli became an early Web folk hero because he stocked the Amazing Fish Cam, a favorite aquarium in cyberspace. Its popularity was no fluke, getting more than forty thousand visits a day.

Clark knew the group's credentials were great. But he wanted to know if Andreessen could be a Pied Piper. "Each of them had a different story," Clark recalls. "But each began with Marc. That was enough for me."

All six got job offers. Clark discussed salaries (roughly $80,000), benefits, and the small matter of stock options—about 1 percent apiece of the new company. It was a generous percentage, given what other companies did. Clark said that if the company took off, the stock options could someday amount to a lot of Skittles. He was right. The Mosaic "others," who didn't get Andreessen's headlines, nonetheless became multimillionaires after the IPO. "We've all revisited that meeting with Jim," says Montulli. "I used to worry how I'd buy a PC every five years."

Montulli and the others accepted on the spot. To celebrate, they headed to Gully's Pool Bar, where Andreessen brought over formal "offer letters." Clark, still in his hotel room, had typed up one on his laptop and faxed it six times to the hotel lobby. "We laughed and laughed at ourselves—that we were going to California," says Mittelhauser. Within weeks, they had packed their bags and bits and were gone, except for Bina, who would commute from Illinois because his wife was a tenured professor there.

After the party, Clark and Andreessen headed back to California to rev up the new Mountain View company. It was clear to both that an online endeavor for a particular business—like Foss's—was small potatoes compared with developing the ubiquitous tool needed for *anything* online. Foss sensed as much, but didn't learn that his real-estate venture was off until Andreessen showed up at Foss's annual Easter Beer Hunt— you search for bottles of Budweiser—and playfully announced that Clark wanted his $50,000 check back. Foss was disappointed enough that he came in last in the hunt.

Clark and Andreessen started out calling their new company Mosaic Communications, but changed the name to Netscape after legal posturing by NCSA. "Mozilla" was considered as a name until grown-ups like fifty-one-year-old Jim Barksdale took over. "Netscape," connoting the Internet and a virtual landscape without limits, was safer, more corporate. Barksdale was the head of McCaw Cellular (which merged into AT&T Wireless Services) and formerly a senior executive at Federal Express; he had twice turned down the chance to be the No. 2 man at Microsoft, deciding that he didn't want to be a first mate, especially on a ship steered by a captain with the name of Gates.

In January 1995, Barksdale joined Netscape as its CEO to ride a corporate stallion that was threatening to run out of control. Clark knew from his SGI experience that he wasn't cut out temperamentally or organizationally to remain as CEO. He was too moody, too impulsive. Foss tells the story of how, early on in Netscape's existence, Clark was in Chicago closing a deal with Playboy's Christie Hefner. Playboy was opening an online mall and Clark offered to route it through Netscape's servers in return for 15 percent of the profits. Clark and Hefner shook hands on the deal. But, minutes later, as he got into his limousine, Clark remarked to Foss, "We can't be in bed with Playboy, can we? It's not the right image." The deal didn't happen. Clark was smart enough to realize his own expectations. "I was looking for a get-rich-quick scheme right now. That's a little crass, but I wanted to get this thing done, get it on the road, and get on with other aspects of my life that didn't amount to running a company. I knew this thing would quickly grow to the point where I needed better management talent."

Barksdale was regarded as one of the half-dozen or so best CEO candidates in the country and John Doerr, working on Clark's behalf, found him through a headhunter. Barksdale was financially set for life, but wanted to hit the proverbial "home run" that only a start-up could offer. Doerr had considered himself for the CEO job—a prospect that made some of his partners guffaw—but he didn't want to give up his position at KP. In return for taking the CEO job of what was then an unknown company, Barksdale—not even a founder—was given a remarkable 12 percent of the company, more than four times Andreessen's stake. Next to Clark, Barksdale was the individual with the largest piece of the pie.

Born and raised in Jackson, Mississippi, Barksdale liked to be known as an aw-shucks country boy dispensing Southern homilies—usually involving a critter. "If you see a snake, kill it," he'd say. "Snakes" were the big problems facing a company. Snakes, dogs, rabbits—they'd all become familiar in Mountain View. "If you can't run with the big dogs, don't get off the porch" was his way of lauding risk-taking. "If I tell people chickens can pull trains, it's their job to hook 'em up" meant he expected his troops to follow orders.

Netscape grew from three employees in April 1994 to 100 by Christmas, and 2,600 in three years. Even Clark in late 1994 had doubts about his company's revenue prospects and curbed its hiring, but Barksdale came roaring in and said the company would steam ahead with expansion and marketing. Under Barksdale, Netscape would be as aggressive as any company in the history of the Valley. But while "Marc" joined "Bark and Clark" in Netscape's executive cubicles, he remained the software guru and media darling. His khaki shorts and plaid shirt with clip-on tie were as remarked on as Larry Ellison's double-breasted suits.

The new company's mission was to start from scratch and come up with a new-and-improved browser. Anything they lifted from the original

Mosaic code would only cause more legal wrangling because the NCSA and the University of Illinois took the position that everything Andreessen and the others did at NCSA belonged to NCSA. But no matter what Netscape did to create a wholly pristine browser, NCSA claimed that its intellectual property had been stolen and demanded fifty cents in royalties per copy. NCSA had decided to enter the browser business itself, licensing the original Mosaic through a company called Spyglass. But NCSA also wanted a piece of the action in Silicon Valley, arguing its case both publicly to the press and privately in ominous cease-and-desist letters to Clark.

In a preemptive strike, Clark had Netscape sue NCSA in California federal court, asking that it be cleared of any copyright infringement and charging the University of Illinois with trade libel. Barksdale had said he wouldn't begin until the legal dispute was resolved. Shortly before Christmas 1994, Netscape settled with NCSA, agreeing to pay close to $3 million in cash. In one final mistake, NCSA rejected taking the settlement as Netscape stock—fifty thousand shares of it. Nine months later, when Netscape launched its IPO, NCSA found out that its decision had cost it a fortune, not to mention, as Andreessen noted, "tens of millions in possible donations from Jim, myself, and the other Illinois alumni at Netscape." A few years after that, the original Mosaic browser was a relic—used by almost nobody in cyberspace.

Netscape released its first browser, called Navigator, in October 1994. Faster and full of doodads, it was an even bigger hit than Mosaic had been—for Unix machines, for Macintoshes, and most important, for Windows-based PCs. The primary means of distributing the software was the Internet itself. Given away like its progenitor—a brilliant and counterintuitive strategy, it seemed—the Navigator immediately ruled the Net, claiming more than 70 percent of the browser market and representing two million people using the software. By the following summer, that number had increased five fold, making the World Wide Web the dominant source of traffic on the various lanes of the Internet. The Web's popularity transcended borders and it took off in Japan in particular. The hundred or so Internet hosts of 1976 had grown to 6 million strong. Netscape claimed that 70 percent of the companies in the Fortune 100—including AT&T, MCI, and Hewlett-Packard—were using its corporate products. "It was this yawning chasm of a market," says Todd Rulon-

Miller, the company's first vice-president of sales. "It was like Oklahoma in 1840, with all the wagons lined up at the border. The marshal went 'Bang,' and all the stagecoaches streamed across the plain."

The browser's *free-ness,* according to Metcalfe's Law, would ultimately *increase* its value. The more people in the network, the more useful it was to get on the network, which led to more people on the network. At the outset, Clark and Andreessen had thought of browsers the way Gillette thought about razors—use them to create a brand and then bring in revenue from other products. Clark and Andreessen just didn't know what their blades would be. Would they make money by taking a cut of transactions made over the Web, or from services that companies provided to other companies, or from traditional advertising? They didn't know. However, they assumed revenues would somehow come from their new software tool; for now, they wanted to seed the marketplace. "My attitude was that we'd generate such a ground swell that we'd figure out later how to make money," Clark says. "With so many people on the Internet, I thought: How can I avoid making money?"

Quickly enough, Netscape decided that businesses had to buy the browsers, along with the more sophisticated server software that acted as a hub for commercial Web sites—products that could cost hundreds of thousands of dollars for a big company. Businesses were quickly figuring out that the Web held promise. From mere postings of annual reports, to "content" sites mirroring existing publications, to truly interactive sites enabling a customer to check a bank balance or track a UPS package—the Web foretold vast efficiencies of scale and speed that no prior network or database dared imagine. (Amazon.com, for example, can offer a range of books electronically that no corporeal store could.) With encryption, the Web made it safe to buy and sell online using credit-card numbers. Even the proprietary online services like AOL succumbed to the openness of the Web and made their services compatible with Web browsers. Netscape would generate revenue by selling to businesses, while at the same time gearing up for battle with another, more established operation called Microsoft that would soon discover the Web.

Andreessen had pushed for the Netscape browser to have no licensing restrictions whatsoever. But the sales staff, supported by Barksdale, insisted that the company couldn't forgo the kind of multimillion-dollar revenue that corporate sales would provide. In return for paying up—

thirty-nine to forty-nine dollars a copy—the companies would get the kind of technical support that the average, freeloading user didn't expect. Netscape critics later said that the decision to charge was a mistake, because Microsoft came along and gave away its own browser. Had Netscape done that uniformly from the beginning, Microsoft might never have gained a foothold. But even ignoring Microsoft's natural advantage of being able to tie its browser to its flagship Windows operating system, Netscape then wouldn't have had the revenue it needed to build a business and make its payroll—revenue that went from $1 million in its first quarter to $40 million in its fifth, from $85 million in its first full year to almost $350 million the next. Browsers made up more than half of revenues those first two years. "Cannibalizing your own product can be a great strategy," Barksdale says, "unless it kills the company"; even with browser revenue, Netscape still wasn't turning a profit most quarters. Given Microsoft's position, Netscape was damned if it sold its Internet browser and damned if it didn't.

Clark put $4.25 million of his own money— roughly a seventh of his net worth—into the new company. For that kind of sum, he regarded himself "as much of a venture capitalist as an entrepreneur," though he thought the latter was more complimentary. But given the burn rate of Netscape spending, he knew he'd need more money, and just as important, he wanted the Valley connections and seal of approval that an established venture capitalist could offer. He also knew Netscape would need an experienced management team and a business plan, and that he wasn't the one to assemble either. That led Clark to the doorstep of John Doerr at 2750 Sand Hill Road. Keiretsus, Zaibatsus, corporate jujitsus—these were for him.

The two had known each other since 1979, when Clark was at Stanford and on his way to starting Silicon Graphics. Kleiner Perkins Caufield & Byers had already committed to Sun Microsystems, so it decided it couldn't also back SGI. But Doerr and Clark stayed in touch and talked of working together someday. They both had frenetic minds and similar instincts. Standing side by side—six inches apart in height—they resembled a sort of Mutt-and-Jeff team. Clark also knew the retired KP partner, Tom Perkins, with whom he had sailed in Fiji.

Clark had little fondness for the VC crowd and had bristled for years about SGI. "If it wasn't for Netscape," Clark says, "I'd have turned out to be a very unhappy guy." He felt the VCs had taken advantage of his naïveté by grabbing 40 percent of SGI and leaving him with a measly 15 percent that got diluted down to 3 percent at IPO time. (In the small world of Silicon Valley, Clark's very first backer wasn't a venture capitalist at all, but Frank Caufield's ex-wife, who lent Clark $25,000; "the worst that can happen is you'll lose $25,000," Caufield advised her.) Clark's discovery that the VCs had gotten the better of him was akin to the calf who just learned where veal came from.

Part of Clark's bitterness may have come from envy. SGI went public the same year as Microsoft. "My stock was worth about $10 million," Clark recalls. "If I had put it all in Microsoft, it'd be worth $10 billion and I'd be one of the wealthiest people in the world. Back then, I thought software was a bad business because it could be so easily copied. I didn't understand intellectual property." Actually, he'd be worth only about $2.5 billion, but we get the point.

In discussing the Sand Hill crowd, Clark rarely uses the word "venture." It's always "vulture capitalists," "velociraptors," or "vampires." The *V*, he says, "is one and the same." At the risk of fighting the last war, Clark was not going to be had again. Even considering his personal regard for Doerr, Clark was suspicious, since KP had once recruited away someone Clark wanted to hire at SGI. This time, Clark's negotiations with the VCs were more even-sided.

Besides Kleiner Perkins, Clark had dangled Netscape before the two venture firms—the Mayfield Fund and New Enterprise Associates (NEA)—that had first backed SGI in the 1980s. KP didn't like to acknowledge that it might not have been Clark's first and only choice: In its promotional materials for the 1995 Zaibatsu Information Sciences Fund, KP said simply that Clark had called Doerr and the two "had a handshake."

To any suitor, Clark's asking price was steep, given that the company barely had a business model, let alone significant revenues. Clark was initially valuing Netscape at close to $20 million; if a VC was going to buy, say, a quarter, it was going to cost $5 million. Valuation of private companies is always an inexact science because there's no stock market to set a number objectively. But whatever the range of possible dollar amounts, Clark's represented the stratosphere. He was charging the VCs three times as much per share as he had paid out of his own pocket. Mayfield and NEA wouldn't bite. Clark didn't even give them the chance to quiz Andreessen, the über-super-wonderboy. (Clark might have owed NEA a little bit more. A few years earlier, Clark was in their offices when a reporter for *Upside* magazine, Nancy Rutter, was there, collecting dish for a critical story on KP. Dick Kramlich, the NEA managing partner, introduced her to Clark and they subsequently got married. "We're a full-service operation," says Kramlich, who in his negotiating with Clark over Netscape had generously suggested bringing in KP as a coinvestor.)

Kleiner Perkins wound up getting the deal by itself, much to the consternation of the other two firms. "I maneuvered KP into getting interested at just the right time," Clark says. "I was careful not to introduce John to Marc too soon. I trusted John, but I was prudent."

Andreessen, no wallflower, remembers thinking that Doerr was from another realm. "He's going 170 miles an hour, while you're puttering along at about 55." Andreessen says the best part was the catered lunches at KP. "I was, like, there must be a lot of money in this business, if they have all this food brought in every day. And they had every conceivable type of soft drink ever made." Andreessen made an immediate impression around the office. Later on, he and his girlfriend went to a movie with Brook Byers and his wife. "Unfortunately," says Andreessen, smiling, "the movie was *The Long Kiss Goodnight,* one of the bloodiest, most violent movies in years." Also one of the biggest flops that year, but Andreessen and his girlfriend loved it. Apparently, the Byerses did not. They never went out with Andreessen again.

With his characteristic fervor, Doerr was impressed by the Clark-Andreessen idea and wanted in for several reasons. Kleiner Perkins was an early investor in AOL—Frank Caufield was still on its board—and Netscape held out the prospect for a much larger audience. KP was coming off a less-than-stellar year and was eager to swing for the fences. And there was little, if any, of the technical risk that KP abhorred. In the fall of 1994, in just forty-five minutes at their regular Monday meeting, Doerr and his partners agreed to Clark's numbers. From Clark's perspective, given how the VCs had treated him before, it was a magnificent concession.

KP invested $5 million for 20 percent of the company. Doerr figured, correctly, that if Netscape was a home run, a few million here or there wouldn't matter. And indeed KP's $5 million became worth $765 million, more than double the entire amount of committed capital in KPCB VII, the fund that put money into Netscape. "We knew it was a high price," recalls Caufield, "especially with what seemed like a twelve-year-old as the technology guru behind it, even with Jim Clark's blessing. But we all wanted to do the deal." When Netscape hit a bumpy period in late 1994, KP tried to renegotiate the deal, but Clark would have none of it. He got to keep about a third of the company, far more than he had retained at Silicon Graphics. It would be the first of many Kleiner Perkins bets on the Web. But Netscape would be the key. There was no doubt that the engine of the Internet now would be business, not academe.

For a time, NEA felt betrayed by KP swooping in and taking the deal by itself. All may be fair in financial warfare, but VCs regarded each other as genteel combatants rather than cutthroat competitors—especially when NEA believed it was partially responsible for bringing in KP. NEA got over it, but a partner at the similarly spurned Mayfield Fund did not.

Glenn Mueller was one of two venture capitalists behind Silicon Graphics (along with Kramlich)—and the man whom Clark blamed for his getting too little stock in SGI. They remained friends, but when Mueller pressed Clark for a piece of the Netscape action, Clark couldn't forget and turned him down. In words that later haunted Clark, Mueller told him: "SGI should be behind us. If you don't let us invest, my partners will kill me." Mueller had few big hits in the 1990s and his wife had become her own entrepreneurial success, founding Nancy's Specialty Foods, the largest maker of frozen quiche in the country. Mueller was going through a range of psychological problems as well—"increasing paranoia," Kramlich later called it—though no one realized how serious it was at the time.

On the day that Netscape formally incorporated—April 4, 1994—Mueller was on his powerboat *Sirena* in Cabo San Lucas off the Baja coast. He had talked with his wife earlier and made plans for her to pick him up the following week at SFO. He gave no hint of any difficulties.

But sometime that evening, fifty-two-year-old Glenn Mueller went into the cabin of his boat and blew his brains out. The police found no note, only the gun beside him.

The suicide staggered the Valley. "We worried that it said something about *us*," remembers Bob Metcalfe, whose 3Com start-up had been funded by Mayfield. "Was there a spiritual emptiness that caused this? Were we just like him and it just happened to him first?"

Clark wondered if he bore any responsibility for Mueller. "I did feel guilty and after the funeral I did visit Mayfield," he says. "They never got back to me. My guilt was gone."

His ability to be tough with the VCs soon paid dividends again. When Doerr was incubating the @Home Network, a company that would provide high-speed Internet access, he asked Clark to help recruit SGI's Tom "T.J." Jermoluk as CEO. "You're a good friend of T.J., right?" Doerr asked.

"Right," said Clark.

"Can you help us recruit him?"

"Sure." If the price was right.

Clark ended up with 1 percent of the company, which translated to tens of millions in its own right.

Whatever business plan it settled on, Netscape had to be up and running quickly. In days gone by, a semiconductor start-up could take years to erect factories and produce a product—this was *building* a company. Even Oracle, one of the software pioneers, didn't bring anything to market in its first year of existence. Netscape would have no such luxury. The economics of the Internet wouldn't permit it. The Internet offered an instantaneous means of distributing software; retail stores and hardware makers would still exist, but the Net added something new. The first company to build a better browser, or otherwise utilize the Web, would have a big advantage, especially in creating a brand and gaining market share. Doerr and Clark wanted to put together a management team in 120 days. Clark could draw from SGI. Doerr could use the legendary KP network; Mike Homer, for example, an alumnus of both Apple Computer and the GO disaster, was brought in to run marketing. And then there was Todd Rulon-Miller, whose whirlwind courtship with Netscape epitomized the new rules of engagement, as well as the moves of John Doerr.

In late September, Rulon-Miller had breakfast with Kevin Compton, an old friend and one of Doerr's partners. "I've heard you can sell ice to the Eskimos," Compton told Rulon-Miller lavishly. Rulon-Miller, who was CEO of a small software company and had been vice-president of sales for Steve Jobs's NeXT, said he wasn't in the job market.

Two days later, Rulon-Miller found a message on his answering machine at home. "This is John Doerr," the message said, at a clip so fast Rulon-Miller had to replay it. "I've just talked to Kevin Compton and he says you'd be great for Mosaic. Jim Clark and I are flying in from Atlanta in two days at 8:32 on Delta Flight 105. Meet us at SFO." Clark and Doerr were coming in that night from the inaugural Internet World conference, where they'd helped run the Netscape booth.

"Who was that?" Rulon-Miller's wife inquired.

"I told her it was somebody from *Mission Impossible* and that the tape was going to self-destruct."

Rulon-Miller had never met Doerr, but decided he couldn't pass up the chance. He went to the Delta gate to meet Doerr and Clark. (In those salad days, they still had to fly commercial.) They all talked in a coffee shop, which meant that Doerr mostly talked and the others listened. The next day, Clark and Rulon-Miller met in Palo Alto. Rulon-Miller thought this would be the "real" interview. But Clark "spent forty-five minutes condemning the repressive tax laws of the United States and explaining how entrepreneurs were creating all this new wealth, goddammit, and had this unique ability to get beyond details." Rulon-Miller then had two more traditional interviews. One was with Andreessen, during which the young programmer played the video game Doom. The other was with Barksdale. They came from similar backgrounds—having started out selling for IBM—and spoke a language distinctly different from the jargon of engineers. It was clear to Rulon-Miller that, whatever technological crusades the programmers had in mind, Barksdale was going to be orienting the company to sales. It was also clear that Barksdale would set a greyhound pace.

"Do you have any heart problems?" he asked Rulon-Miller.

"No. Why do you ask?"

"I like your energy level, but I want to know if you're going to give out in six months."

That night, Clark called Rulon-Miller and formally offered him the job. His equity stake would be 1 percent of the company. "I went to this 1-2-3 Copy Center the next day, called Jim from a pay phone, and gave him the fax number," Rulon-Miller says. "I was nervous about confidentiality and wouldn't use the machine in my office. The fax came, I signed, and faxed it back." Rulon-Miller often compared the experiences of working for both Jim Clark and Steve Jobs. "Predictable people-behavior is not an attribute of intense, passionate, mercurial entrepreneurs," he says. "You don't take it personally when the laser blast comes at you. You depersonalize it—like some thermonuclear material leaked from the shield. You get knocked down, you put back on the lead gloves, and you pick yourself up. I was known at NeXT as the guy who could survive a meltdown of the heat shield." And Rulon-Miller isn't complaining. "These are positive attributes for an entrepreneur. You can't grade them in the normal human way." (As Sequoia Capital's Mike Moritz put it, "You *have* to be a lunatic to start a company.")

In the new accelerated world of "Internet time," Rulon-Miller had been tempted, courted, and signed to run sales in eleven days. "Doerr and Clark invented Internet time," Rulon-Miller says. In the new world, an Internet year was like a dog year, compressing seven human years into one. A pauper today, a billionaire tomorrow. It just depended on your idea—and chance. Even Clark admitted as much. "I was lucky twice," he says, "though you also kind of manufacture your own luck—you stand on a tall building, hold a rod in the air, and wait for lightning to strike."

In 1995, as Barksdale officially came on as CEO and Clark retreated to the chairmanship, Netscape was on its way to becoming the fastest-growing software start-up ever. Due largely to Netscape, the World Wide Web was growing as a medium that left even the industry pundits gasping for hyperbole. Even the Holy See saw the power of the Net and got itself a Web site. The rates of growth—the number of users doubling *every few days*—had never been seen before in world commerce. Nobody knew where the numbers would go, but nobody in Silicon Valley wanted to be left behind.

Companies besides Netscape were staking out their Web claims. Cisco thrived as it kept up the behind-the-scenes hardware business of routing traffic on the Internet. (In the spring of 1999, Wall Street valued Cisco at $175 billion, making it the eighth-most-valuable company in America.) Designers and programmers rented out their services to develop Web sites for those who barely knew what a URL was. Three-dimensional objects and radio broadcasts were introduced to the Web. Most critically, Sun Microsystems invented Java, the Internet programming language that allowed Web applications to work equally well on any operating system. It didn't matter whether you logged on to the Web from a high-end Unix box, a Macintosh, or, most likely, a Windows-based PC; if the Web site you were visiting was Java-based, its pages appeared on your screen with all the bright colors and whirling dervishes its creator intended.

Java was potentially earthshaking because it seemed to threaten the hegemony of Microsoft's operating system. A vice-president of Compaq Computer referred to Java as "letting the world out of jail." Windows, like the MS-DOS it replaced, reigned supreme not because it was good, but because it was, well, supreme. You used it as your PC's central nervous system because you had to—you couldn't even buy a PC without it. Windows was akin to a dial tone. You had the product not because of its quality, but because of its inevitability. The Windows software came with the machine and with good reason—applications like word processing or flight simulators had been written to work with it (and often only it). An operating system served as a computer's "platform," from which all applications were then launched.

Because an operating system naturally tended toward one standard, and because software developers had little incentive to write applications for two-bit competitors, Windows' dominion was locked in (along with the huge number of applications designed just for it). As the ubiquitous operating system, software would be written primarily for it, which in turn made it only more dominant and tended to drive out of the marketplace whatever remained of competing operating systems. It was a vicious cycle, unless you were Bill Gates and his merry band of shareholders. Windows was the checkpoint through which any entrant into PC Land had to travel.

But Java was nondenominational, ministering to all operating systems—it's the preferred language for "applets," tiny applications sent directly over the Web. If it really caught on and the Web became a computing platform unto itself—just as the PC had—then Microsoft would lose its natural monopoly, which amounted to 80 to 90 percent of the 100 million desktop computers on the planet. For the moment, Netscape and Java ran on top of an operating system (like Windows or Mac). In time, they might *replace* the operating system, acting as a computer user's main desktop screen (where a combination of icons and file folders gave the look of the top of a desk). The Net would be the equivalent of a great big hard drive. The era of mainframes and minicomputers had given way to the PC age—now the Internet suggested a world beyond personal computers, a world filled with billions of low-cost information appliances tailored to specific household and business uses, a world without Bill Gates at the digital cash register. The Microsoft credo was: "A computer on every desk and in every home, running Microsoft software." Netscape imperiled that.

The synchronous ascension of Netscape and the Web was exactly what Jim Clark had dreamed of back in 1994 when he left Silicon Graphics and met Marc Andreessen. He had unleashed another revolution in the Valley, but what took two decades for the transistor and chip to accomplish, Clark had done in less than two years. All that remained was to cash in on it.

"Going public" was always part of the core mythology of the Valley—Dennis Barnhart notwithstanding. Founding a start-up was a solitary act, usually accomplished by the formality of a signature on a legal document. "Going public" was the start-up's coming-of-age for all to see—a debutante's ball or a bar mitzvah, depending on your analogical preferences. Steve Jobs made $155 million the first day Apple offered a piece of itself to the highest bidders. Larry Ellison was worth $93.5 million on paper when Oracle went public. Gordon Moore owed his billions to the public's love affair with Intel stock. Genentech's IPO in 1980 put biotechnology on the map. In no other single financial transaction—short of a Powerball lottery—could an individual reap such a fortune so instantly. This was the twentieth-century counterpart to Sutter's Mill. When the riches came, they filled the pockets not just of the founders and the venture capitalists who backed them, but of the "little" people who staffed the mailroom and answered the phones.

Yet, against that historical backdrop, what happened to Netscape Communications was without parallel and came to define the modern Silicon Valley. The company had decided to go public to bring in working capital ($140 million of it), to raise its public profile, and . . . because it could. Revenues, if not profits, were blossoming and the outlook was good for several financial quarters. It didn't matter that the company had been around for only sixteen months—Microsoft didn't have its IPO until eleven years after its founding. For Netscape, there was nothing to lose and everything to gain, not least of which was the money for Clark, Andreessen, Barksdale, Rulon-Miller, and dozens of others. It was one thing to have a paper stake in a private start-up—which had no liquidity—and quite another to own stock in a publicly traded company. SEC rules, as well as company vesting policies, would prevent Netscapers from unloading some of their shares, but only for a matter of months. Apart from the matter of having to open the company's books to public inspection, an IPO—a way literally to print money—was a no-brainer. Whereas Bob Swanson at Genentech fifteen years earlier was dead-set against going public because he didn't want his life to change radically—Tom Perkins had to convince him otherwise—the Valley had mutated so much that a stock offering was seen as a gimme.

On August 8, 1995, Netscape and its underwriters had to set the price for the five million shares being offered to the public the next day. In preceding weeks, the target had been twelve to fourteen dollars, the normal range for most offerings. If the price was too low, the company would look cheesy; if it was too high, it would deter modest investors from getting in the action. But the underwriters were being flooded with calls about the Netscape offering—there were industry press reports that investors were looking to buy 100 million shares—and they decided to double the price to twenty-eight dollars. That figure put the company's market value at a billion dollars—astounding for an unprofitable company that was just a pup. "They wanted to price it at thirty-one," Barksdale says. "My reaction was, 'This is nuts.' "

He was right. The next morning, euphoric employees gathered round their computer terminals to wait for the bell on Wall Street, a continent away. It was 6:30 in the morning, because of the time difference. Clark had been thoughtful enough to arrange for a deli truck to pull in early to serve espresso and bagels in the parking lot. But when the market opened, trading in Netscape did not. There was such high public demand for the stock—with no corresponding supply—that the imbalance delayed trading for ninety minutes. In another time and in another place, it was like waiting to hear on the radio whether Joe DiMaggio had gotten a hit in his fifty-six-game streak. And, then, the Netscape symbol appeared on the ticker—with a "71" after it. The law of supply and demand in action—El Dorado! In a valley that had seen money made before, the IPO still had a fin de siècle quality about it.

Even factoring in greed, fantasy, and the notion of a Ferrari free-for-all, nobody expected Netscape to open at more than two and a half times its offering price. Soon the price climbed to its peak that day of 74¾. The company's ubiquitous teal logo, an *N* soaring into the virtual stratosphere, could just as well have been Netscape's market value. Mom-and-pop investors, who didn't know the difference between a megabyte and an overbite, showed up at Netscape's Mountain View offices hoping to buy in. (They were sent away with a T-shirt.) That same day, up the San Francisco peninsula, the locals were mourning the death of the Grateful Dead's Jerry Garcia. But around the happy Valley, entrepreneurs were passing around a joke: What were Jerry's last words before his heart attack? Answer: "Netscape opened at *what*?"

Just as Wall Street was unfamiliar with the Internet, Netheads knew little about Wall Street. One caller to the Netscape switchboard, looking to get stock, was told to call the underwriters, Morgan Stanley or Hambrecht & Quist. "Is that Hamburger Kissed?" the caller asked. Unfazed, he then asked to speak to "Mr. Stanley." Chris Holten, the Netscape PR executive who kept track of the funny calls, was asked numerous times if she was married.

The delirious Netscape IPO became a marketing tool unto itself, as valuable as the cash it brought to the company. Netscape was selling not just a business but also a dream. Investors bought not for a stake in the profits, but on the assumption that another buyer would one day pay more. This was the Greater Fool theory in full throttle—a wonderful pyramid scheme, except for fools who bought shares at the top. At Charles Schwab, the 800 number got a new recording: "Press 1 if you're calling about Netscape." Even though a few market analysts and TV commentators did acknowledge that the company had rosy prospects— it even posted a small quarterly profit the next month—they compared the stock frenzy to a tulips craze.

By nightfall, the price simmered down to 58¼, which still fixed the start-up's market capitalization at $2.3 billion—double what Apple Computer was worth the day it went public, and ninety-two times what Kleiner Perkins had valued Netscape at a year earlier. That was more than Eastman Kodak, more than Microsoft in its youth, more than some high-tech businesses were worth during a corporate lifetime. Amdahl, Cypress Semiconductor, Intuit, Tandem Computer—none has flown so high. This was the dawning not only of Internet time, but of Internet finance. The *Alice in Wonderland* valuation of Netscape simply made no sense by any traditional market measure. It was a phenomenon more like the once-in-a-generation arrival of a great, big, loopy comet than anything rational.

Jim Clark's stake alone was $544 million (not counting stock options), more than making good on his expressed desire to hit the jackpot. Barksdale was worth $224 million; Kleiner Perkins, $256 million (based on its investment of $5 million); and Andreessen, $58 million. The Wisconsin kid who two years earlier was making $6.85 an hour was hardly hurting, but wasn't his take rather small compared with Clark's windfall? And couldn't Clark be accused of doing to Andreessen what had been done to Clark years earlier when Silicon Graphics rode the rocket? Andreessen laughed off any such suggestion, emphasizing that without Clark he'd never have started a browser company. Clark, too, scoffed at the notion that some twenty-four-year-old rookie, however conceptually adroit, ought to be in his league of compensation. The better question might have been why Tim Berners-Lee, for example, hadn't managed to earn a dime from the new marketplace that his Web unlocked.

The hype from the IPO reverberated far beyond the mother lode of Silicon Valley—around the national media, then through the financial community, then into the media echo chamber. WHY BILL GATES WANTS

TO BE THE NEXT MARC ANDREESSEN! proclaimed *Wired* magazine shortly after it figured out how to spell his Scandinavian name correctly. *Time* magazine put him on its cover, barefoot—the kind of honor accorded Jesus and Gandhi. It was a story everybody loved—instant millions, the boy genius and the adult entrepreneur who discovered him, and a company that might someday topple Microsoft. The presumably calmer heads on Wall Street didn't react any better. In November, Goldman Sachs gave Netscape stock a "buy" recommendation—only days after issuing a "hold" on Microsoft, the first such warning since Gates took the company public. The symbolism of these moves wasn't lost on anyone. Microsoft stock suffered. Netscape's went up, hitting 100 by Thanksgiving, 1995, and heading for lunar orbit at 174 two weeks later. Wall Street had decided the next Microsoft had arrived, even if it didn't have the balance sheet to prove it.

These astronomical numbers, rather than a nifty software product for the Internet, put Silicon Valley back in the public consciousness in a way not seen since the Apple days of Jobs and Woz. Clark was now worth a dizzying $1.6 billion; Andreessen, $174 million. After he started Microsoft, Gates had to wait twelve years before joining the billionaires' club. For Clark, it was sweet vindication of his decision to leave SGI and of the way he'd handled Kleiner Perkins. It also made him feel he measured up to Gates, whose privileged upbringing was in sharp contrast to Clark's Texas childhood. Clark never tired of noting that Gates "had his way paid to Harvard." In this insecurity about his own economic roots, Clark sounded a lot like Ellison and, to a certain extent, Steve Jobs. It was easy to wonder if striking it rich was a kind of psychological compensation, too.

The money necessarily altered the culture at Netscape. You could rationalize the in-house dentist as a subtle way of encouraging the workers never to leave, but $150 car washes in the parking lot—it's called "auto detailing," but steam-cleaning the engine costs extra—suggested certain lifestyle changes. Thirty-one-year-old Marc Coelho was one of the early employees and took care of the various Netscape buildings. Now, with several million dollars in his ledger, he was able to buy a '56 DeSoto, a Range Rover, and six motorcycles, including a Ducati 955SP. His basic transportation was a $65,000 black Dodge Viper RT/10 two-seat hot rod. It was basically a six-hundred-horsepower engine with wheels. "I like to go faster," he told me, as we went from 0 to 120 in seven face-peeling seconds on an entrance ramp to Highway 101. Lou Montulli, one of the founding engineers, bought fewer toys and found himself waxing philosophical on his instant wealth. "I used to say you couldn't make more than a million dollars honestly," he said a few months after the IPO. "It was sort of Marxist." He had ample opportunity to work out any lingering ambivalence about capitalism: A year later, he married Jim Clark's daughter (alas, splitting up after two years).

Barksdale deplored the magical millions and the excess he believed it fostered. "Sure, we hit a home run," Barksdale said at the time, "but the pitch was thrown right to us and we happened to be at bat. My hope is that the people here will not let this largesse influence their sense of self-worth." Coelho was featured with his toys in a *Newsweek* photo that incensed Barksdale. It may have been amusing when employees were asked by friends when they were buying that hillside retreat in Woodside, complete with pachinko arcade and Surround-sound sauna—but the line wore thin the eleventh time round. Barksdale told the fellow with the NSCP license plate on his Porsche to give it back. He also instituted a company rule that employees couldn't talk about the stock price in the office, though that didn't stop one clever programmer from rigging his PC to flash an NSCP price quote every day at four. *Jim said "no talking," not "no looking."* Barksdale's own secretary bought an electronic ticker for her cubicle wall, setting it to blink the stock price all day long. He couldn't believe it.

"Didn't you hear what I said?" he asked.

"Yes," she answered, "but I didn't know you were serious."

Barksdale himself seemed exempt from the rule, wearing a pager that gave him updates on the market. And nobody made him—or Andreessen or Clark—get rid of their private jets. Everybody was drinking the Kool-Aid.

The dark side to Netscape's meteoric debut was that it set the company up to disappoint. The brand-name recognition was great, something not even the best PR machine could generate. But the market and the media created expectations few companies in the history of capitalism had met. The laws of celebrity gravity would soon take over. "There was no way to sustain a $174 stock price," says Barksdale, who joked that Netscape's stock certificates should come with seat belts, for the ensuing roller-coaster ride. "It couldn't be done. We were going to try to build the company step-by-step, but we were always going to be compared to that initial period. The stock price was totally disconnected from reality, a spectator sport that had nothing to do with the business. The guys running General Motors don't get as much press in their lives as we got in a week."

Moreover, in the crucible of Silicon Valley, competitors couldn't wait to take down the newest hero. "People were always bipolar about Netscape," Andreessen told me several years later. "There were always two questions: How big can you get? When will you be out of business? We were always the 'most overrated stock' in the *Barron's* poll—until we were on a list of 'most undervalued.' There always was a downside to the upside." The Netscape offices were typical tilt-ups, but they were notable for another reason. They were built on the same land where another legend of the Valley had faded into oblivion, Fairchild Semiconductor. The land beneath was a Superfund site, full of toxic waste—an irony that amused Andreessen, who had an appreciation of history, if not always the ability to learn from it.

The specter of the Valley wanting to eat its young is simply another way of expressing the evolution of things. The noble Darwinian struggle always bears the strongest offspring. But Barksdale sees the Valley's heart for what it is. "There are only two kinds of stories—all-the-glory-of-it and all-the-shame-of-it," Barksdale says. "We got the first one first. Everybody in this town worships success, but what they really live for is to see others fail." Natural selection isn't what motivates Silicon Valley. It's the German concept of *Schadenfreude*—glee in the misfortune of others.

Chapter IX

Godzilla

In the cult 1969 animated classic *Bambi Meets Godzilla,* we find our fawn playing peaceably in the forest. He frolics, he nibbles grass, he bothers no one. Then, without warning and to the tune of the *William Tell Overture,* the giant foot of Godzilla appears and—*boom!*—squooshes him. Splat, squish, The End. The short film is one gag and two minutes—the credits run longer than the film ("Written by Marv Newland," "Directed by Marv Newland," "Costume Design by Marv Newland," and so on). For three decades, *Bambi Meets Godzilla* has regaled film festivals and college campuses. It also pretty much represents the Microsoft Corporation's relationship with its Silicon Valley competitors, at least in the doey eyes of those competitors. Marc Andreessen alternates between *Bambi Meets Godzilla* and *The Godfather* as his cinematic metaphor for Microsoft and its forty-three-year-old leader, Bill Gates. When Intuit agreed to sell out to Microsoft in 1994—*amicably,* such as it was—cofounder Scott Cook referred to Gates's company as Godzilla. Microsoft had been trying to stomp Intuit for years, so Cook gave up. In a memo to his board of directors, he said that Intuit's "future vision is both vulnerable to and would benefit from Godzilla's strengths." (The federal government killed the deal the next year.)

Godzilla, snake, bear, jackal, eight-hundred-pound gorilla, the Beast of Redmond, Don Corleone, the Great Satan, the Borg, the Leviathan, the Antichrist, the Prince of Darkness, the Dorsal Fin, a twenty-first-century Robber Baron, the "Leona Helmsley of technology," the Chinese

Army, Darth Vader, the Evil Empire, the Death Star—the list of male-dictions grows longer as Jim Clark or Larry Ellison or Scott McNealy look for new ways to demonize the richest company in the history of the world. In the cascade of *Star Wars* references, Clark and McNealy not only bill Bill as the Empire, but they refer to their respective companies as leaders of the Rebel Alliance. Microsoft may be the greatest U.S. corporation since World War II, but according to Jim Clark in interview after interview, it "is fundamentally an evil company." Even the gentle Gordon Moore, when asked what he thought about Microsoft, squirms and smiles before giving the obligatory line about his "great respect" for the company. Moore's not the wealthiest man in California for nothing; if it says "Intel" on your ID card, it's best to say only nice things about Redmond.

Sometimes, the sniping gets trivial—and richly amusing. In the 1980s, Steve Jobs and Bill Gates were forever snapping at each other, about such grave matters as who made the cover of *Time* first. (Jobs did, but Gates overtook him in total appearances.) By the 1990s, from the perspective of business success, Gates had clearly won that rivalry, but you wouldn't have known it. In a 1995 interview with *Fortune,* when Andy Grove was asked who inspired him, he answered: Steve Jobs. "Look at his history," Grove said. "He was the first to see what the PC was about. The first to recognize the value of the laser printer, the graphical-user interface. . . . And now with Pixar, his computer-animated movie and game company, he's pioneering real digital entertainment. Not bad for forty years old. True, he made a few mistakes along the way, but he continues to believe in his own technological vision." When Gates read the piece, he went nuts, calling Grove to scream. "What about all that *I've* done?" Gates protested. In recounting the story to Jim Barksdale, Grove said he was surprised by Gates's vehemence and compared it to the way a child be-haves when a parent brags a bit too much about a sibling.

Gates's reaction may be understandable. Didn't he, too, deserve some adulation? But that is not the primary emotion in the Valley toward his company. "In this industry, people are used to treating Microsoft some-what like the Mafia," Andreessen told *The New Yorker.* "You don't say no to the Mafia, you don't challenge the Mafia, you generally don't fuck with the Mafia." But of course, the Mafia isn't as loaded. Microsoft has a market capitalization that surpassed $425 billion in early 1999, give or take the invoices of some new antitrust lawyers. How can this be, when Microsoft's revenues barely put it in the top 150 of the Fortune 500? It helps to have a monopoly—and the best profit margin in the land.

When Microsoft isn't being attacked institutionally in Silicon Valley, Bill Gates is getting it personally. At one point, his company's own personal-finance software didn't have places to the left of the decimal point to account for his riches. In January 1999, the Gates stockpile reached almost $100 billion—more money than any creature has ever amassed, more than the combined fortunes of Gordon Moore, Larry Ellison, Steve Jobs, Jim Clark, Jerry Yang, Marc Andreessen, George Lucas, Andy Grove, Steve Wozniak, Scott NcNealy, John Doerr, Brook Byers, *and* Tom Cruise. Gates is worth more individually than the entire *corporations* of Oracle, Apple, and Yahoo. If you bought $2,700 of Microsoft stock on the day it went public in 1986, you had $1 million in time for Christmas shopping in 1998; if you put the same money into, say, Intel, you had only $132,000. In the ether of Silicon Valley, those kind of riches don't make you popular (though in a weird way they do earn you a grudging respect). Rooting for Bill Gates is like rooting for the Internal Revenue Service.

The Valley may be a hunter's stew of companies—semiconductors, PCs, peripherals, software, and Internet commerce—but its one unifying ingredient is Microsoft, seven hundred miles due north. Microsoft is like oxygen. It's all around. If you're in business in the Valley, you can't inhale without getting a whiff. Many entrepreneurs hope that "innovative" Microsoft will knock on the door one day and buy them out for a fortune. That's what happened in 1997 to WebTV; its three cofounders walked away with tens of millions of dollars apiece. A big, fat check from Bill Gates is the next best thing to an IPO. "Nobody wants to build the next Hewlett-Packard or Intel or Apple," says Steve Jobs. "Now, people start a whizzy thing and try to get it far enough along to be able to sell it."

If there's no cashier's check forthcoming, the prevalent feeling about Microsoft is fear and loathing. Companies define themselves in *terms* of

Microsoft, as in "We make software that Microsoft hasn't, doesn't, and God we hope won't." Start-ups that have yet to *exist* make the same calculation. For the last several years, partners from Kleiner Perkins and other leading venture-capital firms in the Valley have periodically sat down with a roster of Microsoft's executives to hear about the company's plans—presuming that what Microsoft tells them isn't specifically designed to mislead. In theory, the VCs believe, they'll better be able to stay out of the headlights and thereby off the radiator grille—especially if they can find out what businesses Microsoft *wasn't* going to enter. One of those venture capitalists quipped to *The New York Times* that the only niche Microsoft wanted to carve out was . . . software. "I guess that leaves us washing machines and toasters," Ruthann Quindlen said.

The seeds of such antagonism were sown two decades earlier. Everything in Redmond was about winning the last battle. Microsoft knew nothing of tactical moderation: It never left anything on the table, always looking to take the greatest advantage of lesser combatants. Gary Kildall was dead and gone, but the memory of how Gates treated him lived on. Concerning the similarity between Kildall's CP/M operating system and Gates's MS-DOS, which became the industry standard, Kildall wrote in his unpublished memoir, "Gates's DOS was a blatant misappropriation of proprietary materials—and of my personal pride and achievements." Those were strong words from a martyr.

Neither Microsoft's business tactics nor its in-house culture prizes decency or honor. How did Microsoft's lead lawyer defend the company at the beginning of the 1998–1999 antitrust trial in Washington, D.C.? "The antitrust laws are not a code of civility in business," he said in his opening, which wasn't exactly a ringing endorsement of Microsoft's corporate character. The fact of a monopoly alone wasn't the problem. Intel, whose market share in microprocessors was almost as impressive as Microsoft's in operating systems, never was regarded as the same kind of fiend; indeed, Intel was in a more likely position in high-tech to be villainized, since it was Andy Grove's chips that really dictated the pace of the PC industry. Grove, who stepped down as Intel CEO in 1998, might have been a tyrant at times, but he was revered, a father figure as much to Larry Ellison as to Jim Barksdale. (The latter remembers the first time he met Grove. "Andy acted like I was a little boy," Barksdale remembers. "I told him, 'I don't even let my father talk to me that way.' ")

Once upon a time, Microsoft almost wound up in the Valley. Back in 1978, it became clear to Microsoft's founders, Bill Gates and Paul Allen, that the growing company should leave Albuquerque. The only reason Microsoft was in the New Mexico desert was that its first customer, the maker of the Altair personal computer, was there. As that relationship foundered and Microsoft developed other business, there was little reason to stay, especially with Silicon Valley becoming the capital of high-tech. But Allen preferred to return to the Pacific Northwest and Gates, too, was being cajoled by his family to come home. Nevertheless, the Valley made the most sense because it had a ready labor pool of programmers and it had customers; these were the days before MS-DOS, when Microsoft had to hunt for business, and proximity to customers mattered a lot.

But Microsoft moved to the Seattle area anyway. And it wasn't so Allen could fish. The decision essentially came from Gates, for whom it came down, like everything else, to business. Silicon Valley was just too competitive an area in which to retain talent; even in the late 1970s, there were enough companies around that anyone who was restless and ambitious could trade up and out. Gates believed it would be better to spend the energy convincing programmers to come to Washington State and then not have to worry about them wandering off to another corporate mistress a year later. Back then, there weren't a lot of other technology jobs around Puget Sound. Nor was there the sunshine of the Valley that beckoned one to go out and play.

Like so many other Gates decisions, this one turned out to be astute or lucky, or both. Between its regular sales staff and everyone at WebTV, close to a thousand Microsoft employees today work in Silicon Valley. But fifteen thousand more are around Seattle. And those who leave aren't going to another company. They're retiring. "Even back in the late seventies," says Charles Simonyi, Microsoft's chief software architect, who left the Valley in 1980 to join Microsoft, "Bill anticipated what would happen. In Seattle, there would be stability in the workforce." It was enforced loyalty—the kind that patients in a small town have to a family doctor. (Funny that Microsoft even then understood what happened when people had few choices.)

Had Microsoft moved to California instead of Washington, it's possible that the animosity so many companies in Silicon Valley have toward it might have been muted. It's harder to hate the folks who eat the same burgers as you do at Buck's. But it may also have been that Microsoft, no longer isolated, wouldn't have been able to develop the pathology that has served its bottom line so well. The parallel universe that Redmond, Washington, represents would never have coalesced up the highway from Oracle or down the road from Apple. Even though Microsoft intends to open a Valley outpost in 1999—in Mountain View, near a dump—it will never be part of the spiritual core of the Valley.

Whatever the other enmities in the Valley, Microsoft's rivalry with Netscape in the late 1990s would set a new standard. Microsoft doesn't like competition in general (unless it's Windows 95 vs. Windows 98). Anything that threatened its operating system threatened the corporation itself. The triad of Marc, Bark, and Clark was spouting off at every opportunity about a potential "platform shift" or "paradigm shift" in the software industry from PCs to the Internet. The industry pooh-bah George Gilder predicted that the "desktop imperium" would "pale and wither" in the shadow of "the telecosmic amplitudes of the Internet." In this world, software standards would be open. Cyberspace would be filled with the likes of Sun's Java and Netscape's Navigator. Proprietary software, like the monopolistic Windows, would fade away. Microsoft wasn't fond of that kind of talk. It knew it had to respond. If the Internet gained wide acceptance, a Microsoft senior vice-president, Jim Allchin, testified several years later, Netscape might become "a complete competitor," able to "replace the operating system," instead of just running on top of it. "When it started out, it was just an application," Allchin said. "In a blink of an eye, it became a platform."

But what particularly irritated Microsoftians—it takes one to know one—was the early arrogance of the Netscapers. In speaking appearances, Andreessen delighted in dismissing Bill Gates as "one industry observer." In the summer of 1994, when Netscape's first head of marketing was contacted about its browser by Microsoft, he sent back magisterial "we'll let you know" replies. Netscape acted as if *it* were the dominant company. It had no intention of trying to steer around Microsoft, but wanted to directly cross its path. When the two companies began talking again in the fall about making Netscape's Navigator run smoothly with Windows 95, Microsoft's Brad Silverberg, a senior vice-president, perused his E-mail and found the previous offending items from Netscape. He shrewdly read the letters back to his Netscape counterparts and immediately put them on the defensive.

Ultimately, they all met in Mountain View—and accomplished little. Andreessen, for one, wouldn't answer most of Silverberg's questions. Netscape's executives also resolutely turned down Microsoft's proposal that it assume control of the Navigator code. Netscape was unlikely to give up its franchise under any circumstances, but the flat fee Microsoft had in mind was a pathetic $1 million. When subsequently asked at a deposition why he didn't take it, Jim Clark showed his scornful side. "You must not be a businessman," he told the Microsoft lawyer. (Clark would subsequently joke that he would've been glad to sell to Microsoft—as long as the price was $1 *billion*.)

During the lunch break from the meeting, the Microsoft team got on their cell phones back to Redmond—Microsoft refused to use Netscape landlines—and returned with a Microsoft vengeance. "It was a 180-degree shift," recalls Netscape's Todd Rulon-Miller. "It was like Khrushchev at the table. They told us, 'We're going to bury you. Cooperate with us and we'll consider a relationship.' " Netscape wouldn't and Microsoft didn't.

The awakening Redmond bear was going to go into the browser business at some point, regardless. If it didn't dance with Netscape, it would find another partner. Microsoft was always hungry and the Internet offered a million meals. But Netscape taunted the bear and that just made the bear meaner. Microsoft didn't just like to win; it wanted the other guy to lose. If the other guy was cocky, that just doubled the pleasure of breaking him. This was competition as blood sport.

Microsoft obviously wanted to get a look at Netscape's Navigator code. The way Microsoft did business was to learn as much about its prey as possible—before devouring it. But Netscape had something to gain as well. Microsoft would be releasing Windows 95 the following summer, and it was in Netscape's interest to make sure its browser and server software were compatible. The information stream between Mountain View and Redmond could run both ways. At various times between late 1994 and June 1995, the companies again talked about what to do. Soon after he joined Netscape, Barksdale ran into Microsoft's Dan Rosen at a Hambrecht & Quist conference in Snowbird, Utah. Rosen was Microsoft's director of strategic relationships for all communications technologies, including the Internet. Barksdale intended this first conversation to be a bridge toward a more amicable relationship with Redmond.

But through the first five months of 1995, the companies continued to parry each other. Microsoft had already announced it would release a browser of its own. So Netscape was concerned about ceding any competitive advantage by telling Microsoft about new features of Navigator 2.0. Netscape wanted to cooperate with Microsoft the operating-system company, but not Microsoft the applications company. This dilemma seemed to crystallize Microsoft's remarkable position in the software business.

Despite his concerns, Barksdale agreed to meet again with Rosen, along with Nathan Myhrvold and Paul Maritz, two other key Microsoft senior vice-presidents. On June 2, 1995, Barksdale flew to the Redmond campus. He learned that Microsoft was principally interested in getting him to incorporate into Navigator a range of features that "would enhance Microsoft content," Barksdale recalled in written testimony for the 1998 antitrust suit brought against Microsoft by the U.S. Department of Justice and twenty states. For his part, Barksdale emphasized how important Navigator licensing was going to be in Netscape's revenue plan—particularly a Windows 95 version—even suggesting that Microsoft include the browser in its PC operating system.

Was Barksdale worried about the bear? Not at this moment, Barksdale said, "because Microsoft had not yet announced its intention to displace Netscape from the marketplace by creating its own browser, bundling it with the operating system, and shutting us out of numerous browser distribution channels." Microsoft seemed to want to co-opt Netscape, not to crush it. But Barksdale's view changed dramatically a few weeks later in a follow-up meeting in Mountain View. There was no taming the bear.

On June 21, Barksdale, Andreessen, and the marketing chief, Mike Homer, sat opposite Rosen and five other Microsoftians. In the Wyeth boardroom behind Barksdale's cubicle, on the second floor of Netscape's main building, they went over several technical items before the Microsoft contingent allegedly raised a more strategic question. While it desired to cooperate on some projects, Rosen made clear that Microsoft didn't want Netscape making a browser for Windows 95, which would be released in two months. That huge marketplace, Rosen explained, should be left to Microsoft; Netscape could have what was left. At least this was Barksdale's version, when he testified at length about it three years later in the antitrust litigation.

Microsoft "proposed that a 'line' be drawn between the area in which we developed products and competed and the area in which they developed products," Barksdale said in written testimony, in essence defining a textbook case of collusion in violation of both federal and state antitrust laws. Barksdale quoted Microsoft describing its proposal to rig the market as "a special relationship." "They offered to allow us to continue to develop browsers for *other* operating systems, as long as we did not try to compete with them in developing a browser for the Windows 95 platform." It was like the mob agreeing to let honest garbage collectors have Staten Island while it controlled the busy streets of Brooklyn and Queens. Microsoft would even agree to finally give Netscape the technical data on Windows 95 it had been seeking. Everything depended, as one of the Microsoft personnel supposedly put it, "on how we walk out of this room today."

Barksdale wasn't interested in the offer-cum-threat and, at least later, professed to be in a state of shock. "I have never been in a meeting in my thirty-three-year business career in which a competitor had so blatantly implied that we should either stop competing with it or the competitor would kill us," Barksdale railed to the federal judge handling the antitrust suit. "In all my years in business, I have never heard nor experienced such an explicit proposal to divide markets." Andreessen called the meeting the equivalent of a "Silicon Valley IQ test"—if you agree to partner with Microsoft, you deserve your fate, which amounted to corporate suicide. Consistent with its threat, Microsoft delayed giving Netscape technical data on Windows 95 until after the operating system hit the market. Netscape later even suggested the operating system contained booby traps to cause conflicts with Navigator. (In a celebrated incident in 1996, Netscape discovered that Microsoft's home page on the Web couldn't be accessed with Navigator.) In this complaint, Netscape's hands weren't entirely clean, because Web pages created with *its* electronic publishing tools mysteriously worked a lot better with Navigator than with other browsers. Still, this wasn't the same thing as not working altogether.

At this pivotal June 21 session—which former federal judge Robert Bork, who later signed on to advise Netscape, labeled "the Don Corleone meeting"—Microsoft also proposed buying a 20 percent equity slice of Netscape and getting a seat on the board of directors. This was a tried-and-true strategy for a company like Microsoft that had coffers of cash. Before you expend the energy to beat a competitor, why not just buy it? Gates had delivered that encyclical in an E-mail three weeks earlier to Rosen, Maritz, Myhrvold, and Silverberg. "We could even pay them money as part of a deal, buying some piece of them or something," Gates declared. "I would really like to see something like this happen!!"

Yet he later claimed in a 1998 deposition that "at this time, I had no sense of what Netscape was doing" and played no role in setting strategy for the June 21 meeting. He further said that he had only faint recall of having suggested Microsoft buy a piece of Netscape—even though others recalled that the Microsoftians had talked to him by phone from Netscape on June 21. Gates's memory seemed remarkably dim for a man routinely credited with so much business acumen. Gates was so worried about public perception of his deposition performance that he took the unusual step of calling a press conference to announce that he actually had *told the truth*. Not a good sign if you have faith in your own credibility. (He'd have done better to display some wit, as when he and Barksdale both were headed for the same lavatory during a break from a Senate antitrust hearing in 1998 and Gates, getting to the door first, said, "I've got an exclusive on it.")

Barksdale's account of the June 21, 1995, meeting was supported by contemporaneous notes Andreessen took on his laptop. (In addition to his eating habits, Andreessen was known for his prodigiously fast typing.) But Microsoft countered that the notes manifested a ruse. In a 1997 sworn statement to the Texas attorney general, who was pursuing a state lawsuit against Microsoft, Andreessen said he took such copious notes because "I thought it might be a topic of discussion with the U.S. government on antitrust issues." Moreover, two days after the June 21 meeting, a lawyer for Netscape sent a four-page letter to the Department of Justice, detailing the meeting and attaching a copy of Andreessen's notes. All this, Microsoft says, smacked of Netscape and Justice entrapping Microsoft—the Silicon Valley version of Marion Barry's memorable "bitch set me up" defense.

But so what? Entrapment isn't a defense to collusion. If Microsoft wanted to argue that its tactics on June 21 and beyond were just examples of hard-nosed competition, then it's particularly effective to acknowledge that Netscape had outsmarted them, however underhandedly. What comes around, goes around.

Microsoft also claimed that the June 21 meeting itself arose because of an E-mail Jim Clark sent to Dan Rosen six months before, on December 29, 1994. "We'd like to meet with you," Clark wrote, in a very different tone from what some of his subordinates had conveyed to Redmond. "Working together could be in your self-interest as well as ours. Depending on the interest level, you might take an equity position in Netscape." Clark was writing around the time that Netscape had little revenue stream and was laying off employees. It was a time when he had reason to be desperate, and Barksdale wasn't coming on as CEO until the following week. "Given that worry exists regarding Microsoft dominance of practically everything, we might be a good indirect way to get into the Internet business," Clark wrote, adding that "we have never planned to compete with you."

When Barksdale was cross-examined at the antitrust trial, Microsoft's lawyer suggested that Rosen was merely following up on Clark's message. The problem with that argument, however, is Clark himself. "No one in my organization knows about this message," he wrote, which was totally consistent with his impulsive style. Clark sent the E-mail at three in the morning. Barksdale testified that he first learned of the message during his trial preparation, three and a half years later. That belated discovery was a tactical blessing, but at the time it infuriated Barksdale; in his mind, the Clark E-mail was precisely the kind of reflex that made him unsuited to run a company. It was the same reaction that the people running Silicon Graphics had often had about Clark. "For an engineer and a scientist, Jim is often not very meticulous about how he crafts things when speaking," Glenn Mueller told the San Jose *Mercury News* in early 1994. "People would always cringe to some degree when Jim was loose. He would say whatever was on his mind."

Confronted by that E-mail from Clark, Barksdale was asked by Microsoft's counsel for his personal opinion of the man who served as chairman of Netscape. "Do you consider him a truthful man?"

Pausing for a second, Barksdale dryly replied, "I consider him a salesman," which prompted howls in the courtroom. Somewhere else, Jim Clark probably wasn't laughing. He, too, had something riding on the antitrust case, having gone off for hours to the quiet room of a stereo shop in Palo Alto to rehearse for his deposition. He wasn't exactly optimistic. "If you're drowning," Clark said at the time, "the Antitrust Division isn't the one to count on to rescue you."

In Redmond, part of Microsoft's frustration arose from being so far behind Netscape in the Internet game. When Netscape's Navigator first appeared on the World Wide Web, Microsoft had only a few engineers working on a browser of its own.

Bill Gates was no fool. He was always quick to point out that his company's achievements had been at the expense of IBM. Had Big Blue not been so complacent about the arrival of the personal computer—had it not been so careless in its licensing arrangement with Microsoft—Gates might not be an international icon. No matter how big or proven Microsoft became, Gates vowed that his company would never make the same blunder as IBM. To his credit, he showed an extraordinary willingness to switch gears and change the direction of his corporate battleship when the war demanded. Godzilla, yes—but never a dinosaur. Even with its billions in revenue and operating-system monopoly, Microsoft's psychology required that it see itself as a start-up, perpetually vulnerable to upstarts just as it once was. In 1995, the clear and present danger was Netscape.

The Internet wasn't entirely unknown terrain to Microsoft. During the week that Mosaic Communications incorporated back in April 1994, Microsoft held its first major meeting concerning the Internet. The daylong executive retreat at the old Shumway Mansion wasn't a reaction to Mosaic, since Gates and company didn't even know that Jim Clark and the NCSA refugees had come together. It grew out of a few employees pressing for some Microsoft response to the Net beyond an online service. Gates wasn't against investing in the Internet, but he remained skeptical of making money off it. The hallmark of the Internet, after all, was that it was free—apart from the fees earned by online services like AOL and Prodigy and CompuServe, which had several million customers between them. In addition, any online commerce in which Microsoft became involved would be subject to intense competition. Electronic commerce had almost no barriers to entry—all you needed was a product to sell, some computer equipment, and the ability to design a Web page. This kind of open competition was anathema to Microsoft, whose business model was based on the leverage that a monopoly operating system gave it. Nonetheless, at the April meetings, Gates and his senior staff agreed that the new version of Windows would include some Internet-related technologies.

Later on that year, Microsoft confirmed it would develop its own proprietary online service called MSN (for Microsoft Network) that would, of course, be tied directly to its next operating system—with one click and for a monthly fee, a Windows 95 user could be connected to MSN. Moreover, partially in response to the debut of Netscape Navigator, Microsoft struck a deal with Spyglass and NCSA to license the original Mosaic browser. Microsoft would use the source code for that browser as the foundation for its own—Internet Explorer (IE)—which it would release in August 1995. This would mark the beginning of formal hostilities with Netscape. Still, in the latter part of 1994, it had yet to dawn on Microsoft that the Internet represented a fundamental threat to its business.

The company should have had a whiff by then of the Internet phenomenon—something that could reach so many people so quickly. Two weeks before Christmas, 1994, a Web prankster posted a news bulletin online that Microsoft was acquiring the Catholic religion. In classic Internet fashion, the story spread like a plague, though a very funny one. Datelined "Vatican City" and credited to the Associated Press, the story said Gates would get exclusive digital rights to the New Testament, online confessions would become available, and Pope John Paul II would become a Microsoft vice-president. (No word on his stock options.) "We expect a lot of growth in the religious market in the next five or ten years," Gates explained in the story, sounding just as he might. "The combined resources of Microsoft and the Catholic Church will allow us to make religion easier and *more fun* for a broader range of people." The whole thing was a joke, but not to Microsoft, which released a humorless statement denying the acquisition.

Microsoft's curious slowness to respond to the Internet finally ended the following spring. In a nine-page memo to senior staff momentously titled "The Internet Tidal Wave," Gates mapped out a new corporate direction for what he viewed as the second PC revolution. Whereas the prior two decades in high-tech had been based on "exponential improvements in computer capabilities," which in turn made software "quite valuable," Gates said in his May 26, 1995, memo that "in the next 20 years the improvement in computer power will be outpaced by the exponential improvements in communications networks." This was the triumph of Metcalfe's Law, not replacing Moore's Law, but overtaking it in commercial potential. "The Internet is at the forefront of all this," Gates continued, "and development on the Internet over the next several years will set the course of our industry for a long time to come." At Microsoft, he said, "all work we do here can be leveraged into the HTTP/Web world." In short, the Internet could become just another technology to be assimilated into the Microsoft empire.

Acknowledging that he had "gone through several stages of increasing my views" of the Internet's importance, Gates now was declaring unequivocally he was a believer—and a worrier. Netscape had his attention. He mentioned the company by name, noting its "birth" on the Internet and its ultimate desire to "commoditize the underlying operating system." That would be a death sentence for Microsoft. It was no coincidence that Gates's memo was issued just a few weeks before that key June 21 meeting in Mountain View between Microsoft and Netscape.

"Now I assign the Internet the highest level of importance," Gates proclaimed as lawgiver. "I want to make clear that our focus on the Internet is critical to every part of our business. The Internet is the most important single development to come along since the IBM PC was introduced in 1981. . . . The Internet is a tidal wave. It changes the rules."

In hopping on the network bandwagon and making a range of farsighted observations about what Microsoft's expanding Web presence would mean over the next few years, Gates the businessman was acting as he always had. He hadn't generated the "paradigm shift"—that credit went to Tim Berners-Lee and the Andreessen brigade—but he was now going to make up for it with the legendary Gatesian ferocity. Just as the

Microsoft PC operating system steamrolled over Gary Kildall's CP/M, just as the Windows graphical-user interface paralleled and surpassed Apple's in market share, just as Microsoft's spreadsheet and word-processing applications were Johnny-come-latelies—now Gates set his sights on the Internet. He liked to say that his company was an "innovator," but its best innovation was imitation—identifying what others had created and then, at just the right time, exploiting it. Others blazed the trail. He followed, often tucked in the slipstream.

As part of his GO pen-computing experience, Jerry Kaplan met with Microsoft and discerned its ways. "Rather than focusing on any one thing, the company became adept at identifying promising market niches with weaker competitors," Kaplan wrote in *Startup*. "It would closely study their products and tactics, then launch an attack on their position with a strong product and aggressive pricing. Sometimes, Microsoft would propose some form of cooperation or joint development, to learn about the market before staging its own entry. This was the corporate version of the cheetah's hunting technique: keep a close eye on your prey, sneak up, then outrun it. Like the cheetah, Microsoft became well adapted to its game: lean and mean, fast on its feet, observant, shrewd. And the first warning of its presence was often an unexplained rustling in the tall grass."

The Gates M.O. was to "embrace and extend"—the corporate counterpart to Manifest Destiny. Whether Gates did so ethically or legally was open to debate—and repeated antitrust investigations by the federal government—but there was no mistake that he was good at it. A truism of organizational behavior is that big companies don't switch gears well because, by definition, they've already succeeded at doing things that work. You don't mess with that. This was the basis of a popular book, *The Innovator's Dilemma: When New Technologies Cause Great Firms to Fail*. Microsoft proved otherwise, radically retooling itself, craving the big bet and the next battle. Bob Metcalfe quipped that "embrace and extend" in Microsoftian translated into "draw and quarter."

The problem with Microsoft is that it defies a lot of economic theory. In the very absence of competition that its detractors describe, Microsoft should grow lazy—unable or unwilling to anticipate or respond to the marketplace. But Gates created a continuing start-up environment that thrives on fear. Grove's Law held that "Only the paranoid survive," but

Bill Gates has turned it into a religion—"constructive paranoia," they call it in Redmond. Fear was an even better motivator than greed.

Netscape had its glorious IPO on August 9, putting the company on the front page of almost every newspaper in the country. If anybody in Redmond needed confirmation that the Internet had become part of the culture, this was it. The mere fact that a tectonic development in the computer landscape could take place with barely a reference to Microsoft was reason to fret. Bill Gates's response came in two weeks with the release of Windows 95, as well as an online version of Internet Explorer.

The launch of an updated PC operating system was hardly a revolutionary event. Compared with what the Macintosh had wrought a decade earlier, Windows 95 amounted to little more than a superb makeover of an existing product, Windows 3.1—and it was two years late at that. Yet the circus surrounding the Windows 95 rollout was incredible, a testament to what $250 million in marketing can do. In Redmond, Jay Leno hosted a gathering of 2,500 of Bill Gates's best friends in industry and the media. There were fifteen tents, a Ferris wheel, balloons, food, music, demonstrations, and even Mrs. Bill Gates on display. Other software and hardware companies came with their wares, too; they had millions invested in their own products tied to Windows 95. The entire atmosphere was one of inevitability: How could any PC user in his or her right mind *not* buy Windows 95?

Around the world, the Microsoft hype engine worked in overdrive. In New York, the nighttime Empire State Building was aglow in the logo colors of Windows 95. In Toronto, the 1,800-foot Canadian National Tower displayed a fifty-story banner. In England, Microsoft bought out the full press-run of *The Times* of London, stuffing each of 1.5 million newspapers with a Windows 95 supplement. In Australia, all babies born that day were promised a freebie of Windows 95 itself. And, most prominently, on television everywhere, Windows 95 commercials played to the tune of "Start Me Up," for which Microsoft had paid the Rolling Stones somewhere between $2 million and $12 million, depending on whom you asked. Everybody has a price—even the Stones.

All this . . . for a $109 software upgrade (suggested retail price, prices may vary, some assembly required). The ballyhoo worked. In many stores, people stood in midnight lines to get their Windows 95. *Stood in line?* Wouldn't it be available the next day? You were better off investing in a faster modem with which to connect to the Internet. Even car buffs didn't need the newest, shiniest model the instant it was released. No piece of software was ever better on the first day it was sold—and the opposite was usually true, because bugs get discovered and fixed. The Windows 95 show left little doubt that Microsoft could generate a tidal wave, even if it wasn't about the Internet. Obviously, nobody was buying Windows 95 just to get a look at Microsoft's browser—even if it came completely free, even to businesses. Of all the features that Windows 95 offered, perhaps its strangest—perhaps not so strange—was that it just happened to cripple the Netscape Navigator. If you installed the former on your PC, the latter didn't work very well, if at all. Microsoft denied it had deliberately rigged its operating system and Internet Explorer to harm any competing browser—who could think such a thing?—but acknowledged that there were "compatibility" issues.

The autumn of 1995 saw the continued rise of Netscape and the concomitant criticism of Microsoft for not "getting it." Microsoft was already hard at work to reformulate its strategy—Gates had written his "Tidal Wave" analysis six months earlier. But events still seemed to be directing Microsoft, not the other way around, which is what Bill Gates was used to. All that changed on December 7, 1995, a day intentionally chosen by Microsoft for the historical metaphor it offered.

Fifty-four years earlier, in a surprise attack, the Japanese had bombed Pearl Harbor and drawn the United States into World War II. The man who planned the morning raid, Imperial Admiral Isoroku Yamamoto, warned: "We have awakened a sleeping giant and have instilled in him a terrible resolve." In a carefully scripted address at Microsoft's Internet Strategy Workshop, Gates invoked those words and held out his company as another "sleeping giant." If there had been doubts about his commitment to cyberspace, Gates dispelled them. Speaking to hundreds of stock market analysts and journalists at the Seattle Center—in the best FDR imitation he could muster—Gates outlined multiple lines of attack to defend his corporation's honor.

Far from the Internet being a threat to Microsoft's business, Gates said, the Internet could be his second once-in-a-lifetime opportunity. The centerpiece of Microsoft's strategy would be its new browser, Internet Explorer. Now, Microsoft would make a version of IE available not just for Windows 95, but for prior versions of Windows, as well as the

Macintosh, all of which could be downloaded off the Web. In this way, Microsoft would go head-to-head with Netscape, whose browser was already compatible with all types of PCs. Even more important, Microsoft's IE would be totally free, which, all things being equal, made it more attractive than Navigator—and just happened to threaten a major source of revenue for the fledgling company. To defend the crown jewel—the Windows operating system—Gates was willing to forgo money just to make sure the other guy didn't make any either. Even if his "Tidal Wave" misgivings about turning a buck were accurate, Windows had to be protected.

"When we say the browser's free, we're saying something different from other people," Gates explained, alluding to Netscape. "We're not saying, 'You can use it for ninety days' or 'You can use it unless you're a corporation' or 'You can use it and then maybe next year we'll charge you a bunch of money.'" Netscape could hardly claim unfair surprise. Back in 1995, in its IPO prospectus, the company acknowledged the "high degree of risk" that Microsoft might create a competing browser and "bundle" it with Windows 95, which was "likely to have a material adverse impact on Netscape's ability" to sell Navigator.

By its decision, Microsoft wasn't exactly giving up revenue. Since Windows 95 would soon become the prevailing version of Windows—and since IE would be preinstalled into new PCs sold by companies like Compaq—Microsoft would get some money anyway. It could claim it wasn't charging anything for IE, but because it set the price for Windows 95, the claim was baseless.

Besides being free, IE had another advantage over Netscape's Navigator. Since IE, beginning in October 1996, would be bundled with all copies of Windows 95 (including, finally, the retail upgrade version in a shrink-wrapped box) and then seamlessly integrated into Windows 98—this is part of what Gates meant when he said the operating system would "extend and embrace" the Internet—it was reasonable to assume that average Windows 95 users would simply "choose" IE as their browser. Indeed, the operating system itself would begin to take on the look and feel of the Web. So, why would somebody pay for Navigator in a store or bother downloading it from the Internet (a procedure that could take longer than a Ken Burns documentary)? IE's share of the browser market would have to go up, which in the long run would give it an advantage in electronic commerce and other ways to make money online. In olden days, "nobody ever got fired for buying IBM," and now in the 1990s, nobody got in much trouble for choosing Microsoft.

In his speech, Gates also announced that MSN would no longer be a proprietary online service—it had never caught on—and that Microsoft had reached agreement with Sun Microsystems to license Java. (This was a short-lived marriage in which Sun would go to court and accuse Microsoft of poisoning Java.) "Microsoft is hard-core about the Internet," Gates declared. A year before, only a handful of its programmers were working on browser technology; now hundreds were developing new versions of IE. Thousands more were shifted to other Internet projects—more employees than Netscape or any other emerging company had in toto. And millions of dollars went into a marketing drive for IE. Microsoft may have been late to the party, but when it arrived, it hogged the hors d'oeuvres. Despite his company's size, despite its entrenchment in an "old" technology, Gates wasn't going to allow his Microsoft to be "IBMed," as he had done to Big Blue.

Instead, he was going to eat the upstart for lunch—though never concede that predation had anything to do with it. Larry Ellison joked in *Upside* magazine about the "four stages" of Microsoft's stealing somebody else's idea. Stage 1 was to "ridicule it." Stage 2 was to hint, "Yeah, there are a few interesting ideas here." Then Stage 3: "Well, theirs is better than ours." And, finally: "It was *our* idea in the first place!" The joke was reminiscent of the old cartoon of the beaver and the rabbit beside the Hoover Dam. Says the beaver to his friend: "I didn't actually build it, but it was based on *my* idea."

Wall Street took notice of Gates's Pearl Harbor address. The day before, Netscape hit 174—then its all-time high—before closing at 161¼. After Gates and his lieutenants spoke, the stock plunged 28¾, a harbinger of Netscape's fate (though, at the time, it still left the company with a market capitalization larger than Apple's and that of many other established companies). The only complaint from the Microsoftians was that gravity had not pulled Netscape down more. In a front-page story on Microsoft's conference, headlined GAME PLAN FOR THE INTERNET: CRUSH THE COMPETITION, the *Seattle Times* reported that two Microsoft executives were chortling over Netscape's double-digit price drop when a colleague chimed in, "That's not enough."

Asked about Gates's call to arms in Seattle, Jim Barksdale said he expected "a dogfight." But, he added, "God is on our side." Unfortunately, that still left Microsoft ahead. God had nowhere near the market penetration, a far less monolithic operating system, and, as anybody over forty knows, He didn't offer memory upgrades. Valiant or merely proud, Barksdale conjured up the image of a distant cousin named William Barksdale, a brigadier general for the Confederacy. Barksdale led what a Union officer at Gettysburg in 1863 described as the "grandest charge that was ever made by mortal man." Barksdale was killed in it.

Brave words notwithstanding, Netscape did its best to get out of the way. Despite Netscape's dominant share of the browser market—as high as 87 percent in the spring of 1996, compared with IE's 4 percent, according to one estimate—Netscape had to assume it was going to lose ground once Microsoft attacked. Navigator might have close to 40 million users, but the Windows 95 operating system had double that number within the year and it was reasonable to suppose that a lot of those users would be perfectly content with the accompanying browser.

Licensing Navigator was a primary source of revenue for Netscape, going from $2.3 million in the first quarter of 1995 to $58.5 million in the last quarter of 1996. But that money was going to vanish. Netscape didn't need a premonition: It could see what was already happening to Spyglass and the old NCSA crowd in Illinois. When Gates announced that Microsoft's browser would henceforth be free, he was stabbing Spyglass in the back. Along with licensing its Mosaic browser to Microsoft, Spyglass had sold it to dozens of other companies, including IBM. There would be no revenue from those deals anymore—why buy something that Microsoft was now offering for free?

His company awash in cash, Gates remarked to the *Financial Times* of London that "our business model works even if all Internet software is free," because "we are still selling operating systems." In that cut-throat environment, Gates rhetorically asked, "What does Netscape's business model look like? Not very good." It was barely turning a profit, while Microsoft was making a few billion. So, rather than relying on a business built around a browser, Netscape by 1996 had begun to diversify. Netscape formed partnerships with other companies in the Valley and, flush with capital from its IPO, made several corporate acquisitions to fill in technological gaps. It sold advertisements and E-commerce hyperlinks on its Web home page. And to get away from relying on browser revenue from businesses, Netscape intensively began to sell other kinds of expensive "back-end" Internet software. More than one-third of Netscape's revenues was already coming from software to run corporate "servers," which connected the desktop computers within an organization to the Web.

One of the key markets Netscape hoped to control was for "intra-nets," which were private, internal networks that many companies were starting to use. Just as the Internet allowed the interconnection of computers worldwide, an intranet setup allowed employees *within* an organization to share information and databases in a way that went beyond E-mail— calendars, financials, enrollment in benefit plans. If the data was created in HTML, PC users within a company could navigate around the internal network with the same ease they could cruise the Web. In turn, suppliers and consumers could be connected with this internal network, providing an efficient communications tool. While the Internet was more dramatic in its reach, intranets could be much more practical, because colleagues had more to share than total strangers. Information could be posted quickly, and with certain security features, an intranet could also be programmed to restrict access to specific sites. It was this corporate market that IBM's Lotus Notes software had tapped into and that caused Barksdale earlier to say that he regarded it, rather than Microsoft, as Netscape's chief competitor. Fueled by intranet sales, Netscape's revenues in 1996 hit $346 million. (Microsoft took more than a decade after its founding to hit that kind of number.) Netscape's stable of intranet customers included more than two-thirds of the companies in the Fortune 100.

Netscape also beefed up its products that helped businesses to create Web pages, to customize Internet applications, and to do online commerce. These disciplined efforts by Barksdale to move out of Microsoft's scope made sense. But Barksdale could not ignore the centrality of Netscape's browser to its mission. Quite apart from revenue, it was the browser that gave the company its identity—what had transformed it into a brand name and what helped it to distribute other products. By the summer of 1996, Microsoft and Netscape were both coming out with Versions 3.0 of their respective browsers, and by this round, IE was every bit as good as Navigator in features and simplicity. To the extent that Netscape could heretofore tell the world its browser was superior and justified the special effort to install it on a Windows 95 machine, that argument was gone. Internet Explorer 3.0 also marked the first version bundled with the retail Windows 95.

Meanwhile, Microsoft had decided it wasn't enough to sit back and wait for the natural advantage of Windows 95 to take hold in the marketplace. Maybe that would happen, but maybe it wouldn't; after all, MSN hadn't succeeded in displacing online services like AOL. The "browser war" required an assault on other fronts—at least as strategized by the generals in Redmond. In the descriptive phrase the Justice Department later attributed to Microsoft's Paul Maritz, this all-out campaign against Netscape was designed to "cut off their air supply"; "Netscape pollution must be eradicated," as another Microsoft executive, Jeff Raikes, put it. The prototypical example of Microsoft using its operating system to lever its browser occurred in the summer of 1996. The target was Compaq Computer, which assembled and sold more PCs than any manufacturer in the world, which was another way of saying it sold more Windows-based PCs than anyone else.

Making a straightforward business judgment, Compaq had recognized the popularity of Netscape's browser and chosen to preinstall Navigator in a popular line of PCs. Moreover, even though IE came with the Windows 95 operating system, Compaq decided it would remove the IE icon from the screen that users saw when they turned on the machine. Microsoft responded with its version of a death threat: If Compaq deleted the IE icon, Microsoft would rescind Compaq's license for Windows 95. Compaq was not unimportant to Microsoft's bottom line, but Microsoft was indispensable to Compaq's. Yes, it needed plastic to make the shell for a computer and, yes, it needed microprocessors from Intel or another chipmaker. But if Compaq was denied a license to install the only mass-market operating system, it was out of business.

Compaq got the message. The Microsoft browser remained. Other PC makers reportedly were extended cash discounts on Microsoft software or given marketing tie-ins—just as long as their machines made IE the preferred browser.

There were more subtle attempts to coerce companies. Intuit, which made Quicken, the industry's leading software for balancing a checkbook, was considering a plan in 1996 to sell Navigator along with its product. In return, Netscape would provide Intuit with the ability to embed its browser in Quicken, a function that Intuit badly wanted. But Intuit wound up doing an exclusive deal with Microsoft, for which Intuit got placement on the Windows desktop. The genesis of the deal became clear in an embarrassing E-mail from Gates that came out during the federal government's antitrust suit. In it, Gates recounted a July 23, 1996, conversation he had with Intuit's boss, Scott Cook. "I was quite frank with him," Gates explained in fractured syntax, "that if he had a favor we could do for him that would cost us something like $1M . . . in return for switching browsers in the next few months I would be open to doing that." (As smart as Microsoft was, one might have thought it would've figured out how to delete incriminating E-mail from its computers. Tom Hanks and Meg Ryan in a story of love and deceit: *You've Got Blackmail!*)

Stories within Netscape and in the press about Microsoft's competitive high jinks abounded. According to *The Wall Street Journal*, Gates's No. 2—Steve Ballmer—called the CEO of PacBell Internet Services after PacBell signed a contract with Netscape. Ballmer informed the CEO that he was now "an enemy" of Microsoft. There was pressure brought to bear on bigger companies. Disney was told it would get no Microsoft online promotion if Disney dealt with Netscape. Barksdale said an executive of Hewlett-Packard reported that both Gates and Ballmer had called to vent after Netscape and HP announced a joint-product marketing agreement. Apple Computer claimed it had been forced to bundle IE into its machines because Gates threatened not to update word-processing and other software for the Mac. "The threat to cancel Mac Office 97 is certainly the strongest bargaining point we have, as doing so will do a great deal of harm to Apple immediately," wrote a Microsoft official in an E-mail to Gates. (Gates later claimed under oath he didn't recall receiving the message.)

Gates had presaged these efforts in his 1995 "Tidal Wave" memo, where he urged senior staff to work with companies "who are considering their [Netscape's] product." Even the great and powerful Intel—the only company in the Valley to survive a partnership with Microsoft—was not immune to its strong arm. Intel made the microchips inside most of the PCs that ran the Windows 95 operating system and was as close to a true partner as Microsoft had. A "Wintel" machine was half-Intel. Nevertheless, when Intel's software division pursued its own Internet R&D, Gates got involved. One internal memo at Intel referred to "vague threats" by Gates that, in turn, brought in Andy Grove himself. Intel fell in line. While this didn't hurt Netscape, it did show Microsoft's muscle.

Microsoft's tactics sometimes affected Netscape more directly. In March 1996, Netscape thought it had reached an alliance with America Online—the largest of the companies selling dial-up access to the Internet, and a natural enemy of Microsoft—to incorporate Navigator into AOL's service. But Microsoft also had been negotiating with AOL, which told Netscape that it simply would offer its customers a choice; by no means, according to Barksdale's account, would AOL let either browser exclude the other. On March 11, from its headquarters outside Washington, D.C., AOL announced its licensing deal with Netscape. Now, Navigator appeared to be the main Web browser for AOL's coveted six million customers—and, as Netscape personnel gloated, Microsoft would not. When the stock market closed that afternoon, Netscape was up 15 percent.

But the licensing deal didn't give Netscape exclusive AOL placement. Netscape would get paid only for those Navigator browsers AOL customers actually selected. AOL could ally itself with other browsers if it wanted—and it quickly did. The next day, in a double-cross that would do Bill Gates proud, AOL announced a deal with Microsoft, not only licensing its browser but making IE the *default* choice of AOL users ("force-feeding," Barksdale called it). Unless they specifically chose Navigator, they'd get IE when they went to the Web. Typical AOL customers were computer novices, so they were unlikely to seek out a different browser, assuming they knew what a browser was to begin with. This time, on the Wall Street roller coaster, Netscape's stock went down.

In some ways, Microsoft looked to AOL like a better partner than Netscape, despite AOL's long-standing hatred of Redmond. Microsoft wasn't going to charge for its browser and Microsoft offered better technical support to ensure the software of the two companies was compatible. But the clincher for AOL was that Microsoft agreed to put AOL's icon on the Windows 95 desktop. To subscribe to AOL's service, all a Windows user had to do was click on the AOL icon. This concession by Microsoft gave AOL equal standing on the Windows desktop with Microsoft's own online service, MSN. (Ironically, the year before, the placement of the MSN icon there had so angered AOL that it complained to antitrust regulators at the Justice Department.) Gates was now willing to sacrifice MSN if it got his browser more market share over Netscape.

AOL understood the advantage that Microsoft's operating system offered. That's why it sold Netscape out the day after making a deal. In the law of the jungle, the weasels make out as well as the tigers. Other Internet access providers—including AT&T, MCI, Netcom, and CompuServe—followed AOL in hooking up with Microsoft. Some of the contracts, Netscape later claimed, were written specifically to exclude Navigator as a competitor; others not only provided for free Microsoft browsers, but paid "bounties" to access providers if they enticed users into signing up for other Microsoft services.

The president of one access provider named Global Telecosm sheepishly wrote to Netscape in the summer of 1996 that it couldn't very well distribute Navigator because "Microsoft gave me a deal I couldn't refuse. . . . I know Netscape is better, but $0 vs. $18K is impossible to beat"; another small access provider, on capitulating to Microsoft in 1997, said, "We held out as long as we could, but can no longer bear this tariff," especially "when Microsoft will fly a blimp with our name on it for free." Then came deals with online content providers like *The Wall Street Journal* that gave IE users free access to Web sites that charged everybody else (like someone who had reached the Web site using Navigator). Other Web sites got the velvet-glove treatment from Microsoft as long as the sites removed any reference or hyperlink to Netscape.

The AOL deal gave Microsoft a solid foundation on which to build market share. By the fall of 1996, it had gone from almost nothing a year before to 20 percent. A year after that, it had almost doubled to 39 percent, with Netscape down to 51 percent. There was no mistaking the trend, which seemed inexorable. John Doerr had wishfully predicted "the most titanic battle for market share ever—bigger than Coke vs. Pepsi"— but it was fizzling out.

In Mountain View, the Netscape accountants began to see the effects of Microsoft's sorties. In 1997, Navigator revenue plunged. From its high of $58.5 million in the last quarter of 1996, Netscape revenue plummeted to $18.5 million at the end of 1997—a drop of 68 percent and one of the reasons behind Netscape's disastrous year-end financials (a quarterly loss of $88 million, which led to layoffs of three hundred and a stock price 80 percent below its peak). The collapse was complete in 1998, as Netscape acknowledged it could no longer charge anybody for its browser and even released Navigator's source code to the public. Its 1998 Navigator revenue: zero. During the summer, IE's market-share momentum overtook Navigator—44 percent to 42 percent. That advantage would only grow wider, even with Netscape's belated decision to give away Navigator. As promised, Microsoft had been ruthless—and, critics charged, illegal. Less than three years after Gates spoke on Pearl Harbor Day, his company had officially passed Netscape by. The underdog was just a dog.

The irony was that Microsoft had accomplished its mission with free software. There was a time when free software was the rule. But then, IBM and other companies had the bright idea to sell it. In January 1976, as a very young Bill Gates and Paul Allen were beginning Microsoft, Gates wrote an "Open Letter to Hobbyists" (published in the *Homebrew Computer Club Newsletter* and elsewhere), in which he excoriated hackers who didn't pay licensing fees and compared them to thieves. Besides denying him his cut, Gates said, the failure to pay had the effect of stunting innovation. "One thing you do is prevent good software from being written," he scolded. "Who can afford to do professional work for nothing? What hobbyist can put three man-years into programming, finding all the bugs, documenting his product and distributing it for free?"

Fair point. But then how could he justify his pricing policy for Internet Explorer? Of course, no one was ripping Netscape off, as Gates claimed the hobbyists were doing to him in 1976. But his observation about innovation perfectly fit Netscape's plight. Lucky for Gates he wasn't an enthusiast of irony.

You couldn't argue with Microsoft's Internet success as a matter of realpolitik. Antitrust, however, is another matter. Marc Andreessen may or may not have been thinking of the antitrust posse back on June 21, 1995, when he took such good notes of the big meeting with Microsoft. But over the next year, the executive machinery of Netscape began to gather a record of what Barksdale would call "Microsoft's various exclusionary, restrictive, and predatory actions." The Compaq fiasco, in particular, led Netscape's lawyers to contact the Justice Department again in August 1996. Barksdale was no fan of government regulation—he thought it would be a distraction—yet he became convinced that Netscape had no fair chance in the marketplace against a company that ruled the underlying operating system for PCs. Clark had wanted to intervene earlier and not rely on the government. "If I was running the company," he says, "I would've sued Microsoft myself. It's impossible to go after Microsoft in a business context, so I would've gone to the courts and I would've been just as ruthless as them. It gives me no confidence that the United States government is handling this."

The Justice Department was well versed in claims of Microsoft machinations—from Lotus to Novell (which bought out Gary Kildall's once-proud operating system) to Jerry Kaplan's GO. Between the Justice Department and the Federal Trade Commission, the feds had been investigating Microsoft since the late 1980s, each time reaching some limited agreement that fine-tuned Microsoft's ability to leverage its market position; the only exception was in the spring of 1995, when Justice capsized Microsoft's $2-billion attempt to buy Intuit. (Employees at Intuit threw a party when the deal crumbled.)

In October 1997, the United States sued Microsoft again. The Justice Department alleged the company had violated a 1994 consent decree—settling a prior antitrust dispute—by forcing PC makers to install the IE browser as a condition for getting a Windows 95 license. Two months later, U.S. District Judge Thomas Penfield Jackson issued a preliminary injunction against Microsoft, ordering it to stop arm-twisting computer manufacturers. Three months after that, Microsoft loosened its grip on

Internet access providers by no longer forbidding them to promote Netscape's browser. The move was akin to turning off the faucet after the baby had already drowned in the tub: By March 1998, Netscape was in full market-share retreat in the browser wars.

The skirmish over the earlier consent decree was necessarily narrow. As an appeals court made clear in May 1998, any restrictions imposed by that decree and Judge Jackson's enforcement of it would not extend to the impending release of Windows 98. In short, the Justice Department's lawsuit would soon become moot. For Barksdale or anyone convinced Microsoft needed reining in, a new and broader antitrust salvo would be required.

After considerable braying throughout the software industry and within the Justice Department, it finally came on May 18, 1998, as the federal government sued Microsoft again, this time with twenty states adding claims of their own. The crux of the case was Microsoft's war against Netscape, but included a range of other allegations that Microsoft acted illegally to maintain its Windows monopoly. These included actions against Sun, Apple, and other competitors. Larry Ellison celebrated the antitrust filing by sending his Oracle employees to see the newly opened *Godzilla* movie, which ended with the slaying of the monster, albeit not by government lawyers.

Civil action 98-1232, *United States of America* v. *Microsoft Corporation,* represented the Justice Department's most ambitious antitrust suit since the late 1960s, when it went after another high-tech titan. In that thirteen-year duel with a company called IBM, the government wound up surrendering in 1982 and getting nothing in return. But it didn't really matter. The marketplace had begun to erode IBM's dominance—most impressively because of a start-up in Albuquerque founded by Bill Gates and Paul Allen. Now that start-up was citing its own history as an object lesson in the wisdom of economic competition and the folly of government regulation.

The Microsoft plea: Darwinian conflict, not courtroom analysis, should resolve the whining of weaklings like Netscape. It was the law of the jungle, not of the century-old Sherman Act, that should govern. Why should it aspire to be dinner rather than diner? Microsoft was right that at least some of Barksdale's complaints amounted to bellyaching. "Microsoft's comments about Netscape appeared designed to create doubts about Netscape's ability to compete in the market," Barksdale testified. "It was not a totally uncommon event for a customer to question whether it made sense to do business with Netscape because of Microsoft's public position that it was going to crush Netscape's business." So, instead, Microsoft was supposed to say it wanted to *help* Netscape's business? Venture capitalists in the Valley should've agreed with Microsoft in principle, but it was their financial oxen being gored, so they couldn't very well say anything. John Doerr, in particular.

Months before, Microsoft saw the lawsuit coming. Assuming the Justice Department itself couldn't be acquired, Microsoft responded in the only way it knew—a marketing campaign. Gates himself hit the stump. First he taped an interview with Barbara Walters. She asked no probing questions about Compaq or Intuit or that silly, misunderstood meeting at Netscape. Instead, Barbara wanted a prime-time rendition of "Twinkle, Twinkle, Little Star." Bill was no meanie—he's Barney the lovable lullabyster!

Then, venturing into more hostile territory, Gates visited Silicon Valley itself—not coincidentally, at the same time Netscape was announcing its worst financial quarter ever. Here he tried to put his best foot forward, in a way that, well, Bambi might have appreciated. But even the richest man in the world, with twelve advance-people and additional public-relations staff, can't make a silk purse. Consider the disconnect at the morning photo op—the third-grade class at Cesar Chavez Academy in East Palo Alto. In the shadow of John Doerr's "largest legal creation of wealth in the history of the planet," this was the dregs of Silicon Valley. On the other side of the tracks from Palo Alto, in the city with the highest murder rate of any in the United States a few years ago, Bill Gates came to show he was quite-a-guy for donating Microsoft software to underprivileged kids. (Free browsers, too!)

When enough video footage had been shot by the local news crews, he was off to Stanford, for a heart-to-heart with a huge auditorium full of MBA students. The university's external relations staff had done its best to roll out the cardinal carpet for Gates—wouldn't you, if this was the man who'd already put up one science building for you and might someday build a dozen more? But the staff became exasperated, not just from the sheer number of Microsoftians who had been sent to campus ahead of time—more than the White House sends before Bill Clinton comes to visit Chelsea—but by their imperial demands. Gates insisted that the dean of the business school meet him outside when his motorcade arrived—in the rain—even though the speaking event had been arranged by business students and they wanted the honor of escorting Gates inside. "Even Ellison didn't demand this," noted one of the university's executives. Microsoft also asked that no media representatives be permitted in the auditorium because they would deny seats to more worthy students (raising the question of how the Gates message gets out if the press isn't allowed in to hear it). The auditorium held 1,800. Stanford refused the request.

Inside, donning a Stanford sweatshirt and perched informally on a stool onstage, Gates tried to strike the themes of his company's defense. Yes, "Microsoft is in the best position to succeed," he said, "but there's no guarantee in the software business. One of the great things about coming to work every day is knowing you can destroy the company." After all, the Internet had "caught us by surprise"—who knows what might've happened if Netscape had gotten an even larger lead? And another admitted blunder: "Microsoft can be criticized for not recognizing that some of our competitors were back in Washington trying to turn our success into something negative." Well, that's what passed for self-deprecating Bill, whose sense of humor was second only to Michael Dukakis's. Then, at last came true Bill, as he lashed out at his tormentors at Justice. "One of the privileges of success in this country is government scrutiny," he said. "And that's okay. We have a very sexy industry. If you worked at the Justice Department, which would you rather investigate—bread or software?" Compared with wayward husbands and lying presidents, both sound rather dull, but it's a fair point.

Later, before a gathering of editors and reporters at the San Jose *Mercury News,* Gates's tone was dismissive, as if he had forgotten the public-relations mission that brought him to the Valley. He was gratuitously uncooperative, refusing to let the *Mercury News* put the Q-and-A session up on the Web as an audio file (apparently thinking he read better than he sounded, except when it came to "Twinkle Star"). The whiny voice didn't help, but the problem, as they liked to say in Webspeak, was one of "content." Gates trotted out the Microsoft party line—that Microsoft couldn't possibly be a monopoly because it had a mere 4 percent of software sales. The fact that it was concentrated in the operating system that controlled almost all PCs seemed to be beside the point.

The argument undercut Gates's credibility. It's dumb to pretend there's no monopoly when the facts show otherwise—all the more so when the mere existence of a monopoly is perfectly legal. As long as you don't engage in illegal acts—predatory pricing, price-fixing, exclusionary contracts, tying the sale of a product to the one in which you hold a monopoly—then having a monopoly is fine. Gates knew better a lot earlier. "I really shouldn't say this," he confessed at a May 1981 computer conference, discussing the need for operating-system standards, "but in some ways it leads in an individual product category to a natural monopoly." He said that before he had a pack of lawyers on retainer.

It was performances like his day in Silicon Valley that made Gates look guilty of *something*—which was remarkable, since he had been the consummate tactician his entire business life. If ever there were a company that appeared in need of behavior modification—even discounting Barksdale's allegations and going by what Microsoft acknowledged as "tough negotiating"—it was Microsoft. Though it was hard not to admire a business built in barely a quarter-century from scratch into the most valuable corporate enterprise ever, Microsoft seemed like a monster that needed to be cut down to size. But whether any of Microsoft's entrepreneurial aggression deserved the obloquy of an antitrust lawsuit was a much trickier question.

At its root, antitrust isn't very complicated. After all the economists have testified, after all the high-minded political rhetoric about free enterprise, it is glorified consumer-protection law, resting on the principle that sometimes the free market doesn't know best and that the government must get involved. It's all well and good to demonize the likes of John D. Rockefeller and William H. Gates—to dislike "bigness" and the cowardice of bullies—but that by itself doesn't mean a worthy antitrust prosecution.

As matters of fact, the charges against Microsoft were persuasive. There were essentially three:

• Predatory pricing. Giving away the IE browser would be what old man Sherman and his statutory cousin Clayton had in mind—an overbearing monopolist using its deep pockets to sell goods below cost. This is what gave Standard Oil a bad name. With enough market share, you can scare off any future competitors or buy them out early on, both of which . . . lead to more market share. Microsoft had tried it before. Tom Rolander, who was Gary Kildall's partner at Digital Research and now runs a small software company, tries to remind himself every day not to compete with Microsoft: He keeps on his shelf a copy of a shrink-wrapped box of an old product, "Microsoft Money," intended to compete with Intuit's personal-finance software. The box is priced at "$9.99"—with a $10 rebate.

• "Tying." There's no factual doubt that Microsoft used its operating-system dominance—a monopoly, by any other name—to leverage IE's distribution. It looked at first glance like Kentucky Fried Chicken forcing its franchises not only to buy the spices from them— that was okay, since precise taste was key to the product—but also the napkins and paper bags. No bags, no chicken. That's illegal. Similarly, PC users were forced to take IE as a condition of getting Windows; there wasn't an explicit increase in price, but presumably Microsoft factored its costs into the price. Microsoft could still argue that its leverage was no guarantee: MSN flubbed even with a favorable location on the Windows desktop. But the more interesting debate is legal in nature—whether the tying arrangement was no more than a logical linking of related technologies and not solely intended to squeeze out Netscape.

• Exclusionary contracts. It's possible that Compaq or Apple or Disney or Intuit or America Online made business decisions favoring IE purely on the merits—maybe it worked better, faster, or had prettier colors. However, in the face of threats like the one Bill Gates made to the head of Intuit, and the juxtaposition of AOL's respective contracts for IE and Navigator, it seems rather unlikely that the menacing presence of the Windows operating system was irrelevant. But for its monopoly power, Microsoft never would have gotten all those deals. And while most companies are permitted a certain latitude in making exclusionary deals, monopolies are not.

Microsoft's take on the three charges bordered, at times, on the disingenuous. On predatory pricing, it argued that Netscape itself charged nothing for its browser to many of its nonbusiness customers and by 1998 was giving Navigator away to everybody. The latter point is silly—it only underscores the predation. The former would be reasonable, except for both Microsoft's stated goal of attaining market share at any cost and the millions that Netscape had in fact raked in on browser sales.

On tying, Microsoft said it was simply adding a feature to its operating system, much as car manufacturers included steering wheels and even cup holders. "Integration" became the mantra of Bill Gates, especially when Windows 98 was introduced: You could no more separate out IE from the new operating system than you could force Plymouth to leave an odometer off the dashboard. Wasn't it up to the manufacturer in the first instance to determine what consumers wanted? That's not a bad argument, but it's no better than the one that consumers could quite easily add an Internet browser to their computer after they purchased the machine, or better yet, could decide at purchase time between Navigator and IE if they had the choice. If you wanted to see the hollowness of Microsoft's "integration" argument, you had to look no further than the versions of IE made for the Macintosh, as well as old versions of Windows. These products had nothing to do with Windows 95 or 98. Yet Microsoft made an Internet browser for them, too—and offered it to the public for free. The only reason was to win market share. Free distribution of IE for those operating systems presented no suspect "tying" arrangement, but exposed integration to be a pretense.

On exclusionary deal making, in the face of overwhelming documentary and testimonial evidence to the contrary, Microsoft maintained that its various business partners chose IE because it was best. This was another way of articulating Corleone's decision-making theorem. Assuming, say, Compaq or AOL believed it had no choice but to accept Microsoft's offer, it's obviously going to say it believed Bill Gates's products were superior. It wouldn't look very good to say in the press release, "Our company is pleased to announce today a special partnership with Microsoft not because we like their stuff, but because Bill threatened to put us out of business."

Finally, as a catchall response to any allegations that he didn't play well with others, Microsoft gave it the old backyard whine, "But, Mom, he did it, *too!*" Netscape, Sun, Oracle, IBM—they were ganging up against Microsoft. Divvying up markets? That's nothing more than "exploring the possibility of forging a strategic partnership in some areas," as Microsoft officially described its version of the June 21, 1995, meeting with Netscape. Strong-arming competitors? Golly, everybody in high-tech did that. Bundling products? Well, Netscape included an E-mail application with its browsing software. Such was the level at which Microsoft rested part of its defense. Earlier in the 1990s, Microsoft had kept a low profile in its antitrust travails with the Justice Department and FTC, saying its policy was "not to try cases in the press." But in this latest round, arrogating the role of misunderstood prophet, Bill Gates called the action against Microsoft a "witch-hunt."

If antitrust liability turned on facts alone, Microsoft faced almost certain defeat. Yet on more finespun legal technicalities, the company was on better ground. What harm had befallen consumers? Maybe Windows 95 or 98 might be cheaper without a browser, and maybe the amount saved might be more than the cost of a different browser. But since no one could know, wasn't that a thin reed to base an antitrust case on? Maybe the entire Internet business—going beyond Netscape's little piece of it and extending into electronic commerce and alternative computing platforms—might look very different if Microsoft didn't wield so much power. But who knows? Maybe, despite the natural tendency toward a uniform standard, there would even be different operating systems on the market and they'd all be easier to use than Windows. But who knows?

In an industry that changed so swiftly—where a tiny start-up in Albuquerque supplanted IBM in the course of twenty years—did it make sense for a federal court to step in and restructure the marketplace, particularly when most remedies might be obsolete within months? Judges have enough trouble overseeing the law without adding product design and pricing policy to their dockets.

Even antitrust's version of the death penalty—breaking up Microsoft into a series of "Baby Bills" (either, say, several downsized identical companies, or an applications company and an operating-system company)—wouldn't necessarily stimulate competition. Certainly the permanent separation of the Windows operating-system business from the applications-software business would hurt Microsoft's bottom line. Bill Gates might never accumulate a net worth of a trillion dollars. And instead of the big company having as many as four thousand millionaires, maybe the two little companies between them might only have a thousand or two. But would the breakup necessarily lead to more competition or just less efficiency? Would consumers be comfortable scrapping their attachment to, if not satisfaction with, a family of products that a single company could give them? Would a breakup facilitate the emergence of another would-be monopolist and then another round of antitrust litigation? A Microsoft apologist would say that companies like Netscape were using antitrust actions as a substitute for competitive inadequacies. But honest economists will tell you that determining the outcome of market regulation is a parlor game. More than any strict economic principles, it is the law of unintended consequences that governs the result.

Antitrust law at its simplest level is a paradox. Companies exist to make money. Nobody consciously aims for less than a 100 percent market share—in the fall of 1994, what was Netscape after when it released the first Navigator *other* than market domination? There's a reason Parker Brothers called its most popular board game "Monopoly." In applying antitrust statutes, courts recognize that. There's nothing by itself illegal about securing or keeping a monopoly. Yet, when a company does just that, it is subjected to a set of rules that don't apply to other companies. The kind of abuses ascribed to Microsoft exist in any business. Microsoft's offense is just being better at the abuse.

As virulent as the Microsoft disease seems to be, a court-imposed cure would likely be worse in the end. As successful as the company's anticompetitive actions have been, the facts remain that it has presided over the most successful American industry in generations. Instead of deploring that fact, the normally libertarian and techno-narcissistic companies of Silicon Valley might do better at trying to emulate it. Fight fire with fire. The day may yet come when Microsoft deserves to go the route of Standard Oil and Ma Bell, but the better bet is that someday it will have done unto it what it's done unto others.

In 1998 and early 1999, the freely distributed Linux operating system was making precisely such noises. It's true that a better operating system still needs compatible software applications like word processing and spreadsheets—and that those creations require an enormous initial investment in development and marketing. It's also true that Microsoft, which has cornered the market for those applications and has a $19 billion war chest (as of late 1999), might very well be able to fight back such incursions in a laissez-faire environment. But on the other hand, there's just too much booty built into unseating Microsoft's operating-system dominance to assume it could never happen. This is especially so when there's no guarantee that PCs will even be the computing appliance of choice in five years; Palm Pilots and handheld devices already are making inroads. The No. 1 Microsoft-hater himself, Larry Ellison, notes that "Microsoft has no friends." Next to Windows, its regnant product is antipathy. "No one will rush to help them if they ever have trouble." That's just another reason for the government to let Microsoft do itself in—and stay out of it.

Whatever befalls Microsoft won't matter to Netscape. Around the time the Justice Department was finalizing its lawsuit, Netscape began talks to put itself out of its own misery. Once before, in 1994, America Online's CEO, Steve Case, had been eager to acquire a piece of the company; conveniently, his brother was on his way to running Hambrecht & Quist, the investment bank for Netscape. However, Netscape wanted no part of AOL, fearing that it not only might present obstacles for partnerships with other Internet access providers, but was too lowbrow. AOL was the Chevrolet of the Internet; Netscape was a Ferrari. In the fall of 1995, Case again made overtures to Netscape, this time about working together to "attack the common enemy"—the "Hitler" from Redmond. As Netscape was concerned about its browser, AOL was petrified that Microsoft's MSN might muscle out its online service. "My Dearest Comrade Barksdale," Case wrote to Netscape's CEO, referring to himself as "Franklin D." and Barksdale as "Stalin." In his reply, Barksdale, knowing a bit more history, went with the more respectable "Winston C." The talks produced little, except to show that both men had a healthy respect for themselves.

Then in 1996 came AOL's opportunistic flip-flop with Microsoft and Netscape over their respective browsers, poisoning any possible close relationship. Mutual opportunism, of course, made everybody forget any hard feelings. By 1998, AOL had become the nation's largest online company, providing Internet access and Web content tailored to the masses, including chat-ter rooms on all things topical; a newcomer to the Web might have asked if the country was really a better place for allowing hundreds of freedom-loving Americans to debate whether Calista Flockhart really needed to scarf a few more cheeseburgers. AOL's "carpet-bombing" marketing gambit for signing up new customers—an AOL floppy disk or CD-ROM in every mailbox, and better yet, forty-two of them—made it as ubiquitous as Ed McMahon at Super Bowl time. "You've Got Mail!" indeed.

For its part, Netscape, despite the loss of browser market share, was still a leader in Internet technologies with a long list of plum customers and a popular online "portal" called Netcenter. Portals—most notably, Yahoo—were becoming the gateways to the commercial Internet. They were the starting points for navigating the Web and often they were destinations unto themselves. You could program your browser at start-up to automatically connect to a portal, which, in turn, could be personalized for any user. Portals gathered together on one page all sorts of free information: search engines, E-mail, news headlines (politics, Hollywood, Afghan troop movements), sports scores, horoscopes, weather forecasts, street maps, movie listings, instant stock prices, phone directories, bulletin boards, games, and hyperlinks to other sites (maybe you wanted instantly available an index of gardening sites, or a list of every fan page devoted to Rocky and Bullwinkle). Besides the printed word, there was also audio and video entertainment available. One day soon, the portal might be the place to pay bills, manage personal finances, and make phone calls. The beauty of a personalized all-in-one page was that, unlike a newspaper or TV station, it was specifically tailored to a user's interests.

That's why portals drew Web users—and, inevitably, advertisers, who could better target prospective customers. If millions of people were going to be surfing the Net, they couldn't very well be allowed to do so in peace. They needed to be sold stuff, told to buy things, bombarded with slogans and brands. If it was good enough for highways and rooftops, TV and radio, magazines and newspapers, then it was good enough for cyberspace. Web advertising took off in 1996 and 1997, dancing across the top and along the side of Web pages. When you clicked on an ad for Amazon.com or a more conventional site like General Motors, you were taken to the home page of that company, where you could get more data or make a purchase. Advertising was the revenue model for the portals, along with kickbacks on commerce referred elsewhere. (You buy a book at Amazon.com and Amazon gives a cut to the referring portal.) Because so many users began their Web travels at a portal, advertisers were willing to pay for the chance to lure prospective customers away. Just as there were only several TV networks, a few Internet portals evolved as the major funnels into the electronic universe.

Netcenter (as well as AOL's main site) was consistently among the top five portals. Netscape had maintained a home page from the outset, but at the beginning it was full of dull product information. Its chief function was to be the place where a Web user could download Navigator—all the more so because it didn't come gift-wrapped with most PCs. While Netscape's Web site languished—Barksdale mistakenly told colleagues that he thought advertising revenue would be "incidental"—other start-ups like Yahoo moved in. Had Netscape foreseen that Web traffic would coalesce around a few sites, it might have become the unquestioned portal leader. But that was hardly an intuitive notion in 1994 or 1995, and Netscape had all it could handle running one business without entering another. It seemed a reasonable judgment: Given the Web's potential for unlimited, fluid channels of information—so different from the centralized model of TV and even cable—the idea of a portal violated all that the Internet stood for.

Even so, this failure to capitalize is one of the many mistakes Netscape's critics cite as a reason it didn't keep its IPO luster. Microsoft launched its torpedoes, but Netscape inflicted wounds of its own. A 1998 book by two business professors, MIT's Michael Cusumano and Harvard's David Yoffie, compiled many of them. In *Competing on Internet Time: Lessons from Netscape and Its Battle with Microsoft*, Netscape's own people described sloppy software code and an arrogant business strategy content to rely on past hype. There was a fine line between waiting too long in perfecting a product and delivering it before it was ready. Netscape didn't always strike the balance (much as Oracle didn't with its unsuccessful network computer). NETSCAPE CHAIRMAN POINTS AND CLICKS, swiped a front-page headline in the *New York Observer*,

MILLIONS DISAPPEAR. Barksdale was right: The euphoria of the IPO gave it instant recognition, but made it a media target.

The impudence it first demonstrated in mocking Microsoft especially annoyed customers. Some interviewed for the Cusumano-Yoffie book went so far as to compare Netscape with Microsoft. Solve a customer's complaint? First you had to pick up the phone. According to Andreessen, Steve Jobs visited the company in 1996 to meet with senior staff and "spent about two hours telling us how fucked up we were." Jobs, who well knew the phrase "the arrogance of Apple," then publicly commented that he couldn't stand dealing with Netscape. It takes one to know one.

In the boardroom, Netscape seemed to be constantly adopting new strategies. From browser king to intranet master to pioneer of the "extranet" (business-to-business electronic commerce) to portal power, Netscape seemed to be a new company every six months. It was understandable, given the advance of Microsoft's fleet, but it still gave Netscape the appearance of being adrift. One of the main discussion items among directors, Clark says, "was how to get out of Microsoft's way." During the time frames allowed by securities law, Clark did his utmost to get out of the way himself, selling off three million of his shares. "No insider information," Clark explained. "Just instincts. If I could have sold it all, I would have. . . . It was clear Microsoft wanted to wipe us out." Andreessen did similarly, giving up almost half of his stock in 1997 and 1998.

If brought together, AOL and Netscape could forge a powerful media company—and a further humbling of Microsoft's MSN—particularly if a third company could be found to distribute Netscape's high-end software to businesses. Between AOL's total of 19 million subscribers and Netcenter's 20-million monthly audience, an AOL-Netscape company would control more online revenue—advertising and E-commerce commissions—than any company, including the bear in Redmond, as well as Yahoo. "Scale" meant everything at AOL, and acquiring Netscape meant scaling the company up in a way no amount of carpet bombing could accomplish. AOL would control both ends of the demographic spectrum in cyberspace, from the gearheads wise to the Web and loyal to Netscape, to the "newbies" and families who discovered the Internet last Christmas, right after Santa delivered a new PC (alas, without decipherable instructions). Best of all, Netscape came at an attractive price—far below the ether of the IPO.

Merging Netscape into AOL made so much sense it actually happened. In late November 1998—even as the Microsoft antitrust trial plowed ahead in Washington, D.C.—AOL reached agreement to buy Netscape for about 9 percent of all AOL shares, which wound up amounting to $9.8 billion when the deal closed four months later. In a side deal, Sun Microsystems agreed to handle Netscape's business in intranet and server software.

The tripartite alliance partially undercut the feds' antitrust case, which all three companies supported. Conventional wisdom had it that AOL-Netscape represented a formidable obstacle to Microsoft making Internet inroads. Maybe it would turn out that way, maybe not. The point was that the marketplace, not regulation, had brought three companies together for the purpose of competing against Microsoft. The trial judge noted the possibility of "a very significant change in the playing field" that "could very well have an immediate effect on the market." By year's end, AOL's valuation on Wall Street, months before the Netscape acquisition became effective, was $75 billion—far more than Oracle's and close to the venerable Hewlett-Packard's; only a short time later, by the spring of 1999, AOL's value nearly doubled to $150 billion. Was Microsoft so bulletproof that the marketplace couldn't be trusted to remedy its abuses?

Netscape's answer was that the AOL deal proved its point. It had tried to take on Godzilla and, like the Bambis before it, had been squashed, reduced to the West Coast branch office of a company that the technocracy derisively referred to as the "cockroach of cyberspace." Larry Ellison likened Netscape to "Little Nell strapped to the railway lines." One of the early Netscape employees spoke for many when she said she felt that her four years in the front lines had been for nothing. "At least the old-timers among us came to Netscape to change the world," she says. "Getting killed by the Evil Empire, being gobbled up by a big company—it's so incredibly sad." The fact that it was AOL doing the gobbling made it that much worse. A greater clash of corporate cultures was barely imaginable; a Netscape employee would no sooner admit subscribing to America Online than Jim Clark would admit to drinking Ernest and Julio Gallo.

That may be an understandable sentiment: Netscape had embodied the dream of Silicon Valley—and now it reflected abject defeat. But there's nothing in the Bill of Rights guaranteeing an independent corporate identity, especially when the goals of the company may be better served through a sale. Moreover, while Netscape's stock price in late 1998 was barely half of its all-time high, many of its employees had still hit it big in the IPO. If Marc Andreessen was no longer worth $200 million, he still had enough to feed his bulldogs and fill the gas tank of the S600; he planned to ditch the Valley to go east to work at AOL in Virginia. Jim Barksdale, despite his symbolically frugal one-dollar salary, had a few hundred million dollars left, a board seat with AOL, and none of the hassles of running a desiccated company anymore. Those Netscapers, like Barksdale and unlike Clark and Andreessen, who held on to their stock, actually did more than well. In the days just before the AOL merger closed, Netscape's stock price hit an all-time high, in large part because of the surge of all Internet stocks.

The biggest Netscape winner of all was Kleiner Perkins Caufield & Byers, which helped to put the deal together. At its annual Aspen retreat for CEOs in mid-June, Case and Barksdale had been seen in deep discussion, joined by KP partners, who were understandably interested. The firm had not only multiplied its $5 million Netscape investment more than a hundred times over, but its "Keiretsu" partners—AOL and Sun— now got to pick over the carrion. If Bill Gates was the godfather, John Doerr ran a pretty nice family of his own. And Kleiner Perkins wasn't even subject to the antitrust laws.

In the month before the AOL merger was announced, the man who lost more than anybody in Netscape's downfall—his status as a *billionaire*—was throwing a party. Jim Clark was still chairman of the board—and worth $500 million or so—but emotionally he had moved on. "I knew when the Justice Department wouldn't do anything in 1996 that we were dead," he said, looking back. "In the debris of failure, you find examples of people who said they'd proceed 'come hell or high water.' What you usually found afterward was hell or high water."

It was now Barksdale's company to make or break. And Andreessen, the whelp with whom Clark had blown open the Age of the Internet, was just a memory. But this was no Jobs-Wozniak relationship—didn't there have to be some good feeling left, after the two had achieved so much? "It's not a good relationship or a bad relationship," Clark told me one afternoon, sitting on the veranda of his Palm Beach mansion, Quinta Marina, and lamenting his distance from Silicon Valley. "It's a 'no relationship.' Yes, I exploited him, but we were also pretty good friends at the beginning, when we were having dinner several times a week."

What happened over time, Clark says, is that Andreessen grew too big for his already large britches. "Marc is a very strange character. He's petulant, he's impatient, he made demands on Barksdale that were signs of extreme immaturity. I think he began to breathe his own exhaust—thinking that, by God, the whole thing really did rest on his shoulders. But if it weren't for me, he wouldn't have made it. I knew enough about marketing to know that this kid would be Netscape's star, and that we ought to work to that end rather than do what Larry Smarr did at NCSA, which was to try to take away the glory and bring it on himself. There was no way I was going to do that—I didn't need it and I was looking for the financial lift, not the PR or ego lift. So we pumped Marc up to be a big guy. Now, Marc expects to be running a company someday—he's another super-ego like Jobs or Ellison. Leadership requires intelligence and Marc is very, very smart. But it likewise requires people to admire you and for you to have their respect. Marc came across as an elitist and that backfired among the other engineers."

In some respects, Andreessen couldn't win. Once just an engineer, he was now executive vice-president. The shorts and T-shirts gave way to Ermenegildo Zegna suits. The "golden geek" in bare feet, seated on a throne on the cover of *Time,* was now standing tall in *Business Week* doling out management exegeses. How could it not appear to the programming proles that he was a turncoat?

And then Clark made it more personal. "Marc hasn't talked to his parents since he went to California. I've had his parents call me and ask for help. And I say, 'Look, I don't know what I can do.' They were distraught about it. A while back, Marc's girlfriend contacted me and wanted to know about chartering a boat. It would've been the perfect opportunity for Marc to call me himself. But I think he fancies himself too busy for that kind of stuff." First, Andreessen's parents; then, Clark?

"There is a pattern, if you will, of leaving previous elders behind," Clark says. "I'm really puzzled by it."

Andreessen declines to talk about any of it. Who said business relationships were simple?

In keeping with his reputation as the recidivist entrepreneur of Silicon Valley, Jim Clark was well along in creating yet another start-up called Healtheon—an Internet system aimed at physicians, to manage the trillion-dollar health-care industry's Everestian pile of records. Clark had put $16 million of his money into the company and received the venture-capital seal of approval with an investment from Kleiner Perkins.

Healtheon was scheduled to go public in the fall of 1998, but pulled back because of lukewarm investor response ($73 million in losses over a three-year period didn't help) and a choppy stock market. "Healtheon may yet make me more money than Netscape," Clark says. "Business is about making money and money better be the top objective. It is for me. I grew up poor and I've learned that money can make you a lot happier. No matter how much money you have, there's always the next thing you might enjoy." And he's been generous with the money, giving millions to each of his siblings and to his mother—though not his father, whom he blames for troubles in his childhood. (Clark's ruminations about his father sounded like Ellison talking about his.) "My mother's still in Texas," Clark says. "I'm astonished by it. Who in the hell would want to remain in Texas if they could afford to leave?" Clark's personal staff of twelve—including a cook, a housekeeper, and two pilots—has benefited, too. He asked them to commit to five years of service, in return for which they'll receive a chunk of five thousand Healtheon shares; he tells the story of a friend with a castle in Austria who lost her housekeeper to Dolly Parton. Beyond Healtheon, Clark mused about the far-flung field of biocomputing and the wired brain—"the most intimate form of human-machine interface and nanomechanics, bypassing the clumsy ways that we communicate with each other."

But what of changing the world again—a third act (after Silicon Graphics and Netscape) when most mortals don't get two? "Oh, that, too," he adds. Money was a metric of success in Silicon Valley, and it also bought really cool toys. Like the *boat*—the magnificent 155-foot, 298-ton white cutter *Hyperion* he was building in a little village near Amsterdam at the famed Royal Huisman shipyard. "A giant aluminum computer with sails," *Fortune* dubbed it in its annual issue celebrating "Life Outside the Office"; Clark posed for the cover photo atop the 193-foot mast, stogie in hand. If critics said Jim Clark had sailed too near the wind with Netscape, this boat's technical advancements were evidence that he loved to take chances. *Fortune*'s pronouncement aside, a yachting magazine suggested it might be "a waterborne '1984.'" Clark simply referred to it as "an RV"—"a self-contained, extremely sophisticated vehicle made for a harsh environment" (with a Picasso, a Monet, and a Renoir adorning the cabin). The Boeing MD600 helicopter was a diversion; *Hyperion* would be a fixation.

Three years in the making, *Hyperion* was to be the greatest sailing craft since the *Niña*, the *Pinta*, and the *Santa María*. She came complete with the largest mainsail ever—weighing more than a thousand pounds—and more than twice the total sail area of Tom Perkins's *Andromeda La Dea*. Clark knew the numerical comparisons by heart. *Hyperion*'s mast is so tall that when Clark brings the ship home, it will clear the Golden Gate Bridge by only thirty feet; the mast might have been twenty-nine feet taller, except it then wouldn't fit under the bridges across the Panama Canal. Larry Ellison's *Sayonara* was fast but had all the creature comforts of a Leavenworth cell. *Andromeda* was supremely elegant but lacked the high-tech accessories that Clark wanted. Bill Gates? He just had a big, ugly motorboat—an empty vessel for vulgarians. For $30 million to $55 million, depending on who's doing the accounting, *Hyperion* had to blend luxury, old-world craftsmanship, and state-of-the-art technology in a way befitting a Silicon Valley legend.

The eight-person professional crew for the boat is beside the point—any tycoon worth his sea salt has that. So, too, the forest of Honduran mahogany trees that gave their lives for the cabinetry and trim. What Clark has put together is a system of systems to operate his dreamboat. Not in some telemetric remote-control kind of way—press reports had him sailing the yacht from his cubicle at Netscape, which was ridiculous and also defeated the point of "Life Outside the Office"—but as onboard information tool, providing even virtual instrumentation. From twenty-five liquid-crystal touch screens throughout *Hyperion*, Clark or crew will be able to monitor coefficients on the mast, boom, stays, halyards, and sheets—the better to balance out load and optimize the driving power of the sails. Load is a big deal when the sails are so large and their space-age materials have never been subjected to these kinds of stresses. That enormous, revolutionary carbon-fiber mast? Tom Perkins, who's sailed every ocean, winces at its mention, then makes a snapping noise with his fingers.

Assuming the mast stays together, the computer system will record the boat's ongoing performance and learn what rigging and sail configurations work best. The Internet will allow Clark to see from afar what's going on with the crew aboard *Hyperion*. If there's a coffeepot on electronic display at Cambridge University and that prankster Steve Wozniak has a Web camera focused on his office, why not a cam for the best sailboat in the world? If the Net fails, Clark can always phone the crew— *Hyperion* has its own cellular exchange.

A nautical network unto itself, *Hyperion*'s twenty Silicon Graphics computers and five hundred gigabytes of memory also will keep track of the engines, fuel, drinking water, rudder, navigational instruments, satellite dishes, surveillance video, night-vision cameras, heat and air-conditioning, the home theater, and perhaps the inventory of the saloon, while at the same time establishing a wireless connection to the World Wide Web. Passengers will be able to use the touch screens to choose from a massive CD and DVD library, select wine from the cellar, and activate the heated towel racks in the head.

Beneath the teak deck, packed in the aluminum hull, is the anatomy of the system: thirty-seven miles of fiber-optic cable and conventional wire. Creating the software to run the boat's nervous system required a little company of its own, founded obviously by Clark. Seascape Communications employed a few programmers back in Menlo Park—they work in an unglamorous office that Clark rented above a Jenny Craig Weight Loss Center—and Clark himself spent hundreds of hours writing code to bring his oceangoing network to life. Seascape was likely to be nothing more than a service company for *Hyperion*, but Clark the entrepreneurial Road Runner couldn't help but fantasize about it one day hatching a systems business for commercial buildings, showcase homes, and complex industrial operations like petroleum processing.

All of Seascape's programming information, along with copies of schematics of design and construction, will be part of the boat's electronic log, in case a technician needs to re-create how the database was assembled; of course, a technician would need to do so if the system had a problem, which would mean the system wouldn't be available to provide the desired information, but that's just a detail and I'm not planning a ride on the boat for a few years anyway.

As *Hyperion* prepared for her christening, Clark decided to have a weekend-long party—a real Dutch treat—for three hundred friends who flew in from around the world. Even Tom Perkins and Danielle Steel made it, to get a look at the new maritime queen. "Everyone who came said it was like the Dutch version of a Fellini movie," Clark says. "Which means funny and weird." The menu: Chinese food, and then at the shipyard, a combo French buffet and Indonesian "Rijstaffel" feast. Party favorite: the Oyster Lady. She was the young waitress in leather pants, black boots, and chain-mail gloves with the tray of fresh Dutch oysters hitched to her waist, shucking as she went. The condiments were self-serve.

"If you can find your way back to Silicon Valley with me," a guest named Rich Green told her, "I can make you very wealthy."

"You're the fifth person to tell me that," she replied. Green co-owned The Audible Difference, a high-end stereo store in Palo Alto that he convinced Oracle's Ed Oates to bankroll in 1995. Clark was a big customer and acquaintance, and invited Green to the Amsterdam party. Green designed the audio-video systems for the boat—another shopkeeper earning a tidy living off the gold miners. Clark's entrepreneurial instincts had rubbed off on Green, his failure with the Oyster Lady notwithstanding.

Runner-up party favorite: the bare-breasted mermaid stretched out atop the French banquet table, her body painted in a festive array of greens and blues and sprinkled with glitter. She didn't get any oysters. Neither did her merman. There was also the marching band on bicycles, wearing clogs and playing "Louie Louie" in the pouring rain. Richard Berry's classic was the song of a seafaring man and a Jim Clark favorite.

For a poor boy from the Texas Panhandle, Clark had come a long way in the sailing department. You don't get a lot of water time living midway between Lubbock and Amarillo, especially when much of your high-school energy is spent sneaking in whiskey on band trips and drag-racing on Saturday nights. Clark never outgrew that sort of revelry. One of his buddies is $650 millionaire Bill Koch—oil heir, America's Cup sailor, art collector, and wine connoisseur. Koch and Clark knew of each other through yachting circles. To get acquainted a few years ago, Koch invited him to his summer home on Cape Cod. Clark's wife, who had interviewed Koch for a *Forbes ASAP* story, teased that the two men would be drunk before dawn. She was right.

Koch brought Clark down to his special wine cellar, stocked with $15 million worth of reds and whites, some dating back to Thomas Jefferson. "I was like a kid in a candy store," Clark recalls, to the extent he recalls any of the evening. Koch and Clark went through seven bottles—all from France, most from this century. Before dawn, Clark's wife went looking for her husband, but found only the empty bottles and a trail of clothing from the boat dock adjacent to the house. At three in the morning, Koch and Clark apparently had gone outside to see Bill's collection of old wooden boats, then proceeded to take a Chris-Craft skiff out for a ride. "I want to drive!" Clark announced. They got fouled in a small buoy, whereupon Koch disrobed to his underwear and dove in with a pocketknife to clear the line. It didn't work, and when Koch got back in the boat, he was so cold he lay down on the backseat and Clark lay on top to keep him warm. Clark fell asleep. "Bill said I snored a lot." At sunrise, they were rescued by a fishing boat and brought back to Koch's house. Now they're Palm Beach neighbors and the best of friends.

Clark's interest in sailing began after he dropped out of high school and joined the Navy. That's also where he learned electronics, going on to a Ph.D. in computer science from the University of Utah, teaching jobs at the University of California at Santa Cruz and Stanford, and then into high-tech. When Silicon Graphics went public in 1986, he bought his first boat, then traded up to a ninety-two-foot beauty from the Netherlands, renowned for three centuries of shipbuilding mastery. Clark wanted the best the Dutch had to offer, but it would take more money than an SGI founder could afford. This was the chief motivation to found a company like Netscape. When it hit the jackpot, Clark went back to Royal Huisman with his checkbook open.

And that led to *Hyperion* and the fascination some folks have with it. One of Clark's charms is that he's brutally frank, including on occasion about himself. "Yes, I'm not like other people," he says, describing how he went crazy on a visit to the shipyard upon learning that two steps, rather than three, had been built from the lower salon into the galley. "I told them, 'I don't understand. Did people not listen to what I said the last two times I was here? I want three steps, not two. You're stepping into the galley and carrying stuff and it shouldn't be like climbing down a ladder. I'm paying the bill. This is my boat and I want three steps!'" Why did they build two steps? "I have no idea. Maybe because they're Dutch—space is at a premium. They always build up.

"I guess I'm kind of an eccentric."

But not so much so that he hasn't learned anything from the Netscape experience. "I'll never do anything that gets in the way of Microsoft again," Clark says, afraid even now that it might decide to get into Healtheon's kind of business. "Bill Gates is a megalomaniac."

Good thing he doesn't know how to sail.

Chapter X
Yahoo

The road not taken can be very expensive.

In the spring of 1995, I was completing a journalism fellowship at Stanford University. In addition to convincing me that golf was a game best left to people with names like Skip, the yearlong program gave me an immersion in Silicon Valley that led to this book. I got to meet, among others, an unknown like Jerry Chih-Yuan Yang, a twenty-six-year-old Ph.D. student in electrical engineering, who had recently started an Internet company. I spent time with Yang, an evangelist and tekkie wrapped in one, at his new hole-in-the-wall offices in Mountain View, where he introduced me to his company and how he and his graduate-student partner, twenty-nine-year-old David Filo, had turned down $1 million apiece for the company.

"Isn't that a lot of money to pass up?" I wondered.

"We think we might do better one day if we can figure out a business model," Yang said. "We might go public at some point."

Yang showed me around the windowless work spaces, occupied by leftover meals, sleeping bags, racing bikes, and the few other employees. The only hint of a real company was a darkened room filled with shelves of monitorless computers. This was the place that thousands of users of the World Wide Web visited virtually every day.

As I prepared to leave, Yang casually asked, "Interested in coming to work for us?"

I told him I was returning to my regular journalism job. That seemed wiser than taking the radical step of permanently relocating my family to the Valley and going to work for two graduate students in a company that had no profits, no revenue, and not even a name on the door.

And this is why I'll never make the Forbes 400 or even the Forbes 400,000. In the four years since, Yang has frequently reminded me what my limited world-view cost. Let's assume I came on in some kind of public-relations or creator-of-content role—hell, I wouldn't even have to know HTML for that. Let's assume also that I got a modest one-tenth-of-one-percent interest in the company—which was peanuts in that early period, a year before the IPO.

Of course, Yang's company was called Yahoo, and in December 1999, Wall Street valued the company at $91 billion—eclipsing the entire economy of Kuwait. My teeny-tiny piece would be worth $91 million. Just suppose I had wangled a better deal—say one-half of 1 percent. That would mean $455 million, enough to be Larry Ellison's neighbor in Woodside and still have enough left over to purchase my own baseball team. Nope, I was too smart for that. Easy come, easy go. Yahoo, boo-hoo.

In the spring of 1994, at roughly the same time that Jim Clark and Marc Andreessen were founding Mosaic Communications, Jerry Yang and David Filo were graduate students looking for anything to do except work on their dissertations. In a cramped office trailer on the Stanford campus—behind the Bill Gates computer science building under construction and not far from the labs where Sun Microsystems, Cisco, and Silicon Graphics got their start—Yang and Filo spent their afternoons and evenings hacking away on the Web. Filo had discovered the Mosaic browser soon after it appeared on the Web, and he and Yang were immediately hooked. They each wrote their own home pages, including such useful material as Yang's golf scores and their high mutual regard for Natso Basho sumo tournaments.

To keep track of these and other favorite sites, Yang and Filo began to compile a list of hyperlinks—arranged by subject—and post it on the Web. They called it "Jerry's Fast Track to Mosaic." There was News, Health, Science, Arts, Recreation, Business and Economy, and thirteen other headings. It was an ordinary hierarchy—much like the Yellow Pages. The end of the string of categories and subcategories and sub-subcategories—a series that might extend to a half-dozen levels—would be actual Web sites. The online journey could be both revealing and maddening. En route to the information you wanted, you might take detours to data you hadn't realized even existed.

Under Entertainment, for example, was Humor, then Words and Wordplay, then Palindromes, which contained collections from around the world written by linguists with too much time on their hands. Or under Computers was the wildly popular Camera Devices Connected to the Net, which, in time, would include more than 700 entries, including an ant farm, a hot tub, 8 soda machines, 19 robots, and 613 Web Cams focused on such objects as Earth, Mars, Tulsa, a roll of duct tape, a seismograph in Los Angeles, a toilet in the South African Parliament, and the world's largest tapeworm.

The index wasn't about original content—it was about context. While the integrated circuit or personal computer had been engineering break-throughs, "Jerry's Fast Track to Mosaic" was a simple consumer ser-vice—nothing more than a taxonomic index available to anyone with the guide's address in the Stanford computer network. It was mostly text, devoid of dancing icons and neon graphics. It was put together by hand—time-consuming searches of Web pages, rather than any automated word-specific cataloging system. Even the Mosaic browser created by Andreessen and his Illinois cronies reflected proprietary technology and a degree of design prowess. "It wasn't rocket science," Filo says, looking back. "We didn't have patents or anything like that. Someone smart with resources could've done the same thing." What a contrast to the early days of the Valley, when Bob Noyce and Gordon Moore performed al-chemy in a laboratory.

But the genius of the ever-expanding Yang-Filo directory was that nobody else had assembled one. The secret, as it so often had been in Silicon Valley, was timing. A few years earlier, there wasn't much of a Web to classify; a few years later, Microsoft would've figured it out. Yang and Filo got started at the perfect moment, just as the Mosaic browser took off and just as Netscape was created. The dozen or so Web sites of 1990—soon after Tim Berners-Lee introduced the Web—had given way to thousands. (By the end of 1998, there were close to three million.)

Virtual chaos was setting in and the Web cried out for, if not a traffic cop, at least some helpful road signs. Searching through the Web was akin to a drive through Boston—you might never figure out how to get back to a particular place. There were no directions, no instructions, just seemingly endless possibilities.

"Jerry's Fast Track to Mosaic" was as hip a name as Wal-Mart, and its subsequent incarnations were not much better, "Jerry Yang's Guide to WWW" and "Jerry and Dave's Guide to the World Wide Web." Yang and Filo replaced them all with something more spirited for their directory: Yahoo!, a word that traced its roots to the ruffians in *Gulliver's Travels* (not that Yang and Filo were English majors). Yahoo—to which they added an exclamation point to make it difficult for copy editors ever after—had to stand for something. So Yang and Filo came up with Yet Another Hierarchical Officious Oracle, a parody of sorts. "Yet Another" was in the lexicon of software programming and the directory was hierarchical, but the full name was basically gibberish, and Yang and Filo always went with Yahoo. It turned out to be the best business decision they ever made—a brand name that within two years would rank with any in high-tech. There would be other directories—Lycos, InfoSeek, Architext, WebCrawler, and some that were better suited for certain searches—but none achieved the following of Yahoo, which took on cult status among the digerati.

As their inventory grew—twenty thousand entries in the first year— Yang and Filo took the next step of creating a searching tool, so that they and the legions visiting the Yahoo site didn't have to manually search through endless topics and subtopics. They added a "search engine," which let users type in a word and then get a list of hyperlinks that contained that word. Type in "Baseball" and you'd locate the Cubs, ESPN, and maybe a fan's page dedicated to Sandy Koufax. The Yahoo combination of directory and search engine was invaluable to anyone navigating the Web. Netscape provided the software to get to Yahoo, but Yahoo organized the online content and made it accessible. Yahoo, more than Netscape, held out the potential for "eyeballs," as advertisers liked to call viewers.

Yahoo traffic passed 100,000 "page views" a day in late 1994—up from only a few thousand in May. Yang and Filo could categorically say they were witnessing more Web activity than anyone else in the world—

and with no marketing. (A year later, daily page-views reached a million and, by late 1998, a remarkable 167 million—a higher number than even the daily viewing audience of *Oprah*.) Friends and digital correspondents sent in their own recommendations of Web sites. Yang and Filo added some editorial commentary with headings like "What's Cool" or "What's New." What began as a hobby had become a preoccupation. Yang and Filo wanted to classify not just their favorite Web sites, but virtually any interesting ones. When their thesis adviser returned from a European sabbatical, he was surprised indeed to learn what his students had been doing while he was away.

Yang and Filo had started Yahoo as a lark. But as its popularity soared and the index consumed their time, they started thinking about turning it into business—cashing in, if not selling out. What kind of business? Who could tell? Yang and Filo didn't have a Jim Clark to lead them by the hand to the land of riches. As best they could figure, nobody using the index would pay for the service—not even a few cents a day. Nor would the individuals or companies owning the listed Web sites. Advertising was the most promising revenue model and had long worked for mainstream media like television and radio. But the culture of cyberspace had long abhorred the idea of crass commercialism (AOL notwithstanding), especially when it meant banner ads blazing across the top of a Web page.

Yang and Filo knew they weren't going to remain at Stanford and they knew further that the university's computer resources weren't equipped to accommodate their service. Various venture capitalists had called them, and they finally agreed to meet suitors. Kleiner Perkins was one of them. But KP already was invested in a competitor called Architext—later to become Excite—and the KP partners wanted to fold Yahoo into it. That ended any chance of Yahoo going with Kleiner Perkins. KP's failure to offer a deal to Yang and Filo amounted to one of its great missed opportunities—far more costly than its actual investments in GO or Dynabook.

Another top-tier venture fund, Sequoia Capital, was in the process of scouting out the new Internet marketplace. Founded in 1972 by Don

Valentine, Sequoia represented a completely different style than Kleiner Perkins. Apart from Valentine's 1950 Bentley parked in front of Building 4 at 3000 Sand Hill Road, Sequoia—like its name—was solid and silent. While KP's offices looked like a mock-up for *Architectural Digest*, Sequoia's were ready for the repo man—small and spartan. Andreessen once visited the offices and wondered if the space was new and the partners hadn't yet moved in. They had, but didn't repaint for twenty-five years. KP served a gourmet buffet at lunch; Sequoia put a bowl of Butterfingers in the reception area. But Sequoia didn't pass up the chance to claim that the value of all the companies it had financed was $200 billion—60 percent more than KP's. What they might have lacked in self-promotion and interior decor, they made up for in dollars.

In addition to Cisco—its pride and joy—Sequoia had backed such companies as Atari and Oracle. Now it saw a vast new marketplace in the Internet. KP already had invested in Netscape and the deal made other firms take notice. Sequoia sought industry gossip from Randy Adams, a long-time, small-time entrepreneur whom it had bankrolled before. Adams was president of the Internet Shopping Network in Menlo Park, the first online retailer (eventually bought out by Barry Diller). A forty-year-old Sequoia partner named Mike Moritz made the call to Adams.

Moritz grew up in Wales and possessed an accent that gave him an air of refinement in the hurly-burly of the Valley. He was smart, animated, and dry-witted—with a master's degree from Oxford and an MBA from Wharton as his pedigree—but he hardly had the grand industry reputation or track record of a John Doerr. His most notable attribute was his prior life as a journalist for *Time*. After stints in Detroit and Hollywood—two other company towns—he worked in the 1980s as a high-tech correspondent, which gave him a taste of what the Valley had to offer. He did profiles of Arthur Rock and Tom Perkins, among others. When *Time* decided to make Steve Jobs its Man of the Year in 1982, it sent Moritz to do the interviews. But he disliked Jobs so much—despite Jobs's attempt to win him over, including an unannounced Saturday-afternoon visit to Moritz's home—that the magazine wound up ditching Jobs and making the PC "Machine of the Year." The accompanying profile of Jobs included this memorable quote about him from a former employee: "He would have made an excellent King of France." According to Moritz, Jobs went so far as to call Henry Grunwald, *Time*'s legendary editor, to try to get Moritz fired. Grunwald laughed it off.

Jobs told me that the entire episode was seminal for him in forming a view of his own celebrity and the often skewering treatment he received from the press. "*Time* called me and said they wanted to make me Man of the Year," he recalls. "I was a little younger and had more hubris then. I thought, 'What a wise choice.' But this reporter they sent was the same age as me and had a giant problem with me. So he wrote this hatchet-job piece. When his editors saw the piece, they decided, 'We can't make this guy Man of the Year.' I remember I was off somewhere with a girlfriend and the magazine was delivered, and after I read it, I literally started crying. It wasn't long after that I decided that people like symbols and that I became one. I had to look at it as if I had a famous twin brother—but it ain't me."

Several years later, as Moritz tired of journalism and the newsletter business he'd started, he decided he wanted to be a venture capitalist. He visited Tom Perkins. Moritz remembers the scene. Perkins—smoking a cigar and seated in his glass office, surrounded by ship models—said to him, "So, you want to get rich? Let me tell you. Nobody in the venture business gets rich. *Bill Gates* gets rich." This, says Moritz, was at a time when "$100 million was considered a large net worth." Perkins had nothing for him. But Sequoia gave him a shot.

In early 1995, when Moritz called, Randy Adams told him about a "rinky-dink little service called Yahoo" that helped find things on the Web. Adams said he knew the two Stanford students and could introduce them. So Moritz and Adams paid a visit to the trailer on the Stanford campus where Yang and Filo worked during the day and, since they became Yahoos, slept most nights, surrounded by old clothes and overheating computers. *Fortune* called it a "cockroach's picture of Christmas." *Animal House* meets *Revenge of the Nerds*. "If a script called *Yahoo* had been pitched in Hollywood," Moritz says, "even there it would've seemed far-fetched. How could you possibly invest in a company that gave away its service for free? This wasn't something that we were in the habit of doing. What was the business?" Netscape, at least, looked like it would derive revenue from selling both its browser and server software to businesses. Netscape, too, might create an online index and supplant Yahoo. As Moritz put it, "Wasn't Yahoo a mere 'spell-checker,' compared to Netscape's 'word-processor'?"

There was also that impish name that seemed to have nothing to do with the world of propeller heads. Fred Gibbons, who had founded an early and successful PC software company, happened to be in Sequoia's offices when the partners were considering Yang and Filo. "I can't believe you guys would finance a company with a name like Yahoo," Gibbons said. To which Don Valentine, standing nearby, replied, "A long time ago, we helped finance a company called Apple." Other VCs whom Yang and Filo met were less understanding. They advised: Lose the name.

Despite the absence of revenue, Moritz was intrigued. "We're very good at being able to explain with unerring accuracy why some of our companies have failed," Moritz says, "because we've identified the weaknesses beforehand if we've done our job halfway right. But we always underestimate how successful our successful companies can become. We're always wrong."

In Yahoo's case, Moritz kept thinking about other mass media. "Maybe too simply, I just felt you're in your car listening to radio for free or you're at home watching CBS for free. So why will the Internet be any different? The trick, strategically, was to get an audience and retain the audience and at some point the advertisers would come. That was the line we used to win over the rest of the guys at Sequoia. It doesn't matter if you've got blimps or billboards, radio or TV or cable or the Internet." It's all the same if you have an audience. Yahoo had audience.

Moritz, Yang, and Filo shared the belief that maintaining viewer loyalty was paramount and that turning Yahoo pages into an online Times Square filled with advertisements could wait several months. Yang and Filo liked Moritz's manner—a bluntness and leanness that contrasted with the clubby atmosphere elsewhere on Sand Hill Road. "David and I were duly impressed with KP's offices," Yang says. "But we were more impressed by Sequoia's cynicism. They're the most iron-fisted of VCs. And they squeezed every penny, like David and me. There was a certain amount of desperation on both sides."

Yang and Filo also liked that Moritz was partial to rookies. Moritz's view of the proverbial start-up garage was utilitarian rather than romantic. "Two guys walk through the door. Give us a choice between the one who's built two successful companies and has a house in the hills—or the one who's got a mortgage up to the chimney tops and who can't afford to fail—we'll always finance the hungry guy if we can only write one check." Moritz detested sloth.

First, Yang and Filo had to decide whether to accept any venture capital. The alternative was just to sell the entire franchise—take the money and run. Microsoft, typically, expressed interest. But the two companies that made bids were America Online and Netscape—both of which offered Yang and Filo $1 million each. Netscape would have been a better fit. (The joke was the new company would sound like an Israeli leader—Net-and-Yahoo.) Unlike AOL, Netscape had embraced the Web, and Yang and Filo already had a relationship with it. When they gave up the Stanford trailer in early 1995, Yang and Filo needed a way station. Marc Andreessen persuaded them to use the servers at Netscape as a connection for Yahoo's burgeoning traffic. Andreessen might have had an ulterior motive—Jim Clark wanted Yang and Filo recruited to Netscape—but this was also classic Valley cooperation in action. Yahoo.com—free of the Stanford address—was thus born, the greatest URL of the post-PC era.

The cofounders considered the instant wealth. As engineers, Yang and Filo relied not on intuition but on all the information they could accumulate. Moritz wanted an answer and finally gave them twenty-four hours to choose. Ultimately, says Moritz, "they concluded they didn't want to play second fiddle, or third fiddle, or the occasional bassoon." Instead, they decided to take a risk—and Sequoia's money—and see if they could create a business. They chose to succeed or fail on their own. It wasn't Ellisonian ego, but it was ego nonetheless.

Netscape was sore that it hadn't landed the two of them. Andreessen and the marketing chief, Mike Homer, ordered the Yahoo servers out of the building. But Bill Foss, one of the first Netscape employees, hid the computers in dead space between his floor and the ceiling below—until Yang and Filo made other arrangements.

Had Netscape swallowed Yang and Filo, it might have become a fundamentally different company. But, more likely, it would have continued to concentrate on browsers and let Yahoo languish; for their part, lacking independence and the accompanying fear, Yang and Filo probably would merely have gotten a little richer. Like so many start-ups before it, Yahoo became a significant business because of a perfect alignment of variables—timing and technology, product and luck. Despite being rejected, Netscape did Yahoo the favor of linking its browser to Yahoo's home page—for free; if you clicked the browser's "Internet Directory" button, you arrived at Yahoo. Since almost all new visitors to the Web used Netscape's Navigator—Microsoft hadn't yet entered the browser market—this meant Jerry Yang and David Filo were their tour guides. Netscape's browser was a means to get online; Yahoo actually helped you do something there. (Netscape's generosity lasted for much of the year until it figured out that its Directory button was worth real money and leased it to Excite, which produced subsequent bidding wars among Yahoo and other search engines.)

In the spring of 1995, Sequoia invested $1 million in the company, in return for a 25 percent stake. It was a far smaller price than, say, Kleiner Perkins had paid for Netscape—$5 million for 20 percent. (By those numbers, Yahoo was initially valued at $4 million, Netscape at $25 million.) It would prove to be the best $1 million ever invested in Silicon Valley. After Yahoo went public, the value of that initial $1 million in 1995 rose to nearly $8 *billion* in early 1999. Million to billions—the magic of the Valley.

Within a month of the Sequoia financing, Yahoo moved out of Netscape and into its own offices in Mountain View near the railroad tracks (the place where I scoffed at the notion of a job). Yang and Filo even got around to putting a cardboard company sign on the door.

Yang was a Valley poster boy for immigrant success—Silicon Valley as the Ellis Island of the late twentieth century. He came to the United States from Taiwan when he was ten, with his mother and younger brother. (His father died when he was two.) Raised in San Jose, he had none of the engineering compulsions of a Steve Wozniak, but he was an excellent student and an excellent talker, and he got himself into Stanford. While Yang's Mandarin Chinese is as good as his English, his mediaspeak is even better. He is his company's Chief Yahoo, preaching the Web to all who will listen. If Wall Street believes Yahoo is the next great media company—*media*, not technology—it is in part because of Yang's performances on the financial and press circuits. On his frequent trips to the Far East, he's treated as an icon. When he visited Beijing in the spring of 1998, his limousine was given a motorcycle escort through downtown—an event he considered hilarious; in his native Taiwan, he once had to register in a hotel under an assumed name in order to avoid groupies.

If Jerry was the yang of Yahoo, Filo provided the yin. Raised in Louisiana, the second youngest of six children, he's as laconic as a riverboat captain and not as well dressed. His Yahoo cubicle is so slovenly it would take a shovel, not a vacuum, to find the floor. Some of Filo's friends affectionately nicknamed him the Unabomber because he was so introspective. He was also a bit strange, naming Yahoo's conference rooms after the Ten Plagues. But it wasn't that Filo lacked enthusiasm for or savvy in the Yahoo product. When Princess Diana was killed in August 1997, it was Filo, not Yang, who recognized the notoriety of the event, and he got Di material up on Yahoo's Web site within two hours. "As her pulse was still flickering," says Moritz, the unsentimental Brit. Similarly, it was Filo who lowered the prices on Yahoo's online T-shirt stand.

Both Filo and Yang were smart enough to realize they weren't managers. Moritz brought in forty-three-year-old Tim Koogle to run the company and provide "adult supervision" for the striplings. Koogle, a native Virginian and another Stanford engineer, got his entrepreneurial start by fixing car engines of his wealthier classmates. Since then, in Tokyo, Toronto, Chicago, and Seattle, he helped start various technology companies and worked as an in-house venture capitalist for an established one, Motorola. But he liked the Valley—despite what he calls its "nouveau riche" tendencies dating as far back as the 1970s—and kept at least one house there.

Beyond his experience, Koogle's appeal was that he appeared wholly comfortable with the bizarre new business—or lack of it. He understood that ads were the likely source of revenue, but he also understood that nobody, frankly, had any idea how the company might progress. Maintaining audience was obviously key, but beyond that, he didn't pretend to have a Harvard Business School printout of solutions. He says, "I started with a totally clean slate" (as well as a 5 percent stake in the company). Koogle also had a frugal corporate philosophy—a trait that Moritz and Sequoia particularly liked. In his Day-Glo yellow sneakers, with his collection of zippy cars, Koogle arrived in the fall of 1995, soon after Netscape's tectonic IPO shook the high-tech landscape.

With Koogle onboard, along with a thirtyish Jeff Mallett to run operations, Yang was content to proselytize; Filo, the technical wizard of the two, liked nothing better than to hang out in his littered cubicle, tinkering with code and keeping the servers up to speed. (As Yang liked to be called the Chief Yahoo, Filo identified himself as Cheap Yahoo. Koogle, the grown-up, refused the "Chief Chief Yahoo" designation.)

Advertising revenue did start coming in—from such companies as MasterCard, Wells Fargo, Toyota, and Sony. And Yahoo's Web site started to expand. Rather than just being a bus terminal that routed visitors elsewhere, Yahoo wanted to be a destination resort. This was the beginning of the one-stop "portal" notion. At first it meant just displaying Reuters wire stories through a "Headlines" button on Yahoo's home page. But that would eventually lead to Yahoo-designed, city-specific hierarchies (for example, Yahoo San Francisco Bay Area), classified ads, and a fully personalized home page (www.my.yahoo.com) updated continuously.

Given Netscape's success in going public, it made perfect sense for Yahoo to follow—as a way of raising tens of millions in operating capital and also as a way for the founders and Sequoia investors to cash in. An IPO was scheduled for April 1996; the prospectus was made available on the Web—a securities first. But, first, Yahoo made a private deal that secured its founders' financial futures. Just before the IPO, Yang and Filo sold one-fifth of their respective stakes in Yahoo to Softbank, a large Japanese software-and-publishing company. Five months earlier, attuned to the developing Internet market, Softbank invested $2 million in Yahoo. It then acquired Ziff-Davis, an American computer trade publisher. Now it wanted a much bigger stake in Yahoo—that would give Yahoo $64 million more in capital and a strategic partner in the Asian market. It also would make Yang and Filo instantly rich—$12.5 million in cash for each of them. To make the deal work, Yang and Filo had to sell off part of their holdings because Yahoo, the corporation, simply didn't own enough shares itself. The question was: Would Yang and Filo sell?

It would seem obvious that two men in their twenties would jump at the money, especially because they'd still retain most of their Yahoo shares. But Yang and Filo were ambivalent. "I had a conversation with both of them about taking some money off the table so that they weren't gambling everything they owned," remembers Moritz. It wasn't just about financial prudence, he told them, but about giving them sufficient freedom not to be "too timid." Principal before principle. But Yang and Filo didn't want to sell anything—unless Sequoia sold the same number of shares as each of them. It was an astounding act of faith—or foolhardiness—that Yang and Filo wouldn't sell one share. They didn't want to be viewed by future shareholders as running away from their creation. It was also a sign of their increasing maturity that they parried Moritz— their friend—with a "put your money where your mouth is" ultimatum.

Moritz came back to them and said, "Look, you guys, it doesn't matter to us—we're diversified in other stock. If Yahoo goes south, we won't be hurt too badly. But your entire net worth shouldn't be in this thing." He then told them the story that Tom Perkins had told him about a company called Imagic that made video-game cartridges back in the early days of Atari. Perkins offered to buy some shares from the founder before Imagic went public. The founder said no, explaining that he thought the value of his shares would soar after the IPO. But Atari hit a rough patch and then Imagic never went public. The founder was left with stock certificates as bathroom wallpaper. "Don't make the same mistake," Moritz cautioned Yang and Filo.

They agreed—but stood their ground on Sequoia's selling, too. And that's what happened. Yang and Filo each took his $12.5 million from Softbank—as it turned out, worth roughly a hundred times that at Yahoo's peak in early 1999—but left the rest of the shares on the table (which, between them, amounted to 31 percent of the company). Yang bought a silver BMW 540i with a talking global-positioning navigational system, and a nice house with a view of the Santa Cruz Range many miles from Woodside. He even got married—without a prenup. Filo replaced his 1980 Datsun and continued to share a rental apartment in downtown Palo Alto.

After the IPO on April 12, 1996, Yang and Filo had even fewer financial concerns. Just as had happened with Netscape, the public swooned over Yahoo stock. Initially priced at $13 a share, Yahoo opened at 24½ and zoomed to 43 before closing at 33. "Yahoo! Yeehaw!" the San Jose *Mercury News* wrote. "Or should that be yikes?" Just thirteen months after becoming a business, Yahoo had a market capitalization of $849 million. Yang and Filo were each worth $132 million, in addition to their Softbank cash windfall.

The stock meandered much of the year, falling back almost to its initial price. "Yet Another Highly Overhyped Offering," hooted one online newsletter. Yahoo needed more revenue, if not profits, and that meant more advertising. Netscape—the other Internet darling—was having its own growing pains, caused largely by a bully on the block named Microsoft. Yahoo faced competition from other start-up search engines and directories, as well as AOL. But none of these quite represented the threat of a Microsoft; none of these had the leveraging power of a monopoly and, for the moment, Microsoft had yet to make a direct assault on Yahoo's franchise.

So Yahoo stayed with its business plan for becoming a media company for the millions and potentially hundreds of millions of customers in cyberspace. AOL had already begun to do that with its proprietary services. Koogle expounded three legs of Internet marketing: "content, distribution, and brand." Content was the easy part—just keep adding to the big index (which, by then, had half a million entries) and coming up with new "sticky" services to entice visitors to stay on Yahoo pages.

Key among the new services was free E-mail, which Yahoo provided by buying another company—one in a series of acquisitions that the IPO and Softbank's wallet made possible. These new features served to create more audience, which, of course, was what advertisers wanted. So did Yahoo's widespread distribution through partnerships. The company worked to get its name and hyperlink placed on other Web sites; just as Yahoo pointed people elsewhere, these other digital venues would point to Yahoo as a directory and search engine. These kinds of reciprocal links reinforced Yahoo's position as the most used Web service. The

Field of Dreams notion of the Internet was reasonable in theory—if you built it, they would come. But in practice, only a few gilded gateways—portals—were going to control traffic flow.

Filo had correctly observed that Yahoo reflected none of the technological magic or proprietary knowledge that permitted an Intel or Oracle to thrive over competitors—what the economists referred to as "barriers to entry." But Yahoo's name and the habits of Yahooites were still powerful assets. In the realm of consumerism, brand was king. Logic be damned, it was emotional connection—loyalty—that mattered. "Mindshare," Yang kept calling it. "How do I capture attention and monetize it?" Coke hardly tastes better than RC, but its market share doesn't lie.

Yahoo spent lavishly to flog its brand. While competitors were content to become known only online, Yahoo aimed for the mainstream. It had commissioned a survey to find out what consumers thought Yahoo was. Only 8 percent knew; far more thought this ultra-cool, world-beating Silicon Valley start-up was . . . America's No. 1 Chocolate Action Drink. To change those perceptions, the company early on spent $5 million for TV and radio commercials nationwide. "Do You Yahoo?" was the clever tag line, turning the corporate noun into a verb. The bright purple-and-yellow "Yahoo" logo started to appear everywhere—on ballpark signs and at construction sites; on Ben & Jerry's lids, parachutes, kazoos, and the Zamboni at San Jose Sharks home games; and in great big letters on the car of any adventurous employee (Yang and Filo demurred). TV's *E.R.* showed the Yahoo home page on the computer at the nurses' station. Microsoft was one smart company, but never showed this kind of whimsy.

Yahoo's efforts surely boosted its profile. Yang and Filo, and Koogle on occasion, graced the covers of *Newsweek, Time,* and *Business Week,* and every industry publication. Marc Andreessen's coverboy image had been regal. By contrast—Yang on a surfboard, Koogle in dark glasses, all of them waving from a Yahooified car (there were at least six different versions of this shot)—Yahoo always oozed California cool. (Talk about good "brand management": When journalism joins the cause, in a way not seen in the Valley since the 1980s heyday of Apple, you know you're hot.) And the message was always the same: Yahoo was the digital place to get connected to anything.

In 1997 and 1998, as Netscape's fortunes began to swirl down the drain, Yahoo's rebounded. The actual financials—on which stock market valuations were based in days of yore—were modest. Yahoo's revenues in 1998, for example, were $203 million (mostly from several thousand advertisers), but that was loose change for a major corporation. Microsoft took in that much in four *days;* Yahoo's biggest direct competitors—AOL and Netscape, between them—took in that much in a month.

Similarly, Yahoo's 1998 profits were only $50 million (excluding acquisitions). That was better than 1997's loss of $425,000 (excluding acquisitions), but under normal circumstances still barely enough to warrant a mention in the Fortune *5,000*, let alone the elite Fortune 500. But these were not ordinary times in the marketplace. No profits? No problem. That was just part of the romance of the Internet. (At the 1998 soapbox derby on Sand Hill Road, one start-up with a sense of humor made money-losing part of its marketing campaign: It drove around in an armored car, distributing real dollar bills emblazoned with its logo.)

Even with Yahoo's numbers, investors eagerly rolled their dice. They were not buying a piece of a current business—in 1999, Yahoo had only five hundred employees and $500 million in the bank—but the dream of what might come. Three national newspapers, four big networks, why not only a few big Web portals with Yahoo in the lead? The inflated market values of companies like Yahoo enabled them to acquire smaller competitors, so that the prophecy of a few dominant portals started to become self-fulfilling. Advertisers spent billions and billions in print media and TV—why not a chunk of that online someday, especially if more eyeballs were gazing at computers than newspapers or the tube? According to some estimates, by the year 2000, nearly half of American households will be connected to the Internet. It wasn't just the potential numbers. Internet eyeballs could be targeted by local demographics and interests, since Web users left an electronic trail of crumbs in their journey around cyberspace. If you visited a lot of Web sites on cars, your Yahoo pages might come with a banner ad for Ford; click on it, and you might get price quotes from local dealers.

The projections through the year 2003 were a runaway train: 300 million Web users worldwide, half a billion devices connected to the network, $3.2 trillion in Internet spending. Good-bye, mall. Hello, little shopping-cart icon. Who wouldn't prefer clicking on an electronic mouse to dealing with a gum-chewing, put-upon salesclerk? Now Yahoo was a presentable business. Soon it would be the media company of the twenty-first century. Millions, billions, trillions—hey, why not a few zillion for investors?

In 1997, Yahoo shares went up 511 percent; the next year, another 584 percent. What sold for $13 on IPO Day went for $1,335 in January 1999 (not factoring in stock splits). If you had the chance to pony up $9,738 in Yahoo back on April 12, 1996, you had $1 million three years later. Apple and Oracle and even Microsoft never had it so good. Netscape, at the top of its ride, had a market capitalization of around $8 billion. In January 1999, Yahoo's was $44 billion; Intel had taken nearly a quarter-century to earn that valuation. Yahoo's market cap made it the seventy-second most valuable company in the world— almost double CBS's value, and more than the New York Times Company, the Washington Post Company, Gannett, and Dow Jones *combined*. For a while, wags called Yahoo the Buzz Lightyear of Stocks—to Infinity and Beyond! Yahoo's price fell back to earth, but nowhere near a reasonable valuation by terrestrial standards. Yahoo had become the marvel Netscape had wanted to be. Both had been first out of the Internet gate, but Yahoo had yet to blow its lead.

The riches of Yang and Filo apparently showed that. Moritz had prevailed upon them to take $12.5 million from Softbank prior to the public offering. But, unlike Marc Andreessen at Netscape or the typical entrepreneur, they didn't unload any more. Nearly three years after the company was founded, apart from some charitable donations and gifts to family, they kept everything—like the Larry Ellison of the late 1980s. "If Yahoo goes down," says Yang, "I'll go down with the ship."

Not that he or Filo has to worry. At Yahoo's apogee, Yang was worth $5.1 billion, more than the combined salaries of every player in Major League Baseball. Filo has $5.2 billion. That put them both among the sixty richest people in the world—pretty good for two electrical engineers who in essence became salesmen. Someday, their company's profits might even catch up with them.

Some of the other rock stars of Silicon Valley couldn't help casting a jealous eye. Yang found himself meeting them and getting various entertaining reactions. Andy Grove told him, "I used your product, I loved

your product, but your product is shit—you could make it a lot better."
Larry Ellison offered some insults about Yahoo's being "overvalued";
Yang just laughed, later noting, "I'm sure the women were impressed."
Maybe Ellison wasn't thrilled that the bank accounts of both Yang and
Filo were approaching his.

Market purists will tell you that a stock is worth, by definition, "whatever price somebody's willing to pay" for it. That's what Tim Koogle
says. He's right, obviously. But the founders seemed to understand the
absurdity of the dollars. Yang says Yahoo has created "more wealth for
others than ourselves—VCs, brokers, shareholders, and management."
But he knows that's just brave talk. "I have no idea most days why the
stock goes up or down. When I do know something, the price usually
goes down. I've had such good luck I can't believe I'm alive. I realize it
could turn around at any time."

In particular, he worries about Bambicide. Will the Beast of Redmond
come in search of his little company and squoosh it? When Microsoft
smells money, it arrives soon thereafter. Its efforts at creating an online
portal have largely failed, but Microsoft's history is that it tends to get
things right on the third or fourth try. A monopoly affords that luxury.
"Microsoft can crush us in an instant," Yang says. "I'm terrified of them.
When Bill Gates has three hundred of his people go to Madison Avenue
to sell ads, we're in trouble. But I think he's concentrating on other
markets right now." According to Koogle, Yahoo has gone out of its way
not to taunt the bear—twice declining to join Netscape, Oracle, IBM,
and Sun in their alliance against Microsoft.

Bill Foss, one of the early Netscapers, says whenever he has dinner
with Yang, "I ask Jerry what he's worth and Jerry always knows—though
he then always points out that the number is ridiculous." When I had a
fancy French dinner with Yang, we didn't talk much about money—
except when the tab came and he suggested with a smile that my ethics
required I pick it up. I did. (Next time, he paid for the $6.95 noodle
lunch special.) "I know we're not a $6 billion company," Yang told me
in the spring of 1998. He was right—indeed off by $38 billion, if you
believed the stock market of January 1999.

Mike Moritz shows the same kind of bemusement as Yang. "I find it
hard on occasion not to detach myself from everything that takes place
between San Francisco and San Jose," Moritz says. "I'm sort of a twin

foreigner. I grew up in Britain and parachuted in here. And I came from another profession, not in the servitude of Intel or HP before I went into the venture business. So every now and then you take the opportunity to walk to the top of the stands and see the whole game."

And how's the view? "The whole environment is preposterous. It's our game and we like playing it. I can't imagine being anywhere else that has a richer collection of talented, driven, oddball characters." But he says the notion of Yahoo or other companies changing the world is wildly overblown. "Jerry and David don't produce that many jobs. One of the dirty little secrets of the Valley is that all the jobs-creation we like to talk about is probably less than the Big Three automakers have laid off in the last decade. One of the best ways to have a nice Silicon Valley company is to keep your head count as low as possible for as long as possible.

"Look at our companies. Maybe they've produced 100,000 jobs or 150,000. But what kind and for whom? Jobs that 250 million people in this country aren't qualified to apply for. Jobs for guys out of MIT and Stanford. Jobs that in many ways gut the older industries in the Midwest and on the East Coast. Are the benefits of these jobs passed down? I don't know. What are we investing in? Companies that enable people to work harder and longer—anyplace, anytime. You can be reached on a ski lift, on a beach, or on a plane. Why is that good for people's lives?" Moritz's cold eye is downright refreshing amid the hype.

But, Mike, what about the "largest legal creation of wealth in the history of the planet"? "No argument from me," Moritz says. "But, in the end, so what?"

It almost makes you wonder why he and Jerry and David aren't offering to give some of the billions back.

Epilogue
Lincolnville
04849

I t is about as far away in North America as you get can get from Silicon Valley, unless you count Newfoundland.

From Portland, Maine—already 3,200 miles from the land of mega stock options and eighteen-dollar-a-pound ostrich salami—you drive an hour north to Augusta, the state capital, and then an hour east into the Maine outback, full of fields and forest and New England cottages. At Belfast, you head southwest on State Route 52, past the Faith Temple Church of God and country store, and in a few miles you cross into the town of Lincolnville, Maine, the heart of Waldo County. Population: 2,100, except in the dead of winter, when folks fly south. Just out of town, on the left, is rambling Kelmscott Farm—150 acres overlooking Pitcher Pond, Ducktrap Mountain, and Penobscot Bay in the distance. This is where Bob Metcalfe and his family now live, along with a hundred rare Cotswold sheep; three Nigerian dwarf goats; five Gloucestershire Old Spots pigs; an evil-eyed, spitting llama; several dozen geese, chickens, and ducks; the giant lead horse from the Coors Beer team; two Kerry heifers; two barn cats; a pigeon; and a border collie named Tess.

Pioneer of the computer network, inventor of the Ethernet networking standard as an early employee at Xerox PARC, founder of 3Com in 1979, father of Metcalfe's Law, and only fifty-three years old—Metcalfe came here for a different sort of life. He was no latter-day Luddite trying to escape the intrusions of machinery and technology. From Maine, using the Internet that he helped to create, he was still plugged into the wired world. As a featured speaker at big industry conferences and a sardonic columnist for the leading computer trade publication—not to mention appearances on *60 Minutes* and in magazines and newspapers—he remained as visible a member of the digerati as anyone. (His famously incorrect prediction in 1995 that the Internet "would go spectacularly supernova and in 1996 catastrophically collapse" led him to literally eat his words. Turning a mistake into a publicity stunt, he put his column in a blender with some water, and then ate lustily.) Baseball had Bob Costas. High-tech had Bob Metcalfe.

Greeting me on a brisk, snow-robed afternoon in January 1999, in his down coat and lumbering boots, Metcalfe said, beaming, "You're not in Silicon Valley anymore."

Once upon a time, Metcalfe lived in the Valley—in Woodside, of all places. "I was Old Money—from the 1980s," he says. "Yang, Andreessen—they're New Money." Metcalfe had one of the grand old properties in town, up the hill from Mike Markkula's on King's Mountain Road. "Brookside," as the Metcalfes called it, was a four-acre estate carved into the hill. The huge Mediterranean house was designed by Bernard Maybeck and built in 1906 on the rubble of another dwelling that had been destroyed by the great San Francisco earthquake the same year. William Greenwood, a wealthy ship chandler, was the first of a handful of owners through the rest of the century. Expanded, landscaped, accessorized, the property became one of the jewels of Woodside, right up there with the splendid house of Nolan Bushnell and the one Larry Ellison was building.

It had a black-bottom pool, spa, cabana, carriage house, guest house, games court, fountain court, brick-laid motor court, tennis court (with phone), bath garden, bog garden, herb garden, redwood grove, and barbecue pit. The beige stucco mansion itself was classical, with a tile roof, great gables, gold-leaf ceilings, marble floors, and a long colonnade in front—along with five bedrooms and eight bathrooms. There was also a caretaker's residence on the property, which made sense, since somebody had to take care of this little colony. One frequent contractor says visiting the house was like "going to a big public library, except I never could find the front door." For the Metcalfes, the 9,250-square-foot house was a mixed blessing. They liked the comforts, but didn't always like being the object of architectural envy (or the annual $40,000 tax bill). "People who came to the house for a party or dinner," Metcalfe recalls, "only wanted to talk about the *house*." Whoopi Goldberg spent three days there making *Kiss Shot*—one of her worst movies.

By 1994, Metcalfe had left 3Com. He had founded the networking company, but had been passed over twice by the board of directors for the job as CEO. It was time to do something else, and it didn't have to be in an office building in Silicon Valley. He certainly didn't have to work, having racked up "a significant fraction of a milli-Gates" at 3Com. (A "milli-Gates is one-thousandth of Bill Gates's net worth," which translated to around $100 million in early 1999.) And, then, he had what he later described as "an epiphanal moment," on the way to the elementary school. "I was in my Mercedes dropping off my daughter at school," he says. "Then I saw all the other children being dropped off in their Mercedes—by the family nannies, all of whom seemed to be blond. I decided right then I didn't want my kids to grow up as trust-fund hippies. I wanted our two young kids to be raised differently than they would be in Woodside." He had the luxury of $100 million to indulge his misgivings as well as to subsidize his wife's dream.

Robyn Metcalfe was looking to start a farm to conserve endangered breeds. She made the mistake of trying it in Woodside. She bought six sheep—a pilot flock—and leased an adjacent parcel as a paddock. One rainy afternoon, a truck with flashing lights across the top showed up at the Metcalfe house. The man was from the county's Animal Control department.

"A neighbor is complaining about 'llama abuse,'" reported the officer, as if there was in fact a specific local regulation governing the South American ruminant mammal.

The Metcalfes used a llama to guard the sheep from such predators as indigenous mountain lions. (Border collies are great at herding, but don't protect.) Robyn Metcalfe further explained to the officer that the llama and sheep had their own barn with an open door, and had made the independent decision to go outside to munch. And, she said, llamas in the wild actually stood out in the rain.

To this day, no one knows why the neighbor only complained about the llama, not the sheep.

There were other Woodside misadventures. Behind the Metcalfe house, on the hillside, were hundred-foot-high redwood trees. Running through them was Union Creek, which led to an old dam that gave the Metcalfes a swimming hole. It was dry most of the year, but during the winter rainy season, the creek ran so swiftly it began to imperil the redwoods. In the early 1990s, the Metcalfes proposed to shore up the banks with a stone wall, at a cost of tens of thousands of dollars. They hired a civil engineer and went before Woodside's always-entertaining planning commission, which told them that they needed an "inspecting engineer." It just so happened that the town engineer was available for private consulting—and this way, by golly, he could perform both roles simultaneously. The Metcalfes hired him.

Nonetheless, the town ruled that the stone wall could be erected only if the dam was removed. It seemed that the dam was preventing trout from getting upstream (and presumably had been doing so since Ulysses S. Grant was president). Trouble was, besides the minor matter of destroying the swimming hole, removing the dam would cause the creek bed to erode for miles all the way upstream, which concerned all nearby property owners. "We were caught between a rock and a hard place," Bob Metcalfe says. And he had already spent $25,000 on a project that had been conceived simply "to protect some old, beautiful trees—not to put up a chemical plant."

He finally convinced the town to approve his plan and not require elimination of the dam. Just one catch: Now he had to build a fish ladder, which would cost thousands more. Metcalfe told the town: "The hell with it. I give up."

Maybe he should've held a charity auction for the fish ladder.

The Metcalfes didn't leave Woodside to flee this kind of silliness, but it was a nice side benefit. "There were two kinds of lives we could lead," Robyn Metcalfe says. "One was the common set of Valley values—symphony, Stanford Shopping Center, Woodside auction. The other was more of an adventure—more isolated and uncomfortable, but more alive. The kids would learn that every birthday party isn't catered." And that a year has four seasons. Well, okay, Dad was worth more than almost anybody in the state of Maine—this wasn't exactly subsistence farming—but who could argue that the Metcalfes' hearts weren't in the right place?

Married to Bob Metcalfe for nearly twenty years, Robyn Metcalfe knew better than most the utility of the digital universe—computers, networks, the World Wide Web. It had changed the way people live and it had made her family wealthy. Her remote farmhouse in Maine would soon be fully wired and have its own Web site—Kelmscott.org—complete with an Ethernet, which her husband had invented. "But with all that stuff, I thought you had to stay grounded in something real—the warm breath of a horse in your face," she says, sitting in her small office, filled with such books as *Managing Your Ewe*. "My friends in Woodside couldn't understand why I'd give up poolside for 'poopside' on a farm."

Whereas Silicon Valley used to be about adventure, now it represented complacency and extravagance—a citadel of greed. Gearheads once were proud to be social misfits: banging away on their keyboards, tinkering in the garage, unaware of the benefits of hygiene. Today, they're lining up to learn about napkins, stemware, and why you don't hold that bottle of Chardonnay between your knees when removing the cork. Etiquette courses are the rage in the Valley. Marc Andreessen, meet Miss Manners. How dull.

The Metcalfes held out hope for something fresher for their children—something of the Woodside that used to be. The children now attend a K-through-9 school with just seventy-five students, named after the *Life of Riley*. And the children seemed to adapt quickly to farm living. The mission of Kelmscott Farm, which attracted thirteen thousand visitors in 1998, was "to create public awareness of the importance of agricultural biodiversity." The hundred Cotswold sheep are the jewel of the farm, descendants of a bloodline stretching back two thousand years to the Roman occupation of England. Many from the Metcalfe flock are used to replenish other herds in North America. But there simply isn't a demand for every last lamb. So about a dozen every year are sent to slaughter, providing a delicacy for specialty markets—and the Metcalfe dinner table. The children had been told some sheep went to "camp," including the two they named Lucky and Geordi. One night, the family sat down for lamb kielbasa. Max, then five, had a taste and then asked matter-of-factly, "Is this Lucky, or is it Geordi?"

Even Bob Metcalfe, who becomes "Farmer Bob" only in staged photos for glossy magazines, is philosophical on the charms of watching his son chase his daughter around with a frog from the pond. Farmer Bob is on a first-name basis with all the pigs and, under duress, even feeds them some days.

With their wealth, the Metcalfes could have begun their farm anywhere in the United States that provided sufficient acreage. Napa and Sonoma counties—north of San Francisco—were possibilities. But they still were within the gravitational pull of Silicon Valley. Maine represented a more complete break. The Metcalfes had been spending summers there for more than a decade. They owned various seaside homes, as well as part of an island ten miles out in Penobscot Bay—with no electricity or running water and only a cabin. In the 1980s, Bob Metcalfe came in on weekends from the West Coast. During one of his first summers in rural Maine, he found himself walking his two dogs on Main Street in Camden, a few miles from Lincolnville. It was early morning and he was jet-lagged from the long flight the night before. He was miserable. "What am I doing in this jerkwater town?" he asked himself. Then, up the block he saw John Sculley jaunting along the sidewalk in his chinos and Top-Siders. "This was back when Sculley was running Apple and was king of the world," Metcalfe recalls.

"Hello, Mr. Sculley," Metcalfe said back then. "What are *you* doing here?" Like Metcalfe, Sculley had a home in Woodside, so why was he on the wrong coast?

"I've been coming here since the early nineteen-seventies," Sculley said. "Who are *you*?" Sculley's still there—and remains the closest thing to a Silicon Valley neighbor that Metcalfe has.

Metcalfe grew fond of the region soon enough. Maine is only a short flight to Boston, where he was heavily involved in MIT alumni affairs and the computer publishing business. Once he was an entrepreneur. Now he had climbed the evolutionary ladder into the realm of journalism—as both columnist on the editorial side and publisher on the business end.

So, after twenty-two years in the Valley, the Metcalfes in 1994 chose Maine as their new home. They sold the Woodside estate for roughly $4 million to J. Taylor Crandall and his family. Crandall was a financial adviser to the Bass family—and the guy at the 1998 Woodside school auction who paid $125,000 for a cruise aboard Larry Ellison's super-yacht. It was altogether fitting that Metcalfe's self-proclaimed Old Money would be superseded by the dazzle of the young and the new.

The scale of success has changed," Bob Metcalfe says. "People always say, 'Things have changed since I was a boy.' But two things have changed—things and you. I'm not certain which it is for me. But at the beginning of 1998, Meg Whitman was product manager for Mr. Potato Head at Hasbro—and six months later she was worth half a billion dollars as the president of eBay [the online auction company]. You come up with a product and instantly you have 40 million users. I can remember at 3Com when we shipped only fifty units a month. People are leading faster, different lives. I now know how the people at IBM felt when 3Com came along." The hardware companies of another era were doing their best to keep up. Compaq, founded in 1982 and today the largest PC manufacturer, is nonetheless busy transforming itself into an online-commerce company that will sell not only computers but all kinds of gadgetry; it sold 30 million PCs in its first fourteen years in business, but those numbers paled compared with the potential of the Internet.

Metcalfe laments the new pace and new Internet ethos that make the age of semiconductors and PCs seem quaint. "I'm a traditionalist. I think it's immoral for a company to go public without profits or reasonable projections thereof. But I was brought up at a different time."

It's no surprise, then, that he finds himself thriving with a life beyond the Valley, a continent away. He visits the Valley every few months, he E-mails with the denizens every day, his professional punditry requires him to follow its every hiccup. But in Maine he's learned a new freedom. The rhythm is slower—far from the madding crowd and its fast cars and super egos—where discoveries in the piggery and poultry shed seem as marvelous as a new billion-dollar business to market Kathie Lee Gifford on the Internet. Metcalfe delights in introducing Pete the two-thousand-pound Shire horse, who lives in the barn attached to the big, gray family residence. Or describing his sixteen-foot sailboat, *Flash,* designed by the late Joel White, son of E. B. White, who used to live just up the midcoast. (The Metcalfes were admirers of the father. Their prior boat was named *Wilbur,* some pig from *Charlotte's Web.*) Or noting that Kelmscott cost "barely a tenth" of what he sold his Woodside property for. In its simplicity and timelessness, Kelmscott holds out the promise of the best of both worlds.

"And, given the traffic out there in the Valley now, my commute back is only slightly longer."

Nobody appears to be having quite as good a time in Silicon Valley. Passions have become mere professions. Impulsiveness is now compulsiveness. Steve Jobs himself—the closest that Silicon Valley has to a god, even with his company now a bit player—rues what the Valley has left behind. At forty-four, he calls himself "an old man," someone who remembers not only Bob Noyce and Dave Packard—he befriended them both—but the apricot orchards and the beat generation and the times when people cared that Jerry Garcia had attended Palo Alto High School. "There used to be something magical here—scientific and cultural," Jobs told me. "You could smell it, feel it. When I was in high school, I would ride my bicycle to the Stanford artificial-intelligence lab on weekends and hang out. You could feel the magic in the air. I miss that time. People care more about material things now. When I was twenty-three, I was worth a million dollars. When I was twenty-four, I was worth $10 million and, at twenty-five, over $100 million. But I never really cared that much about the money."

The rebels of another era have been replaced by MBAs and accountants; yesterday's tinkerers and thinkers have given way to the moneymen and predators. If Jobs were "young" again, would he not pursue computers and networks and the digital revolution all over again? "Never," he says. "I'd get into molecular biology."

While Steve Jobs may not be Mr. Congeniality, he embodied a Valley of dreams that in time became Babylon. For every Steve Wozniak romantic or Gary Kildall idealist, there are now ten velvet-tongued mercenaries like John Doerr ready to capitalize on them. Larry Ellison and Jim Clark want to author the future, so long as the Nasdaq can tally the revolution. Hype counts as much as product. Hype *is* product. Arthur Rock, the money behind Intel, is as tough a financier as the Valley has ever seen, yet he has little taste for the game anymore. "It's about money: Go public, sell shares as soon as you're allowed, get out, go on to the next thing. Most of the guys come in and tell me how much money I'm going to make if I do their financing. My attitude is if it's just money you're interested in, go someplace else." They do.

The workaholic pace robotizes people—they *have* to get rich. Call it affluenza. No time for fun, no time for family, no perspective on life outside the bubble. Is it only coincidence that there is so little philanthropy in the Valley relative to "the largest legal creation of wealth in the history of the planet"? What shimmered in the Valley now corrodes. The frontier is gone, replaced by a racetrack. Wozniak is counting the days until his children are grown and he can move to the Sierras.

By one estimate, the recent divorce rate in the Valley is 80 percent, and that's not factoring in Larry Ellison's marriages. The birth rate is low—who has time for children when you're going public next week? Ron Wiebe, a family therapist in Los Gatos who counts many CEOs as patients, says the Valley's biggest disease is fear of vacations. "People are afraid that if they go away for a week, they'll return to find themselves obsolete." Wiebe acknowledges what Silicon Valley Inc. has accomplished, but what he sees is the debris of dreams. The trouble with so many people making so much money is that they're always keeping score—and most are making more. The party continues, but the partygoers grow weary.

The Valley once was a new machine. It changed the world. It may do so yet again. But the machine has no soul anymore.

Acknowledgments

I spent most of 1998 in and around Silicon Valley, immersed in the culture and currency of the place. This book is the product. But its beginnings go back to 1994–95, when I spent a year at Stanford University as a John S. Knight Fellow. That wonderful program allowed me the sustained introduction to high-tech culture that was essential to this project. My thanks to the fellowship and to the two journalists who run it, Jim Risser and Jim Bettinger. My thanks also to Professor Gerry Gunther, who first urged me to consider the fellowship; I'm lucky to have him and his wife, Barbara, as friends.

I could not have written this book without the enlightened employment policies of *Newsweek*. My special thanks to the late Maynard Parker, as well as the other editors, for giving me the time I needed. I'm also grateful to the magazine for permitting me to use the fruits of my journalistic efforts in this book.

Other colleagues and friends tolerated my queries and frequent pleas. Thanks to: Jerry Adler, Sharon Begley, Ellis Cose, Bob Costas, Storm Duncan, Carl Falcone, Stephen Gillers, Neil Goteiner, Andrew Joskow, Alex Kozinski, Patricia King, Jon Newman, Allen Reichman, Robina Riccitiello, Adam Rogers, Michael Rogers, Paul Saffo, John Schwartz (the phrase "Valley of the Dollars" is his), Gary Simon, Allan Sloan, Marcy Tiffany, Rich Turner, Mark Vamos, and Stephanie Vardavas. Thanks especially to the amazing Mike Wilson, who knows Larry Ellison better than Ellison does. For translating the mumbo jumbo of venture capital, I was fortunate to have the help of Craig Canine, Mike Curry, Joe Schoendorf, Marty Sklar, Jim Steinberg, Peter Tufano, and Jack Wilson.

In Silicon Valley, a group of individuals—some of whom I've included in a list of interviews at the end of the book—went above and beyond the call of duty: Al Alcorn, Marc Andreessen, Roger Bamford, Frank Caufield, Jim Clark, Chuck Dietrick, Stuart Feigin, Bill Foss, Marge Gianetto, Rich Green, Joyce Higashi, Heidi Johnson, Jamis MacNiven,

Gordon Moore, Mike Moritz, Ed Oates, Tom Perkins, Arthur Rock, Tom Rolander, Todd Rulon-Miller, Jeff Suto, Steve Wozniak, and Jerry Yang. Richard Brandt at *Upside* and Tony Perkins at *Red Herring* were generous with their journalistic resources. Most important of all in Silicon Valley was Jonathan Kaplan, along with Asad and Zura, who gave me a place to stay.

There are bad public-relations professionals, there are good public-relations professionals, and then there are the best ones. In the last category are Katie Cotton at Apple, Howard High at Intel, Chris Holten and Suzanne Anthony at Netscape, and Pam Alexander and Dawn Whaley at Alexander Communications.

Thanks also to: Bob and Robyn Metcalfe for their Maine hospitality; T. J. Rodgers, for the finest bottle of wine I had in 1998; Marv Newland for a print of his *Bambi Meets Godzilla;* Charles Simonyi at Microsoft; Henry Lowood and Margaret Kimball at Stanford Libraries and Archives; Denise Harvill and Cindy Boris at United Airlines; www.dogpile.com; *Law and Order* reruns at three in the morning; Ben & Jerry; Steve Cerutti at the University of Oregon Law School; Catherine Aman at the Columbia University Journalism School; the Churchill Group; and Jean Brown, the master of transcription.

Justice demands that any book about technology be afflicted with a few technological gremlins along the way—it's not *Revenge of the Nerds*, but payback from their machines. For rescuing various audiotapes, my thanks to Miles Perkins and Degi Simmons, and Stacey Moran and Joel Gilbertson at Sonic Foundry. For resolving intermittent Macintosh headaches, my thanks to Damian Rieger and the Woz.

Near its completion, the manuscript benefited greatly from comments by Audrey Feinberg, Barbara Kaplan, Hank Gilman, and John Riley. Katie Hafner, David Howell, and Paul Saffo read portions of the manuscript and offered valuable suggestions. All errors and unpardonable puns, of course, are mine.

I am indebted to two people for their research assistance throughout this project. Nadine Joseph is an extraordinary reporter who never once protested when I asked her to go up in flying machines that I wouldn't even go near. Dana Gordon knows her way around libraries and electronic databases better than anyone; she found needles in haystacks I didn't know existed.

I can't imagine a finer editor or friend on this venture than Henry Ferris at William Morrow and Company. Without his skills and forbearance, this book would not have happened. My thanks, too, to Ann Treistman at Morrow.

I also could not have done this book without the inestimable Esther Newberg, who is among literary agents what Ted Williams was among hitters—the best. I salute her assistant Jack Horner and promise no more questions for a while.

Finally, I want to thank my wife, Audrey, and my two boys. Their love, patience, and indulgence got me through. Yes, Joshua and Nathaniel, we can go outside and play now.

Sources and Bibliography

The primary sources for *The Silicon Boys* were interviews I did during 1998, and in a few cases dating to 1995, with the following individuals:

Al Alcorn
Marc Andreessen
Robert Andrews
Roger Bamford
Jim Barksdale
John Seely Brown
Brook Byers
Gerhard Casper
Frank Caufield
Jim Clark
Bud Colligan
Scott Cook
Dave Corbin
Wilf Corrigan
Roger Craig
Harvey Dale
Chuck Darrah
Jeanne Dickey
Chris Dickson
Chuck Dietrick
John Doerr
Rich Draeger
Esther Dyson
Larry Ellison

Bill Erkelens
Stuart Feigin
David Filo
Mike Foley
Tom Ford
Bill Foss
Sue Fox
Marge Gianetto
Mark Goldstein
Rich Green
Andy Grove
Zara Haimo
Hugh Hempel
Rick Herns
J. S. Holliday
Chris Holten
David Howell
Lyndy Janes
Steve Jobs
Heidi Johnson
Jerry Kaplan
Gene Kleiner
Randy Komisar
Tim Koogle

Dick Kramlich
Jamis MacNiven
Valeta Massey
Bob Metcalfe
Max Metcalfe
Robyn Metcalfe
Jon Mittelhauser
Gordon Moore
Mike Moritz
Dennis Muren
Michael Murphy
John Nesheim
Charles Nesson
Ed Oates
Jenny Overstreet
Tom Perkins
John Quayle
Roger Rickard
Mike Roberts
Arthur Rock
T. J. Rodgers
Tom Rolander
Todd Rulon-Miller
Joe Schoendorf

Carter Sednaoui	Bob Spicer	Ron Wiebe
Dick Shaffer	Kevin Starr	Jack Wilson
Tom Siebel	Ted Turner	Gary Wozniak
Russ Siegelman	Jim Valentine	Steve Wozniak
Charles Simonyi	Jim Warren	Jerry Yang
Larry Sonsini	Jim Wickett	Pierluigi Zappacosta

There were roughly a dozen other people I interviewed who asked not to be identified. Most of them weren't critical to the narrative, with the exception of material in the Prologue concerning Woodside; Chapter VII concerning the venture-capital funds of Kleiner Perkins Caufield & Byers; and particularly in Chapter IX concerning Microsoft. Well into this project and around the time it was sued by the federal government, Microsoft withdrew its cooperation; most Microsoftians who talked to me thereafter understandably asked for anonymity. In all but several places, however, I have not used unattributed quotations.

In addition to interviews, I relied on numerous accounts, both current and historical, from secondary sources: newspapers, magazines, industry publications, and in a few instances, material published specifically for the World Wide Web. Where relevant, I have cited sources directly in the text. In particular, among newspapers, I frequently used *The New York Times, The Wall Street Journal, The Washington Post,* the *San Francisco Chronicle,* the *San Francisco Examiner,* the Woodside *Almanac,* and the San Jose *Mercury News;* among magazines, I often referred to *Newsweek, Time, Business Week, Forbes, Fortune, Vanity Fair, The New Yorker, The Economist, Upside, Red Herring,* and *Wired.*

Books were crucial resources in filling in pieces of the narrative, especially in Chapters I through IV. For Chapter I, I relied on *The World Rushed In,* by J. S. Holliday; *Californians: Searching for the Golden State,* by James D. Houston; *Assembling California,* by John McPhee; *Historic San Francisco: A Concise History and Guide,* by Rand Richards; and *Gold Rush: A Literary Exploration,* edited by Michael Kowalewski.

For Chapters II and III: *Inside Intel: Andy Grove and the Rise of the World's Most Powerful Chip Company,* by Tim Jackson; *The Big Score: The Billion Dollar Story of Silicon Valley,* by Michael S. Malone; *The HP Way: How Bill Hewlett and I Built Our Company,* by David Packard;

The Chip: How Two Americans Invented the Microchip and Launched a Revolution, by T. R. Reid; and *Crystal Fire: The Birth of the Information Age*, by Michael Riordan and Lillian Hoddeson.

For Chapter IV: *Apple: The Inside Story of Intrigue, Egomania, and Business Blunders*, by Jim Carlton; *Fire in the Valley: The Making of the Personal Computer*, by Paul Freiberger and Michael Swaine; *Hackers: Heroes of the Computer Revolution*, by Steven Levy; *Insanely Great: The Life and Times of Macintosh, the Computer That Changed Everything*, by Steven Levy; *The Big Score: The Billion Dollar Story of Silicon Valley*, by Michael S. Malone; and *Fumbling the Future: How Xerox Invented, Then Ignored, the First Personal Computer*, by Douglas K. Smith and Robert C. Alexander.

For Chapter V: *The Difference Between God and Larry Ellison: Inside Oracle Corporation*, by Mike Wilson.

For Chapters VI and VII: *The New Venturers: Inside the High-Stakes World of Venture Capital*, by John W. Wilson.

For Chapter VIII: *Where Wizards Stay Up Late: The Origins of the Internet*, by Katie Hafner and Matthew Lyon; *Speeding the Net: The Inside Story of Netscape and How It Challenged Microsoft*, by Joshua Quittner and Michelle Slatalla; and *Architects of the Web: 1,000 Days That Built the Future of the Business*, by Robert H. Reid.

For Chapter IX: *Gates: How Microsoft's Mogul Reinvented an Industry—and Made Himself the Richest Man in America*, by Stephen Manes and Paul Andrews; *Overdrive: Bill Gates and the Race to Control Cyberspace*, by James Wallace; and *Hard Drive: Bill Gates and the Making of the Microsoft Empire*, by James Wallace and Jim Erickson. I also relied heavily on court documents in the federal government's antitrust case against Microsoft.

For Chapter X: *Architects of the Web: 1,000 Days That Built the Future of the Business*, by Robert H. Reid.

This is a complete list of books used in this project:

Amelio, Gil, and William L. Simon. *On the Firing Line: My 500 Days at Apple*. New York: HarperBusiness, 1998.

Auletta, Ken. *The Highwaymen: Warriors of the Information Superhighway*. New York: Random House, 1997.

Barich, Bill. *Big Dreams: Into the Heart of California.* New York: Pantheon Books, 1994.

Bolt, Bruce. *Earthquakes.* New York: W. H. Freeman, 1993.

Brockman, John. *Digerati: Encounters with the Cyber Elite.* San Francisco: HardWired, 1996.

Bronson, Po. *The First $20 Million Is Always the Hardest: A Silicon Valley Novel.* New York: Random House, 1997.

Carlton, Jim. *Apple: The Inside Story of Intrigue, Egomania, and Business Blunders.* New York: Random House/Times Books, 1997.

Cringely, Robert X. *Accidental Empires: How the Boys of Silicon Valley Make Their Millions, Battle Foreign Competition, and Still Can't Get a Date.* Reading, MA: Addison-Wesley, 1992.

Dillon, Pat. *The Last Best Thing: A Classic Tale of Greed, Deception, and Mayhem in Silicon Valley.* New York: Simon & Schuster, 1996.

Dyson, Esther. *Release 2.0: A Design for Living in the Digital Age.* New York: Broadway Books, 1997.

Freiberger, Paul, and Michael Swaine. *Fire in the Valley: The Making of the Personal Computer.* Berkeley, CA: Osborne/McGraw-Hill, 1984.

Gates, Bill, Nathan Myhrvold, and Peter Rinearson. *The Road Ahead.* New York: Viking, 1995.

Gibson, William. *Neuromancer.* New York: Ace Books, 1984.

Grove, Andrew S. *Only the Paranoid Survive.* New York: Doubleday, 1996.

Hafner, Katie, and Matthew Lyon. *Where Wizards Stay Up Late: The Origins of the Internet.* New York: Simon & Schuster, 1996.

Hall, Mark, and John Barry. *Sunburst: The Ascent of Sun Microsystems.* Chicago: Contemporary Books, 1990.

Hanson, Dirk. *The New Alchemists: Silicon Valley and the Microelectronics Revolution.* Boston: Little, Brown, 1982.

Holliday, J. S. *The World Rushed In.* New York: Simon & Schuster, 1981.

Houston, James D. *Californians: Searching for the Golden State.* Santa Cruz, CA: Otter B Books, 1992.

Jackson, Tim. *Inside Intel: Andy Grove and the Rise of the World's Most Powerful Chip Company.* New York: Dutton, 1997.

Jager, Rama Dev, and Rafael Ortiz. *In the Company of Giants: Candid Conversations with the Visionaries of the Digital World*. New York: McGraw-Hill, 1997.

Kaplan, Jerry. *Startup: A Silicon Valley Adventure*. Boston: Houghton Mifflin, 1995.

Kidder, Tracy. *The Soul of a New Machine*. Boston: Little, Brown, 1981.

Kowalewski, Michael, ed. *Gold Rush: A Literary Exploration*. Berkeley, CA: Heyday Books, 1997.

Levy, Steven. *Hackers: Heroes of the Computer Revolution*. New York: Dell, 1984.

————. *Insanely Great: The Life and Times of Macintosh, the Computer That Changed Everything*. New York: Viking Penguin, 1994.

Malone, Michael S. *The Big Score: The Billion Dollar Story of Silicon Valley*. New York: Doubleday, 1985.

Manes, Stephen, and Paul Andrews. *Gates: How Microsoft's Mogul Reinvented an Industry—and Made Himself the Richest Man in America*. New York: Doubleday, 1993.

McPhee, John. *Assembling California*. New York: Farrar, Straus and Giroux, 1993.

Moritz, Michael. *The Little Kingdom: The Private Story of Apple Computer*. New York: William Morrow, 1984.

Nesheim, John L. *High Tech Start Up*. Saratoga, CA: John L. Nesheim, 1997.

Packard, David. *The HP Way: How Bill Hewlett and I Built Our Company*. New York: HarperBusiness, 1995.

Quittner, Joshua, and Michelle Slatalla. *Speeding the Net: The Inside Story of Netscape and How It Challenged Microsoft*. New York: Atlantic Monthly Press, 1998.

Reid, Robert H. *Architects of the Web: 1,000 Days That Built the Future of the Business*. New York: John Wiley & Sons, 1997.

Reid, T. R. *The Chip: How Two Americans Invented the Microchip and Launched a Revolution*. New York: Simon & Schuster, 1984.

Reynolds, Terry S., and Stephen H. Cutcliffe, eds. *Technology and the West*. Chicago: University of Chicago Press, 1997.

Richards, Rand. *Historic San Francisco: A Concise History and Guide*. San Francisco: Heritage House, 1991.

Riordan, Michael, and Lillian Hoddeson. *Crystal Fire: The Birth of the Information Age*. New York: Norton, 1997.

Rodgers, T. J. *No Excuses Management*. New York: Doubleday, 1992.

Rogers, Everett M., and Judith K. Larsen. *Silicon Valley Fever: Growth of High-Technology Culture*. New York: Basic Books, 1984.

Rogers, Michael. *Silicon Valley*. New York: Pocket Books, 1982.

Saxenian, AnnaLee. *Regional Advantage: Culture and Competition in Silicon Valley and Route 128*. Cambridge, MA: Harvard University Press, 1994.

Sculley, John. *Odyssey: Pepsi to Apple . . . A Journey of Adventures, Ideas and the Future*. New York: Harper & Row, 1987.

Smith, Douglas K., and Robert C. Alexander. *Fumbling the Future: How Xerox Invented, Then Ignored, the First Personal Computer*. New York: William Morrow, 1988.

Smolan, Rick, and Jennifer Erwitt. *One Digital Day: How the Microchip Is Changing Our World*. New York: Random House/Times Books, 1998.

Stegner, Wallace. *Angle of Repose*. New York: Doubleday, 1971.

Stoll, Clifford. *Silicon Snake Oil*. New York: Doubleday, 1995.

Stross, Randall E. *Steve Jobs and the NeXT Big Thing*. New York: Atheneum, 1993.

———. *The Microsoft Way: The Real Story of How the Company Outsmarts Its Competition*. Reading, MA: Addison-Wesley, 1996.

Swisher, Kara. *aol.com: How Steve Case Beat Bill Gates, Nailed the Netheads, and Made Millions in the War for the Web*. New York: Random House/Times Books, 1998.

Wallace, James. *Overdrive: Bill Gates and the Race to Control Cyberspace*. New York: John Wiley & Sons, 1997.

Wallace, James, and Jim Erickson. *Hard Drive: Bill Gates and the Making of the Microsoft Empire*. New York: John Wiley & Sons, 1992.

Weston, J. Fred, and Eugene F. Brigham, ed. *Essentials of Managerial Finance* (5th ed.). Orlando, FL: The Dryden Press, 1979.

Wilson, John W. *The New Venturers: Inside the High-Stakes World of Venture Capital*. Reading, MA: Addison-Wesley, 1985.

Wilson, Mike. *The Difference Between God and Larry Ellison: Inside Oracle Corporation*. New York: William Morrow, 1997.

Index

emoticons, 222
encryption, 239
Engelbart, Doug, 221
ENIAC, 43, 66, 71, 99, 147
Enterprise Integration Technologies, 231
"entrepreneurialogists," 14
EO, 199–200
Equitable Life, 46
Esquire, 47, 51–53, 84
Ethernet, 190, 323, 327
Eubanks, Gordon, 211
European Laboratory for Particle Physics
     (CERN), 223
Excel (Microsoft) spreadsheet, 148
Excite, 26, 208, 213, 307, 312
extranets, 293
Exxon, 206

Faggin, Federico, 66
Fairchild, Sherman Mills, 50–51, 52, 57,
     60, 83, 165
Fairchild Camera and Instrument, 51
Fairchildren, 58, 62, 72
Fairchild Semiconductor, 50–61, 62, 63,
     64, 67, 83, 84, 95, 97, 163, 164,
     165–166, 200, 254
   buyback of, 57–58
   founding of, 50–51, 55
   growth and decline of, 55, 59–60
   integrated circuit developed at, 56–57
Fairchild v. *Baldwin*, 59
Fantasia, 35
Farnsworth, Philo, 32
Farr, Bruce, 127
Federal Communications Commission
     (FCC), 181
Federal Telegraph Company, 30–31, 32,
     38
Federal Trade Commission (FTC), 281,
     288
FedEx, 186, 236
Feigin, Stuart, 130–131, 133, 135, 138,
     139–140, 142, 143, 150, 173, 220
Filo, David, 303, 304, 305, 306, 307,
     310–311, 312, 313, 314–315,
     316, 317, 319, 321
Financial Times, 274
Financial World, 68
Fireman's Fund Insurance, 129
Fitzgerald, F. Scott, 107
Flash, 330
floppy disk drives, market for, 174
Folger, Abigail "Gibby," 90

Folger, James, III, 90
Forbes, 229, 301
Forbes ASAP, 187
Forbes 400, 17, 99, 145
Ford, Henry, 57
Ford, Tom, 172–173
Ford Foundation, 206
Ford Land Company, 173–174
Fortune, 63, 123, 196, 212, 214, 256,
     298, 309
Fortune 100, 238
Fortune 500, 99, 147, 203, 257
Foss, Bill, 232, 233, 236–237, 311, 320
4004 microprocessor, 66, 67, 69, 108
Fowler, Jim, 153
Freiberger, Paul, 85
FTC (Federal Trade Commission), 281,
     288

Gale, Grant, 46
Garcia, Jerry, 250, 330
Gates, Bill, 11, 18, 20, 29, 41, 68, 108,
     129, 132, 133, 147, 161, 189,
     211, 226, 248, 251–252, 255,
     256, 260, 266–267, 295, 298,
     302, 304
   Ellison and, 9, 11, 119–120, 141, 147–
        148, 149, 150, 273, 282
   Internet Explorer browser and
        Netscape and, 255–290
   Microsoft cofounded by, 259
   operating system licensing and, 111–
        117, 118, 137
   wealth of, 113, 116, 119, 145, 252,
        257, 325
Gates, Mary, 115
Gateway Computers, 191
Gehry, Frank, 18
Genentech, 179, 198, 248
General Electric (GE), 34, 43, 45, 70,
     71, 101
General Motors, 206, 292
General Radio, 35, 168
gene-splicing, 17, 178–179
genomics, 191
Georgia Tech, 181, 207
germanium, 39, 40, 41, 56
Gibbons, Fred, 310
Gibson, William, 225
Gilder, George, 229–230, 232, 260
Global Telecosm, 279
GO, 197, 199–200, 208, 244, 281, 307
Godbout, Bill, 111